# *ISS*CAPADES

## THE CRIPPLING OF AMERICA'S SPACE PROGRAM

DONALD A. BEATTIE

All rights reserved under article two of the Berne Copyright Convention (1971).
We acknowledge the financial support of the Government of Canada through the
Book Publishing Industry Development Program for our publishing activities.
Published by Collector's Guide Publishing Inc., Box 62034,
Burlington, Ontario, Canada, L7R 4K2
Printed and bound in Canada
ISScapades: The Crippling of America's Space Program
by Donald A. Beattie
ISBN 978-1-894959-59-9
ISSN 1496-6921
©2006 Apogee Books

# ISScapades

## The Crippling of America's Space Program

### Donald A. Beattie

---

An Apogee Books Publication

## Table of Contents

Preface ................................................................. 7
Acknowledgments ......................................................... 9
Abbreviations and Acronyms .............................................. 10
Introduction: ". . . the next logical step" (1982 - 1984) ............... 14
    Space Station Task Force Initial Studies ........................... 15
    Into the Lion's Den – The President Decides ........................ 15
    The Plan to Proceed ................................................ 16
    Space Station Design Drivers ....................................... 20
    International Partners ............................................. 22
    Moving On – Management Decisions ................................... 24
    Early Budget Controversies ......................................... 26

Chapter 1: First Reality Check – Problems and Solutions (April 1984 - December 1985) ....... 29
    Man-Tended or Permanently Manned? .................................. 29
    Space Station Program Office Established – Phase B RFPs Released ... 31
    Automation and Robotics Studies .................................... 32
    Phase B Schedule ................................................... 33
    Budget Battles Begin ............................................... 33
    Work Package Contractors Selected .................................. 34
    Space Station Operations Review .................................... 37
    International Partners Sign Up ..................................... 37
    National Commission on Space ....................................... 38
    Reference Update Review ............................................ 39
    Potential Users Review Design ...................................... 41
    Beggs Resigns – Management Turmoil ................................. 42

Chapter 2: The Baseline Configuration – Everyone on the Same Page?
    (January 1986 - April 1987) ........................................ 45
    The Challenger Accident ............................................ 45
    Management Changes and Design Reviews .............................. 46
    James Fletcher Returns as Administrator ............................ 48
    The Space Shuttle – A Potential Space Station Single Point Failure . 49
    Program Continues to Change ........................................ 51
    Phillips Report – Andrew Stofan Selected as Space Station Manager .. 52
    Astronaut Gordon Fullerton Critical of Design ...................... 52
    Fletcher Calls Time Out – CETF Formed .............................. 54
    Stofan Organizes Washington Offices ................................ 57
    Phased Program Task Force .......................................... 59
    Conflicts Continue with OMB and Congress ........................... 62
    STS Modifications and Study of a Mixed Fleet ....................... 63

Chapter 3: Replacing Challenger – More Studies – More Paper (April 1987 - January 1989) ..... 67
    GAO Disputes Runout Cost Estimates – Progress Made ................. 68
    Ride Report – Mission to Planet Earth .............................. 69
    Mood in Congress Deteriorates ...................................... 72
    Space Station Operations Task Force ................................ 72
    NASA Courts the Private Sector ..................................... 74
    Phase C/D Studies Begin – Cost Reductions Studied .................. 74
    President Reagan's Space Vision – Impact on Space Station .......... 76
    Rescoping – Rescheduling ........................................... 78
    James Odum Replaces Stofan – Progress Continues .................... 79
    New Administration – Same Congress ................................. 84

Chapter 4: More Management Changes – Design Continues to Evolve
(February 1989 - December 1990) .................................................. 87
    Fletcher Resigns – National Space Council Established ........................... 88
    Truly Makes Management Changes – Problems Persist ............................ 90
    President Bush Announces a Long-Range Space Program ......................... 92
    New Space Station Management Team ........................................... 93
    Program Rephased ............................................................. 95
    Amount of EVA Raises Concerns ................................................ 96
    New Problems Accentuate Congressional Oversight .............................. 101
    Management Adjusts to Congressional Demands ................................. 102
    Augustine Committee Report .................................................. 103

Chapter 5: Restructuring the Space Station (January 1991 - December 1992) ..... 105
    Effects of the Restructuring .................................................... 108
    Space Exploration Initiative and Science Community Issues ...................... 109
    More Management Changes – Progress and Problems ............................ 111
    SSAC Panel Reports on ACRV and Program Verification ........................ 113
    Truly Resigns – Search for New Administrator .................................. 114
    Daniel Goldin Takes Charge – Another New Direction .......................... 115
    Russian Cooperation Explored ................................................. 117
    Space Station Budget Survives in Congress – More Progress ..................... 117
    New Administration – New Program Changes .................................. 119

Chapter 6: New Administration – Back to Square One (January 1993 - March 1994) .... 121
    New Administration Guidelines – Redesign Ordered ............................ 121
    Redesign Team Reports ....................................................... 124
    President Clinton Selects Design Option – Congress Objects ..................... 126
    Redesign Challenged – Goldin Appoints New Managers – JSC as Host Center ..... 127
    Administration Courts the Russians ............................................ 129
    Russian Partnership Changes Program – Congress Confused .................... 130
    New Space Station Implementation Plan ....................................... 132
    Congress Passes FY 1994 Budget – Questions and Concerns Rampant ........... 133
    Cooperative Agreement with Russia Signed – Critics Denounce Terms ........... 135
    Management Changes – Budget and Program Debates ........................... 137
    System Design Review Completed – CBO Forecast .............................. 139

Chapter 7: The International Space Station (Alpha) (April 1994 - December 1996) ..... 143
    International Partners Express Concern with Program Changes .................. 144
    SDR Results in New Cost Estimates – Congressional Reaction ................... 144
    Russian Agreement Modified – Program and Budget Concerns Continue ......... 147
    Russian Funding Transfers Become a Problem ................................. 149
    Elections Bring Changes to Congress – Partners Add Changes ................... 150
    Progress, But Old Problems Persist ............................................ 151
    O'Connor Resigns – More Management Changes ............................... 155
    Cost Runout Concerns – GAO Issues New Report – Program Changes Anger Experimenters .. 157
    Inability of Russia to Deliver their Elements Continues – Spies in the Woodwork? .......... 159

Chapter 8: Assembly Begins – A New Star in the Evening Sky (January 1997 - January 1999) ... 160
    Russian Service Module Continues to Slip – Assembly Schedule Changed ........ 161
    EVA Time and *Mir* Crew Safety Remain as Problems ............................ 164
    *Mir* Decommissioning Creates New Problems – Added Costs .................... 167
    Assembly Sequence Changes – Schedule Delays Continue ....................... 168
    FY 1999 Budget Hearings – Runout Costs Debated ............................. 169
    Phase 1 Completed – New Launch Dates Announced ........................... 171
    Runout Cost Estimates Unresolved – Hardware Delivered ....................... 172
    First Elements Finally Launched – Budget Battles Continue ..................... 175

Chapter 9:  The Story Continues (1999 - 2005) .......................................... 177
    2000 ............................................................................ 179
    2001 ............................................................................ 181
    2002 ............................................................................ 185
    2003 ............................................................................ 188
    2004 ............................................................................ 190
    2005 ............................................................................ 193

Chapter 10:  Who Was in Charge? – No One! ............................................. 197
    Modern Times in Washington ...................................................... 197
    Program Oversight ............................................................... 199
    NASA Top Management Turnover .................................................... 200
    Early Space Station Management Problems ......................................... 201
    NASA Attempts to Improve Program Management ..................................... 201
    Congressional and White House Control ........................................... 202
    NASA Administrators' Space Station Legacies ..................................... 205
    What Are the Lessons to be Learned? ............................................. 209
    NASA's Future ................................................................... 212
    Privatize NASA Programs? ........................................................ 218

Selected Bibliography ................................................................ 221

Index ................................................................................ 222

## – PREFACE –

Space stations, orbiting overhead, shining like bright new stars from reflected sunlight in the morning or evening sky, have been the subject of dreamers and engineers for generations. Numerous books and articles have explored and explained how and why space stations would one day fulfill those dreams. Many different concepts have been proposed and a few have been constructed. The first operational versions, Salyut, Skylab, and *Mir*, have already been placed in orbit, completed their short lives, and reentered Earth's atmosphere in flaming last acts.

The Salyut program, begun by the Soviet Union in 1971, became an accumulation of small spacecraft that Soviet cosmonauts occupied on and off during its 15 year lifetime. Space endurance records were set with two cosmonauts living in a weightless environment for 185 days while conducting a number of experiments. What was learned during those long periods in weightlessness was not shared by the Soviets with the outside world until the 1990s.

Skylab, launched in 1973, was a hurry-up program to place a U.S. space station in orbit to compete with the Soviet Union. It was based, in part, on studies conducted during the Apollo Applications Program that proposed utilizing upgraded Apollo hardware for a variety of space missions including studying Earth from low Earth orbit. Skylab used modified parts left over from Apollo that provided a large volume living quarters and laboratory. The astronauts during their time in orbit were able to conduct a variety of experiments and used the Apollo Telescope Mount to take high resolution photographs. It also allowed U.S. astronaut crews to live in space, continuously, for 28, 59, and 84 days, greatly expanding our knowledge of the effects on humans of long duration space flight beyond the twelve days of the longest Apollo mission.

*Mir*, the next Soviet space station, with the first elements orbited in 1986, most closely resembled early ideas of how a space station would be constructed. Modules were added over time, all connected to a core unit. *Mir* permitted Soviet cosmonauts to set new endurance records and to perform numerous experiments while living in space for very long periods. One, Valery Polyakov, remained onboard *Mir* continuously for 438 days, a time selected to equal a roundtrip to Mars if such a mission would be undertaken. *Mir*, far outliving its original expected lifetime, after fifteen years in low Earth orbit was deorbited, reentered the atmosphere and burned up over the Pacific ocean in 2001.

*Mir*'s final years overlapped the first assembly flights of the International Space Station (ISS), the latest version of the old dreams. The Space Station program, approved by President Reagan in 1984, gradually evolved into a multinational effort and is designed to eventually provide a much larger and more capable laboratory than *Mir*. Under U.S. leadership the program, if completed as planned, will ultimately include contributions from fifteen international partners. The ISS has undergone major changes from the program originally envisioned in 1984. These changes were directed by several administrations and Congress. After many redesigns and delays, the first element was launched in November 1998. Additional elements have been added as they became flight qualified and U.S. shuttle and Russian launchers were available to lift them into orbit.[1] Much still remains to be done to complete the program. It is a work in progress with a uncertain future as a result of the *Columbia* tragedy and its aftermath, and President Bush's new space vision. Further changes can be expected but they are not predictable.

The story that follows is based, in large part, on close involvement with the Space Station program. One of the projects completed by my NASA office in 1980 was a study of the Satellite Solar Power System (SSPS) first proposed in 1974.[2] The SSPS, if ever approved, would place huge satellites in geosynchronous orbit, collect energy from the sun, and beam it back to Earth in the form of microwave

---

1. If you are interested in observing the International Space Station as it orbits the Earth, go to NASA website: www.nasa.gov; Human Space flight –International Space Station – Realtime Data; click on Sighting Opportunities; click on your city or the nearest location listed, and you will find the next observing times and viewing direction in either the evening or morning skies. With 10-power binoculars or a small telescope on a clear night, you may be able to see some detail of the structure.

power. With this background of studying how to place large payloads in Earth orbit, in the summer of 1982 I participated in Space Station planning that had just begun at NASA Headquarters under the direction of John D. Hodge. John and I had worked together in the 1960s and early 1970s when he was at the Manned Spacecraft Center (now named the Lyndon B. Johnson Space Center) and I was at NASA Headquarters. During our collaboration we attempted, unsuccessfully, to sell a long-range lunar exploration program to NASA management and the Nixon White House. My participation in the early Space Station planning was limited as I left NASA in 1983. At the end of 1987, while working as a consultant on government programs, NASA Chief Scientist Dr. Noel W. Hinners recommended that I be appointed to the newly chartered Space Station Advisory Committee and I served on the Committee until 1994.

This history covers the development of the Space Station from 1982 to the launch of the first two elements in 1998. Brief, annual, summaries carry the program from that point until the submission of the manuscript for publication. It is not always a flattering account of how the Space Station evolved. Rather, it tells the story of how the program struggled to survive in a political environment that continually modified its goals and content. It also describes how powerful antagonists toiled through the years to have the program terminated. The final chapter summarizes how Congress, several administrations, and NASA management, failed the test to wisely spend taxpayer funds. The result is a Space Station now on orbit, six years later than originally projected, operating in a impaired and incomplete condition, and faced with a very uncertain future.

To provide a complete history of the evolution of the Space Station Program would result in a book of enormous length. I have been selective in the details and the people I chose to write about based on my judgment of which events would be most informative for the reader in order to understand the programmatic climate in which NASA managers operated. You will find extended discussions of budget issues, hopefully not too extensive. In the last analysis funding, or the lack thereof, shaped the program as much or more than any other events. Following the money for the Space Station is a particularly difficult task. Funding for various pieces of the program moved back and forth between different NASA line items. Some years Congress "locked up" or "fenced off" funds. When NASA was required to began full cost accounting in FY 2004, that further complicated the ability to compare and discuss funding levels.

If the story comes across as too negative it is because from its initiation there was no period during which the program was not dogged by controversy. Some who have been intimately associated with the program will undoubtedly disagree with the approach taken and emphasis. Additional histories will undoubtedly be written about this program. There exists a rich collection of untapped source material available for those that might desire to write about other aspects of the program or cast it in a different light. And, of course, the International Space Station is an ongoing saga that will continue to add to the record, perhaps for the next ten years or more. Based on the history described in the following pages you can decide: Will the end-product of the effort be worth the cost and provide real return on the huge investment, or will the new star in the evening sky end up as a slowly dying ember?

In order for the reader to track the program through time, each chapter heading includes the chronological dates covered. There are small overlaps in time from chapter to chapter but the chronology is generally continuous. My apologies to the reader for the use of so many acronyms. Attempting to describe a major NASA program is always difficult because of the many elements involved, all identified by acronyms in order to save time when discussed. One acronym list in my files for the combined Space Station – Earth Observation System programs consists of 45 pages with 70 entries on each page. Because it has become second nature to resort to their use, is it any wonder that NASA has a difficult time explaining its programs during presentations to the Congress and the public? This can be very annoying to the uninitiated and, I must confess, while at NASA I also fell victim to the same affliction sprinkling my daily speech with acronyms. I was unable to arrive at any other mechanism with which to discuss the Space Station to avoid repetitive, very long, names and other references. I hope it will not detract from the story and with early usage you will not have to refer, too often, to the attached acronym list.

---

2. P.E. Glaser, O.E. Maynard, J. Mockoviak, Jr., & E.L. Ralph, Feasibility Study of a Satellite Solar Power System, NASA Lewis Research Center, Cleveland, Ohio, CR-2357, 1974.

## — Acknowledgments —

In writing the history of the Space Station, I was able to use the extensive files I maintained from 1982 to 2005, seven of those years (1987-94) included material collected as a member of the Space Station Advisory Committee. However, no matter how complete one might think his records are, it was necessary to contact many former colleagues and NASA officials to fill out details and obtain first person accounts of what happened during the program's development. All with whom I spoke graciously shared their remembrances and in several cases material from their personal files. Writing a history is always easier if those who lived it can be questioned in order to clarify and supply details.

The first source for information and background I contacted was Phil Culbertson, a former boss and at one time Space Station Associate Administrator and NASA General manager. Phil had maintained a small collection of material from his Space Station days and allowed me to review his files and borrow some important documents as well as supplying personal accounts of what happened on his watch.

Before I put fingers to keyboard, I interviewed John Hodge and took copious notes of his experience as the director of the Space Station Task Force and as acting Space Station Associate Administrator. John graciously reviewed the progress and the problems he encountered while he was in charge and responded to several telephone requests for more information as writing progressed.

Bill Raney, who was the Space Station Advisory Committee secretary for seven years, also provided some of the program reports he preserved. Bill, in his role as secretary, was responsible for the fact that my files from 1987 to 1994 were so extensive as he provided the background documents we received and scheduled our many briefings and meetings.

Granville E. Paules, who held several management positions in the Space Station program from 1985-1993, had the most extensive collection of material that I was able to find covering the formative years of the program and made it available. Granville also reviewed a late draft of the manuscript and made a number of useful suggestions on how to improve the story.

Terry Finn was able to furnish behind the scenes background on the events leading up to the Space Station briefing for President Reagan. Terry orchestrated the briefing material following reviews with other senior NASA officials as well as the briefings given during the visits to foreign heads of state in 1984 when Administrator Beggs invited their participation.

Marcia Smith, Specialist in Science and Technology Policy, Congressional Research Service, Library of Congress, provided useful material on the Space Station budget as it evolved over the years and responded to requests for additional background. Accurate budget numbers are among the most difficult data to correlate for large government programs. Malcom Peterson, former NASA Comptroller, also was most helpful in clarifying NASA budgets.

In addition, I interviewed a number of former Space Station managers or individuals closely associated with the program, either in person or by telephone, including: Jim Beggs, Bryant Cramer, Dave Criswell, Mike Duke, John Dunning, Peggy Finarelli, Ed Frankel, Danny Herman, Sam Keller, Richard Kline, Ron Larson, Frank Martin, James Miller, Bob Moorehead, Tom Moser, Jim Odum, John Sheahan, Andy Stofan, Bill Stoney, John Talone, Ron Thomas, and Tom Young, all of whom graciously responded to questions.

The NASA History Office was always helpful in assisting while reviewing its Space Station files. In particular, Colin Fries quickly found requested documents, photos and drawings. Local colleagues, Martha Dysart and Roger Van Ghent, were especially helpful in converting my files to the format requested by the publisher.

Finally, my thanks to Rob Godwin and CG Publishing for accepting the manuscript and placing it on a fast track to publication. Hopefully, their confidence will be rewarded.

## — Abbreviations and Acronyms —

| | |
|---|---|
| AA | Associate Administrator |
| ACRV | Assured Crew Return Vehicle |
| AIAA | American Institute of Aeronautics and Astronautics |
| APM/COF | ESA attached pressurized module or *Columbus* Orbital Facility |
| A&R | Automation and Robotics |
| ARC | Ames Research Center |
| ARP | Automation and Robotics Panel |
| Ariane | ESA launcher |
| ASAP | Aerospace Safety Advisory Panel |
| ASI | Italian Space Agency |
| ASRM | Advanced Solid Rocket Motor |
| ASTP | Apollo-Soyuz Test Project |
| ATAC | Advanced Technology Advisory Committee |
| ATV | Automated Transfer Vehicle – ESA furnished vehicle |
| AXAF | Advanced X-ray Astronomy Facility |
| CAIB | *Columbia* Accident Investigation Board |
| CAV | Cost Assessment and Validation Task Force |
| CBO | Congressional Budget Office |
| CCACS | Center for Commercial Application of Combustion in Space |
| CDG | Concept Development Group |
| CDR | Critical Design Review |
| CERN | Centre European por Recherches Nucleaires |
| CERV/CRV | Crew Emergency Return Vehicle / Crew Return vehicle |
| CETF | Critical Evaluation Task Force |
| CEV | Crew Exploration Vehicle |
| CMG | Control Moment Gyro |
| CNES | Centre National d'Estudes Spatiales (French space agency) |
| CODEL | Congressional Delegation |
| COF | ESA's Columbus Orbiting Facility |
| CofF | Construction of Facilities |
| COTS | Commercial Off-The-Shelf |
| CPI | Continuous Product Improvement |
| CRS | Congressional Research Service |
| CSA | Canadian Space Agency |
| CSM | Apollo command and service module |
| CWG | Concept Working Group |
| DCR | Design Certification Review |
| DDT&E | Design Develop Test and Evaluation |
| DMS | Data Management System |
| DOE | Department of Energy |
| EADS | European Aeronautic, Defense and Space Company |
| ECLSS | Environmental Control and Life Support System |
| EDO | Extended Duration Orbiter |
| EIS | Environmental Impact Statement |
| ELV | Expendable Launch Vehicle |
| EMTC | Extended Man-Tended Configuration |
| EMU | Extravehicular Mobility Unit |
| EOS | Earth Observation System |
| ERDA | Energy Research and Development Agency |
| ESA | European Space Agency |

| | |
|---|---|
| ESS | Environmental Support System |
| ETCO | External Tanks Corporation |
| EVA | Extra Vehicular Activity |
| EXPRESS | Expedited Processing of Experiments to Space Station |
| FEL | First Element Launch |
| FGB | Russian functional supply / energy block vehicle |
| FRR | Flight Readiness Review |
| FTS | Flight Telerobotic Servicer |
| GAO | Government Accounting Office later named Government Accountability Office |
| GFE | Government Furnished Equipment |
| GMSR | General Management Status Review |
| GNP | Gross National Product |
| GSFC | Goddard Space Flight Center |
| HAB | Habitation Module |
| HEI | Human Exploration Initiative |
| HQ | NASA Headquarters |
| HUD | Department of Housing and Urban Development |
| ICM | Naval Research Laboratory interim control module – possible SM substitute |
| IG | Inspector General |
| IGA | Intergovernmental Agreement |
| IOC | Initial Operational Capability / Configuration |
| IPT | Integrated Product Team |
| IRR | Interface Requirements Review |
| ISF | Industrial Space Facility |
| ISS-ISSA | International Space Station – International Space Station Alpha |
| ITAR | International Traffic in Arms Regulations |
| JEM | Japanese Experiment Module |
| JPL | Jet Propulsion Laboratory |
| JSC | Lyndon B. Johnson Space Center |
| kbps | 1000 bits/second |
| KSC | John F. Kennedy Space Center |
| LAB | Laboratory |
| LaRC | Langley Research Center |
| LDEF | Long Duration Exposure Facility |
| LDO | Long Duration Orbiter |
| LEO | Low Earth Orbit |
| LeRC | Lewis Research Center |
| MDAC | McDonnell Douglas Astronautics Company |
| MEIT | Multiple Element Integrated Testing |
| *Mir* | Soviet space station |
| MMPF | Microgravity and Materials Processing Facility |
| MMU | Manned Maneuvering Unit |
| MOL | Manned Orbiting Laboratory |
| MOU | Memorandum of Understanding |
| MPLM | Italian / Alenia multipurpose logistics module named Leonardo |
| MSC | Mobile Servicing Centre (proposed by Canada) |
| MSA | Microgravity Science and Applications |
| MSFC | George C. Marshall Space Flight Center |
| MSS | Mobile Servicing System, used interchangeably with MSC above |
| MTA | Man-Tended Approach |
| MTBF | Mean Time Between Failure |

| | |
|---|---|
| MTC | Man-Tended Configuration |
| MTFF | Man-Tended Free-Flyer |
| NAC | NASA Advisory Council |
| NACA | National Advisory Council for Aeronautics |
| NASDA | National Space Development Agency (Japan) |
| NEO | Near Earth Object |
| NGO | Non-governmental Organization |
| NLS | New Launch System |
| NOAA | National Oceanic and Atmospheric Administration |
| NPOP-1 | NASA Polar Orbiting Platform |
| NRC | National Research Council of the National Academy of Sciences |
| NRL | Naval Research Laboratory |
| NSC | National Space Council |
| NSDD | National Security Decision Directive |
| NSF | National Science Foundation |
| NSTL | National Space Technology Laboratory |
| NSTS | National Space Transportation System – the complete shuttle stack |
| OAST/OAET | Office of Aeronautics and Space Technology – name changed in 1990 to Office of Aeronautics, Exploration and Technology |
| OCP | Office of Commercial Programs |
| OMB | Office of Management and Budget |
| OMU | Orbital Maneuvering Unit |
| OMV | Orbital Maneuvering Vehicle |
| ORU | Orbital Replacement Unit |
| OSS | Office of Space Station |
| OSSA | Office of Space Science and Applications |
| OSP | Orbital Space Plane |
| OSTP | Office of Science and Technology Policy |
| OSV | Orbital Servicing Vehicle (Japan) |
| OTA | Office of Technology Assessment |
| OTV | Orbital Transfer Vehicle |
| OWG | Operations Working Group |
| PDR | Preliminary Design Review |
| PHC | Permanent Human Capability |
| PI | Principal Investigator |
| PLM | Pressurized Logistics Module |
| PMA | Pressurized Mating Adapter |
| PMC | Permanently Manned Capability / configuration |
| POP | Program Operating Plan |
| PRD | Program Requirements Document |
| PRR | Preliminary Requirements Review |
| PSIA | Pounds per square inch absolute |
| PV | Photovoltaic |
| RCS | Reaction Control System |
| R&D | Research and Development |
| RD&D | Research Development & Demonstration |
| RFP | Request For Proposal |
| RID | Review Item Discrepancy / disposition |
| RSA/RAKA | Russian Space Agency / Russian Aviation and Space Agency |
| RMS | Canadian remote manipulator system, also SSRMS and Canadarm |
| RUR | Reference Update Review |

| | |
|---|---|
| SAAC | Space Applications Advisory Committee |
| SBIR | Small Business Innovation Research program |
| SD | Solar Dynamic |
| SDR | System Design Review |
| SDV | Shuttle Derived Vehicle |
| SEB | Source Evaluation Board |
| SE&I | System Engineering and Integration |
| SEI | Space Exploration Initiative |
| SES | Senior Executive Service |
| SET&V | System engineering, test & verification |
| SIG (SPACE) | Senior Interagency Group for Space |
| SII | Space Industries Inc. |
| SM | Russian service module |
| SOW | Statement of Work |
| SPP | Science Power Platform |
| SPS | Solar Power Satellite |
| SRB | Solid Rocket Booster |
| SRM&QA | Safety Reliability Maintainability and Quality Assurance |
| SRR | System Requirements Review |
| SSAC | Space Station Advisory Committee |
| SSB | National Academy of Sciences Space Science Board / Space Studies Board |
| SSCB | Space Station Control Board |
| SSE | Software Support Environment |
| SSFPO | Space Station Freedom Program Office, also SSF and SSFP |
| SSME | Space Shuttle Main Engine |
| SSOTF | Space Station Operations Task Force |
| SSP | Space Station Program |
| SSPO | Space Station Program Office |
| SSSAAS | Space Station Science and Applications Advisory Subcommittee |
| STS | Space Transportation System – space shuttle |
| TA | Teleoperator Assembly |
| TBD | To Be Determined |
| TCS | Thermal Control System |
| TDRS | Tracking and Data Relay Satellite |
| TFSUSS | Task Force on Scientific Uses of Space Station |
| TIP | Technical Integration Panel |
| TMIS | Technical and Management Information System |
| TMS | Teleoperator Maneuvering System |
| TQM | Total Quality Management |
| VA | Veterans Administration |
| VAB | Vehicle Assembly Building – called Vertical Assembly Building during Apollo |
| WBS | Work Breakdown structure |
| WETF | Weightless Environment Test Facility – also called Neutral Buoyancy Facility |
| WP | Work Package |

## — Introduction —

## ". . . THE NEXT LOGICAL STEP"

### (1982 - 1984)

When President Ronald Reagan announced in his State of the Union address in 1984 that he was "directing NASA to develop a permanently manned space station and to do it within a decade," he closely echoed the words of President John F. Kennedy, some twenty years earlier. Kennedy had promised we would land men on the Moon and return them safely "before this decade is out." The two announcements had more than a similar ring. Kennedy's commitment to a manned lunar landing included only the barest of details of how the landing would be accomplished. The space station that Reagan proposed to develop was little more than a concept. And perhaps most importantly, what use would be made of such a space station, if it were built, was still being hotly debated. NASA and its contractors had been studying requirements and different options during the two previous years, however, the studies had not progressed beyond what might be called preliminary designs. Understanding how large it would be, how it would be equipped, crew size, and many other critical parameters were still at an early stage of study. With such a meager understanding, the biggest question of all – how much would it cost? – depended on who you talked to. By early 1984, the cost number that was being used was eight billion dollars (FY 1984 dollars) for the "initial space station." What was or wasn't included in this number would generate considerable controversy in the years ahead as NASA attempted to defend the number. Shuttle launch costs during assembly, for example, were definitely not in the $8 billion.

The story of how Reagan came to his decision to approve a Space Station program has been reported several times but deserves another brief retelling to set the stage for what will follow. After many false starts in the decade leading up to Reagan's decision, NASA had finally gotten serious about studying a space station program. James M. Beggs, Reagan's appointee as NASA Administrator, established a Space Station Task Force in May 1982. He chose John D. Hodge to lead the effort. Hodge would report to Philip C. Culbertson, NASA Associate Deputy Administrator, who had joined NASA as a senior manager during the early days of the Apollo Program. The creation of the Task Force had the effect of formalizing and increasing in scope the work Hodge had begun with several cohorts a few months earlier. Hodge, also a NASA veteran with Apollo management experience, decided that the Task Force should abide by a few simple rules as it went forward with its work. Chief among them, and contrary to the usual NASA approach for new programs, was that the initial efforts would not attempt to provide an early design. He reasoned that NASA should first define the "missions" for a space station and that the configuration(s) would follow naturally from these needs.[1] As it turned out, this was a good working approach as the design would undergo many metamorphoses over the next few years as NASA attempted to respond to powerful, conflicting interests and budget constraints.

Although Hodge's Space Station Task Force was a small group of engineers and scientists, he was able to bring many other NASA staffers into the early planning to participate in working groups and a Technology Steering Committee. Members and staff for the working groups and the Steering Committee came from NASA Headquarters and the NASA centers. The names of these working groups are too numerous to mention here but included such titles as Program Definition, Data Management, Environmental Control and Life Support, and Structures and Mechanisms.

---

1. The arguments, pro and con, for this decision are discussed in detail in The Space Station Decision, Incremental Politics and Technological Choice, Howard E. McCurdy, The Johns Hopkins University Press, 1990, 286 pps. In an interview with the author in October 2002, John Hodge stated that during the first year he did not allow anyone on the Task Force to even draw a picture of a space station.

## Space Station Task Force Initial Studies

At the end of 1982 Hodge and his small team published an initial set of program description documents.[2] Book 1 summarized the studies that had been done to date based in part on mid-term contractor briefings. Eight, $787,500 contracts[3] had been awarded in August 1982 to analyze the science, applications, technology development, national security and space missions that a space station program could support. Book 1, not unexpectedly, contained many TBDs (items to be determined). During 1983, studies continued at both contractor facilities and NASA Centers. From October to December Hodge's office compiled these efforts in a new series of documents and appendices and distributed a Baseline Issue of seven documents.[4] Parallel studies were also undertaken at this time by potential international partners. The European Space Agency, Japan, Canada, Germany, France, and Italy conducted mission requirement studies for elements of interest to each of them with data exchanged at intervals with their U.S. counterparts.

Despite Hodge's determination to concentrate on "the missions," and avoid drawing pretty pictures of what a space station would look like, the trade press and contractors could not resist hiring artists. During the presentations at the AIAA / NASA Symposium on the Space Station in July 1983,[5] the screens were filled with many different versions of how a space station would look. Ford Aerospace showed a "Candidate Space Station Complex" made up of nine interconnected modules and large solar panel wings. McDonnell Douglas showed three different drawings of how a space station could evolve through time that were not far off from the eventual selected design. On a wrap-up panel at the same symposium moderated by Hodge and Robert F. Freitag, George E. Mueller, the former NASA Associate Administrator for Manned Space Flight made, perhaps, the most prescient comments of all the panel participants when he said, "there are several challenges. One is maintaining the support of the public. In my definition, the public support can be spelled OMB and, to some extent, Congress . . . The second challenge is the cost of logistics . . . The third thing is that . . . if the space station is to be useful at all, there must be more than one . . . I would guess the first economically attractive space station will be . . . really for establishing the basic parameters on how to survive over long periods of time in interplanetary flight because some day we're going to Mars!"

## Into the Lion's Den – The President Decides

Thus, with a background of two years of studies, as final discussions were being held on the Administration's FY 1985 budget submission, Beggs decided to make his pitch to the President. He had enlisted some supporters on the President's staff but was also well aware of strong opposition led by David Stockman, Director of the Office of Management and Budget (OMB). A meeting of the Cabinet Council on Commerce and Trade was scheduled for December 1, 1983, to discuss the Space Station. Accompanied by Deputy Administrator Hans Mark, Culbertson, Hodge and Task Force member Terence T. "Terry" Finn, Beggs made his presentation to the President and the assembled supporters and naysayers.

---

2. Space Station Program Description Document, Book 1, Introduction and Summary, 1st Edition – December 13, 1982, SSPDD – Book 1. In the Introduction, Hodge writes that the purpose of this initial document is "to stimulate debate on the critical issues so that the decisions made over the next several years will rest on a firm foundation of knowledge."
3. The eight contractors were Boeing, General Dynamics, Grumman, Lockheed, McDonnell Douglas, Martin Marietta, Rockwell, and TRW. The funds for the contracts were transferred to Hodge's office for management from the Associate Administrator for the Office of Space Flight, under the direction of Major General James Abrahamson. Mid-term briefings were held in November 1982, with the final reports issued in April 1983.
4. Space Station Program Description Document, Books 1-7, Prepared by the Space Station Task Force, December 1983, Final Edition.
5. Proceedings of the AIAA / NASA Symposium on the Space Station, Arlington, Virginia, July 18-20, 1983, Edited by Mirellie Gerard and Pamela W. Edwards, AIAA, October 15, 1983, AIAA, 1633 Broadway, NY, NY 10019.

Finn remembers working over the Thanksgiving holiday to develop the first draft of the briefing and then reviewing the presentation several times with Beggs, Hodge and other senior management.[6] The briefing was purposefully short (eleven vugraphs). In 1984, an overhead projector and vugraphs were still the visual presentation technology of the day. Hodge changed the vugraphs and Beggs provided the accompanying commentary. The presentation consisted mostly of pictures of NASA's successful programs and artist's renderings of a space station concept. The major points made were that the Space Station would demonstrate the country's continuing space leadership and that a decision to go forward, similar to Kennedy's Apollo announcement in 1961, required a presidential-level determination. What seems surprising is that none of the vugraphs mention international involvement or partnership as it was well known that Beggs had begun discussions with potential international partners some two years earlier. This omission perhaps reflected the continuing concern in some circles, especially in DOD, about technology transfer. In any case, it was not discussed.

Hans Mark had brought along an impressive large-scale model of a Space Station that he had the Langley Research Center (LaRC) fabricate especially for the briefing. Mark used it to describe to those entering the room how a Space Station might look and to give the impression that planning was well along. Although there was not unanimous agreement by those present that NASA should start such an ambitious program, the briefing convinced the President that the program should proceed leading to his State of the Union announcement one month later.

After receiving Reagan's blessing, the next open question, as important as what the design would be and how much it would cost, was how and where in NASA would the program be managed? This was not a trivial question as all of NASA was trembling under the fear of major budget cuts and potential downsizing as the Reagan Administration struggled with the larger issue of what should be NASA's future role. Important voices in and out of the Administration, led by The Heritage Foundation, were advocating that many of the nation's civil space programs should be privatized in keeping with the Administration's overall philosophy that government's functions should be reduced in many areas. Without a major new start that would justify utilizing the shuttle, NASA's manned space flight future looked bleak. The National Academy of Sciences recommendations in 1980 regarding a Solar Powered Satellite program that would use the shuttle had killed or, at the very least, put the program on the back shelf for the foreseeable future. With all of these uncertainties, once the Space Station was endorsed by the President the race was on within the agency to claim major pieces of the action. Space Station planning was still under the direction of NASA Headquarters; however, the management structure was still to be decided.

## THE PLAN TO PROCEED

On January 20, 1984, a few days in advance of the President's State of the Union address, a Space Station Program Briefing[7] for NASA staff and support contractors was held at the Johnson Space Center (JSC). In his opening remarks, Hodge reviewed the planning that had taken place over the past two years. He described how the space station would support the vision of a national space strategy outlined by the President during an earlier speech in October while commemorating NASA's 25th anniversary. In that speech, Reagan had endorsed the slogan "the next logical step" but had not applied the phrase specifically to the space station.

Hodge outlined eight space station functions that had evolved from the studies: (1) An on-orbit laboratory for both science and technology; (2) A permanent observatory(s); (3) A transportation node; (4) A servicing facility for free-flyers and platforms; (5) A communications and data processing node; (6) A manufacturing facility; (7) An assembly facility; and, (8) A storage depot (Figure 1). He went on to call for maximum agency-wide participation and predicted that the station would be up and operating by 1991. In addition to the above generic functions, Hodge listed a number of other benefits to developing a space station including commercial activities that would result in technology transfer to the private sector. A DOD member was also included in his Task Force to explore possible DOD

---

6. Based on an interview with Terry Finn in August 2003.
7. Space Station Program Briefing at the Johnson Space Center, January 20, 1984. A compilation of vugraph charts used during the full-day briefing.

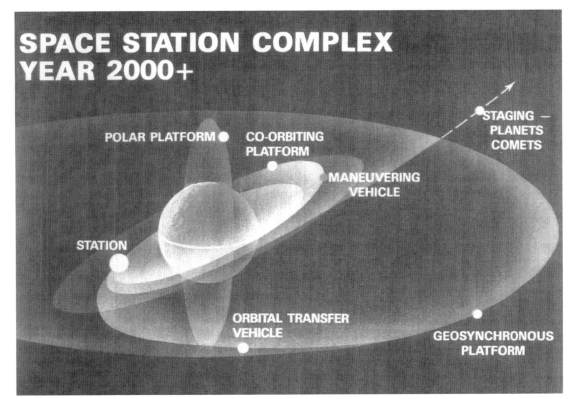

Figure 1. Early simplified schematic of Space Station Complex, Year 2000+. (NASA graphic)

experiments that might be conducted on the Space Station. A DOD presence on the Space Station would lead to many questions and controversies in the years ahead and became of special concern to the future international partners.

Luther Powell, a transferee from the Marshall Space Flight Center (MSFC) to NASA Headquarters who chaired the Task Force Concept Development Group, then described six basic "geometries," each with an accompanying artist sketch. They were named: "Raft, Spinning Array, Power Tower, CDG 1, Streamline, and Triangular" (Figure 2). (Notice that Hodge's early edict still held, they were not called designs, but "geometries.") After some two years of study by NASA and its contractors, each geometry was stated to have advantages and disadvantages but the level of understanding of each varied greatly. Some "common threads" were identified for each geometry. For example, all would use solar arrays for power, and would be supplemented with unmanned platforms operating in the same orbital inclination as the main station or in polar orbit. All of the geometries would support the basic functions described earlier by Hodge.

In addition to the work done at the NASA centers, many private companies participated in these early studies. The list included essentially all the major aerospace companies in existence at that time before consolidation in the late 80s and 90s reduced their numbers to a few. Each company was attempting to keep its foot in the door for what they hoped would be future larger roles and the big plums, building hardware and large engineering support contracts. Some companies, in order to stay abreast of what was developing, did not receive any NASA funding for their efforts and conducted their own internal studies.

NASA funded work at Martin Marietta, Boeing, and Battelle Northwest Institute to study operations and logistics. Boeing, Lockheed, and Rockwell studied the design of a eight-man habitation module. Lockheed and McDonnell Douglas studied maintainability. Chance Vought, General Dynamics, Grumman, Martin Marietta, and Lockheed studied satellite servicing of free-flyers, one of the fundamental functions foreseen for a space station. And McDonnell Douglas, Lockheed, Boeing, and Rockwell, studied the concept of common modules. Reaching outside the aerospace industry, Booz-Allen & Hamilton and Coopers & Lybrand were brought in to develop a list of potential commercial users and provide preliminary requirements for these users.

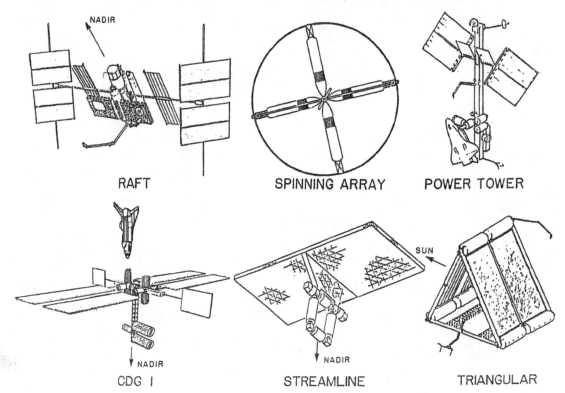

Figure 2. Basic Space Station Geometries. A chart included in the Concept Development Group's briefing given by Luther Powell at the Space Station Program Briefing, Johnson Space Center, January 20, 1984.

Designing the space station around common modules and standard hardware in order to control costs were major objectives for all participants in the early studies. Other engineering guidelines included that the space station would be evolutionary but technology transparent, maintainable, designed for continuous habitation, and dependent on the space shuttle during construction and servicing. In 1983 and 1984, this last guideline was the only option available. As a result, all studies considered that the space station would be placed in a 28.5 degree orbit. Two considerations drove this orbital inclination: the majority of user missions studied could be accommodated, and the shuttle could deliver the maximum payload at this inclination. Any missions requiring higher inclinations would be carried out by the unmanned polar orbiter satellite or other free-flyers. However, some users interested in making observations and measurements of the Earth's surface and atmosphere would have preferred a higher inclination.

Selecting solar cells (planar array photovoltaic cells) as the preferred power source for the initial studies was made after evaluating many potential choices including concentrator arrays, Brayton cycle and nuclear power systems. Solar cell technology was well understood and experience with Skylab was positive in spite of a few deployment problems. Solar systems could be evolutionary in the sense that more and more arrays of the same or similar design could be added as needed to supply what was sure to be a continuously growing demand for more power. An early surprise resulting from the mission studies was the amount of power that housekeeping and potential users were projected to need. Initial requirements indicated a need for 55 kW, potentially growing to 150 kW, the equivalent power usage of approximately 50 average homes. Power storage options studied included batteries, regenerative fuel cells, and inertial systems.

Thermal control was one of the more interesting and challenging design problems. Heat rejection from the many onboard heat sources, such as electric power generation, would require a robust but flexible system. It would have to be modular and able to increase its capacity as the station grew in size and complexity. Orientation of the station in free space to satisfy some potential users would subject the external structure at times to solar baking, especially if it was required that a given orientation had to be maintained for long periods. To satisfy these and other requirements, body-mounted radiators were selected as they could also provide meteoroid and debris impact protection for the habitation modules, a concern that would grow in future years as space debris, left over from hundreds of launches, was carefully inventoried. The thermal control system could also be designed to provide hot water for washing and food preparation.

Propulsion systems for maintaining orbital altitude and orientation was another critical study area. It was desired to design the systems such that there would be commonality between the station and the proposed free-flying platforms that would need to be refueled during their lifetimes. Safety during assembly and maintenance, and possible contamination of external surfaces from thruster gas, especially surfaces associated with potential experiments, were major concerns. Monopropellants, $CO_2$, oxygen / hydrogen, and cold gas systems were evaluated.

Tracking and communication would present a unique set of requirements. Early studies indicated that the station should serve as a communication hub for data generated onboard as well as that produced from the free-flyers. The number of communication links might be very large and security of some of the links might be important. Colocation of transmitters and receivers was considered to be desirable for certain functions. A tethered, or co-orbiting satellite antenna, was studied as a possible way to improve antenna coverage. Although the goal was to design the space station to operate as autonomously as possible, uplink and downlink communication and data traffic would dwarf the needs of any previous space program.

Although the Shuttle was to be the primary logistics vehicle, additional means of supplying and servicing the space station and free-flying satellites were considered necessary. An Orbital Maneuvering Vehicle (OMV) and Orbital Transfer Vehicle (OTV) were two concepts that were studied. As a satellite servicing hub, free-flyers launched from the space station would be placed in geosynchronous orbits or as co-orbiting platforms carried by either the OMV or OTV. These vehicles would also be used to service the satellites and extend their lifetimes. The ability of the OMV / OTV to service various space station elements was critical to the programs success. Both the OMV and the OTV had been identified very early by NASA as candidates for technology development with far reaching potential and application. To address the many new technology problems these systems would require, a large task team from the NASA centers and industry was assembled, led by Hodge's office.

A Remote Manipulator System (RMS), attached to the space station external structure, was studied by the Canadians. This system once in place, operating either independently or under astronaut control, would assist in constructing the station, maintenance and repair functions. Preliminary designs called for the RMS to have the ability to be relocated at various places on the external structure as needed. In addition to the RMS, a Teleoperator Maneuvering System (TMS) was conceptualized to complement the many systems that would be available to service the space station and co-orbiting platforms. Also, Manned Maneuvering Units (MMU's), small, individual astronaut donned systems that would allow the astronauts to independently and safely conduct tasks outside the space station, were already under study for use with the shuttle and would have application to the space station.

Because of a concern for an event that would occur far in the future, and remembering the worldwide anxiety when Skylab was deorbited with parts impacting in Australia, studies were conducted for end-of-life operations. Preliminary design considerations were given to how this huge mass of space hardware could be decommissioned. Two scenarios were studied, boost it to a higher orbit and maintain it there, or disassemble the structure and deorbit in non-hazardous pieces. To put the space station, as a single entity, into a higher orbit that would be safe for many hundreds of years, would require some type of propulsion system that did not yet exist. Being able to eventually break the space station into acceptable size pieces would not be a trivial design consideration. If disassembled at module interfaces, individual modules could weigh many tons and include many types of material that would survive reentry just as did parts of Skylab.

Operations studies were conducted in parallel with the early requirements studies. A large Operations Working Group (OWG), chaired by Frank Bryan of KSC, and consisting of members from nine centers and NASA Headquarters, had been working since July 1982. In November 1982, they published their first report containing an operations philosophy and guidelines identifying critical issues. Over the next ten months, studies were made of the many identified design goals and a set of "white papers" was published in September 1983 for review by the Space Station Task Force. After review, a consolidated Operations Requirements and Baseline was issued in December and a "Final Edition," called the "Yellow Books" after the color of their covers, was issued in March 1984.[8] Designing the Space Station to be "user friendly," regardless of the user, became a design mantra.

In order to bound their studies, the OWG assumed the space station would have a minimum of generic elements: 1) A manned station in an approximate 28.5 degree orbit consisting of various modules serviced by the shuttle and the OMV, 2) One or more free-flyers including one in near-polar orbit, 3) An ability to service the station as needed, and 4) Ground systems to support the flight elements. In keeping with Hodge's dictum to avoid specific designs this early in the study phase, it included a disclaimer that "The assumed generic configuration does not constitute NASA or Space Station Task Force endorsement of a final baseline configuration."[9] NASA was still playing the game enunciated earlier by Beggs to disarm critics, especially those concerned with potential cost, that you could buy a space station "by the yard."

### SPACE STATION DESIGN DRIVERS

The objective of the Operations Working Group was to identify the operational "drivers," those activities that would affect design requirements and operational protocols during the station's lifetime. The drivers identified were safety, maintainability, automation, operations philosophy, prelaunch processing, and customer interfaces.

Automation was a particularly complex issue. How much automation? Should there be shared onboard and ground participation for some tasks? How should the displays and controls be designed? How should the embedded software be designed, and should there be provisions for onboard software modification? This latter question would be a continuing problem throughout station design as it was self evident that computer and software development would rapidly improve. No one wanted to launch a space station that would be one or two generations behind the times in these technologies and unable to adapt to new advances.

To address all the safety issues that would be involved during assembly and operation, the OWG conducted a "First Level" study. The ground rule established, as for all activities involving the astronauts, was that: "Safety for the flight crew and operation of the systems will be a prime consideration in the design of the Space Station System."[10] Among other requirements that grew out of this ground rule was the need for various levels of system and subsystem redundancy, and caution and warning systems to alert the crew to failures. In the event of a major accident, three different capabilities were studied: (1) escape to orbit, (2) escape to Earth, and (3) provision of a safe haven inside the space station. After study, escape to orbit or Earth were considered costly and unreliable and thus onboard safe haven became the approach of choice (Escape to Earth was later revisited and became the prime choice and safe haven became a secondary design criteria). The early safe haven decision then evolved into considering both a single dedicated safe haven destination or several distributed sites as the space station grew in size and complexity. An initial resolution of all of the "drivers" mentioned above needed to be agreed upon by all involved as NASA was preparing to issue

---

8. Space Station Program Description Documents – Baseline Issue: Book 2 – Mission Description; Book 3 – System Requirements and Characteristics; Book 4 – Technology Options and Advanced Development; Book 5 – System Definitions; Book 6 – Space Station Operations; Book 7 – Program Planning. Books and appendices were published during the period October 1983 through December 1983. Appendix C of Book 6 – Summary of Operations Studies (First Level White Papers) compiles the studies initiated in November 1982 for a conceptualized space station and covered Maintainability, Automation, Operations Philosophy, Customer Interfaces, Safety, and Prelaunch processing. The Yellow Books were issued in March 1984 with the titles Book 1 – Introduction and Summary; Book 2 – Mission Description; Book 3 – System Requirements and Characteristics; Book 4 – Advanced Development; Book 6 – System Operations; Book 7 – Program Plan. The original Book 5 – System Definitions, was not issued as a separate book, but sections were summarized or incorporated in the other books.
9. Ibid., Book 6 – Systems Operations, p. 2-1.
10. Ibid. Book 6, Systems Operations Document, Appendix C, Summary of Operations Studies (First Level White Papers), Baseline Issue, October 1983, p. C-V-1.

the next round of Request for Proposals (RFPs) inviting industry to participate in the Phase B definition and preliminary studies.

At this early date, customer operational requirements were spelled out in some detail emphasizing "simple, standard, stable requirements and interfaces for future customers."[11] Those in the user community who had begun to take a serious interest in utilizing a research facility that was still in an early concept phase were already on record asking for clarification as to how NASA would ensure "user friendliness" and control the cost of experiment integration.

Working under a NASA contract, the National Research Council – Space Applications Board of the National Academy of Sciences, published in 1984 the results of their study on the practical applications of a space station.[12] Five user-oriented panels had been established to examine how the Space Station could support study of: Earth's resources, Earth's environment, ocean operations, satellite communication, and materials science and engineering. A sixth panel examined design factors that could result in commonality requirements for users. The five user-oriented panels provided guarded support but generally concluded that a space station operating with a separate platform in near polar, sun synchronous orbit would have many industry applications. Among their conclusions: "a space station development program should make provision for technology advance or growth during the development period. Construction of a space station using off-the-shelf technology, although less costly initially, may not provide adequate capability to support requirements for onboard data processing, user needs, growth potential, or sufficient operational utility." NASA, of course, fully supported this approach but as budget restrictions became the dominant concern, off-the-shelf technology was looked upon as an attractive way to control costs.

Although industry support, as represented by the Space Applications Board, was favorable, many in the space sciences community were concerned as to whether their support for Space Station would result in a reduction in funding for traditional space science and applications programs. During the first round of hearings on NASA's FY 1985 budget, Thomas M. Donahue, Chairman of the Space Science Board of the National Academy of Sciences, stated, "the overriding issue in FY 1985 is the request for a start on the space station and the impact that initiative will have on space science." He went on to say, "the Board will cooperate in every way in its power to maximize the usefulness of the space station to science . . . But we are concerned lest the bad experience space science has had in coexisting with the space shuttle . . . be repeated with the space station."[13]

To alleviate this concern, the Administration's FY 1985 budget included an increase of 21% over the FY 1984 budget for space science, much larger than the increases of a few percent in previous years. In addition, Office of Space Science and Application (OSSA) Associate Administrator Bert I. Edelson established a Space Station Working Group to identify a combined set of science and application needs extending to the year 2000. The Working Group was chaired by Friedrich von Bun, on loan from GSFC. The report[14] issued by the Working Group stated that "The disciplines which will obtain the major benefits from the Space Station program during the early 1990's are Micro-Gravity Science and Applications and Life Sciences." However, the overall strategy would be that until the Space Station complex attained operational status, OSSA would fly individual spacecraft and shuttle payloads as originally planned. Thereafter, OSSA would examine the programs planned for new starts in the late 1980's and early 1990's to determine if they could make use of available Space Station support. In spite of these efforts to calm the fears of the science community, the squabble over the internal allocation of NASA funds (between science and manned space flight) would be used against the Space Station in the years ahead by those in the Congress opposed to the program.

Also in 1984, the NASA Advisory Council (NAC) was requested to establish a subgroup of its Space and Earth Science Advisory Committee to study the scientific utility of the Space Station. The

---

11. Ibid., Book 6, p. 10-14.
12. Practical Applications of a Space Station, Space Application Board, Commission on Engineering and Technical Systems, National Research Council, National Academy Press, Washington, 1984, 89 pps.
13. ASF News, Vol. II, No. 2, Spring 1984. A publication of the American Space Foundation.
14. Future Use and Needs of the Space Station Complex for Science and Applications, Draft, Office of Space Science and Applications, NASA Headquarters, June 22, 1984.

subgroup, with the title Task Force on Scientific Uses of Space Station (TFSUSS), held a summer study at Stanford University in August 1984 under the direction of its chairman, Peter M. Banks.[15] A number of important recommendations came out of the study in which over 150 scientists participated, covering all the major scientific disciplines. Some dealt with the use of the shuttle as a test bed for the improvement of scientific operations that would take place on the space station and continuing use of the shuttle and Spacelab for science after the Space Station became operational. Other interesting recommendations included establishing the Space Station as a national facility based upon experience gained from the operation of ground-based facilities, and the development of standardized modular platforms to meet the needs of polar and co-orbital research facilities.

In describing the summer study with the House Subcommittee on Science, Technology and Space in March 1985, and the implications for conducting science on the Space Station, Banks noted the dramatic change of opinion in the science community toward using the Space Station for research. He stated that as the summer study progressed, "it became clear that this new facility will be the focus of exciting, new research in space."[16] However, he cautioned that their were still concerns about various aspects of the program. The study recommended that when initial operating capability (IOC) was achieved, the total crew size should be at least eight, six dedicated to science and two for general space station operations. It was the opinion of the study members that the proposed co-orbiting platforms were too large and that smaller satellites would better meet the needs of science. The TFSUSS also recommended that NASA proceed with the concept of the Extended Duration Orbiter (a modified shuttle) so that the scientific productivity of the Spacelab could be increased in preparation for similar operations on the Space Station.

## INTERNATIONAL PARTNERS

Backing up a bit and turning to a new subject, some two years prior to President Reagan's endorsement of the Space Station, NASA Administrator Beggs and Associate Administrator for International Affairs Kenneth S. Pederson began courting other nations to participate in a potential U.S. led Space Station program. They believed that foreign participation in the program was essential to assure approval from the Administration. There was also Congressional interest at this time to include other nations in U.S. space activities, especially Japan, that recently had announced it was about to initiate an aggressive space program. As the avowed leader in all matters concerning space, the Administration and NASA, along with its Congressional allies, were determined to channel overseas efforts in ways that would guarantee continued U.S. dominance. The NASA Space Station Program History: 1981-1987,[17] in describing "Reasons Why" the need for a Space Station, included "Challenge Soviet lead in space stations," and, bottom line, "Assure free world leadership in space during the 1990's and beyond." Initial contacts with the European Space Agency (ESA), Canada, and Japan, elicited strong interest in forming partnerships with NASA in the lead as the program developed. Assuring that the U.S. would maintain world leadership in space would always remain a primary goal of the program.

In January 1982, potential international partners met with NASA at the Johnson Space Center to discuss how they might join the program and under what terms. Bob Freitag, at that time NASA Director of Advanced Programs, and Pedersen coordinated these efforts. The result of the meeting was to invite the international attendees to start planning how they would become partners in the program.[18] ESA member countries, Germany, France, and Italy, as well as Canada and Japan, began their own requirements studies. However, it wasn't until Reagan's formal announcement of the Space Station program that NASA was on firm ground to pursue such partnerships.

President Reagan's State of the Union address invited international participation and in March 1984, he sent Beggs to follow-up on NASA's preliminary contacts and attempt to obtain firm commitments of cooperation. In advance of the trip, the President sent a letter to the heads of state who would be

---

15. SESAC Task Force on Scientific Uses of Space Station, Space Station Summer Study Report, February 12, 1985.
16. Statement of Peter M. Banks, Chairman, NASA SESAC Task Force on Scientific Uses of Space Station before the Sub-Committee on Science, Technology and Space, March 8, 1985.
17. Space Station Program History: 1981-1987, Dr. Terence T. Finn, Deputy Director, Policy Division, Office of Space Station, February 12, 1988.
18. McCurdy, Chapter 11 "International Participation."

visited, giving his strong endorsement to the Space Station program and renewing his invitation to partner with the U.S. in the effort. Vice President Bush arranged to have one of the President's fleet of Air Force 707s (whichever aircraft carried the President would have the designation Air Force One) available for the trip.[19] John Hodge remembers this trip as one of the highlights of his NASA career. Being pampered on the equivalent of Air Force One is a life style that few experience but many covet. Hodge remembers that there were two full crews of pilots and personnel that provided around-the-clock service. Culbertson remembers that the one small bedroom onboard was commandeered by Mr. and Mrs. Beggs while the rest of the NASA staff had to "endure" the equivalent of deluxe, first class dining and sleeping accommodations in reclining seats.

Leaving from Andrews Air Force Base outside of Washington, Beggs and some twenty staffers flew to London, then on to Rome, back to Paris and over the top of the world, by way of Alaska, to Tokyo. The presentation given at each destination, almost two hours in length, had been coordinated by Terry Finn. Putting the final touches on this global, nation hopping, at the London Economic Summit in June 1984, the President reviewed the Space Station program with the attendees (Figure 3). A framework for collaboration was proposed and agreed upon by ESA, Canada, and Japan. It would take almost a year of further negotiations before all parties formally signed MOUs containing the terms of the partnerships.

Figure 3. President Reagan shows the Space Station to Prime Minister Thatcher and other attendees at the London Economic Summit, June 1984. NASA HQ S84-1663D.

As mentioned, leading up to the formal signings, Canada, France, Germany, Italy, and Japan had funded and conducted their own mission requirement studies. The ESA countries and Japan studied potential common modules and free-flying platforms. The French conducted a conceptual study for a small crew transporter, given the name Hermes, that could be launched on an Ariane-5, a rocket still on the drawing boards. The Canadians, capitalizing on their experience of developing the remote manipulator carried on the shuttle, began to study the design of other types of robotic manipulator systems for the station. Germany and Italy combined to study a self sufficient habitation module that they called *Columbus*. And, in addition to studying the *Columbus* module, Italy began a study to incorporate tether systems for the future station.

---

19. Based on a conversation with Beggs, March 7, 2005. Beggs believed that the letters sent to the heads of state who received him at each stop were instrumental in permitting him to easily accomplish his objective of convincing the countries visited to join the program.

## Moving On – Management Decisions

In keeping with standard NASA practices, the studies that the Space Station Task Force commenced in August 1982, leading up to the "go ahead" in 1984, constituted the Phase A preliminary definition that all NASA programs observed. To avoid raising the program's profile so high that it might lead to intense, premature, scrutiny in Congress, funding for these studies had been kept to a minimum, amounting to less than $22 million. Funding came from several different NASA offices.[20] Those who are familiar with the present Space Station program will recognize from the above discussion that by the end of FY 1984 essentially all elements of the program had been defined to some level of understanding. In some cases, more elements had been studied than currently comprise the program. To move on to the next phase, still only paper studies but providing more detail and the inclusion of both NASA centers and contractors, would require substantially larger sums. The effort could no longer remain below Congress' horizon.

It was clear that a program of the magnitude projected would require choices to be made on where the day-to-day management would reside and how the various NASA centers would participate. Management of the Space Station would not be the same as that of a typical NASA program such as sending a probe to a distant planet. Nor was Apollo as complex in terms of all the pieces that would have to work together to assure success. Culbertson, following lengthy debate within the Task Force and discussions with Beggs, decided to break the work into three "work packages." JSC, the Marshall Space Flight Center (MSFC), and the Goddard Space Flight Center (GSFC) were declared the winners.[21] JSC and MSFC would have the lead for the space station per se while GSFC would be responsible for the free-flying platforms.

Three weeks after The State of the Union address, Beggs made the decision to establish three management levels. The Level A Program Office, responsible for establishing program policy, budget, and schedule guidelines, and coordinating with external organizations, would be at NASA Headquarters. JSC would be the Level B Program Management Center responsible for daily management. Level C identified how the other NASA center project managers would be involved. Specific technology assignments were given to JSC and MSFC. JSC would have the lead for environmental control and life support systems, data management systems, and thermal management systems. MSFC would be the lead center for attitude control and stabilization systems, auxiliary propulsion systems, and space operations mechanisms. Initially, the electric power system was unassigned with several centers involved during the definition phase but, eventually, the Lewis Research Center (LeRC, now named Glenn Research Center) was given the lead and became the fourth "work package" center. At Level C, the centers would manage the contractors selected to develop the hardware, systems, and subsystems (Figure 4). However, some of the above assignments would be changed before the next round of RFPs were released.

In April 1984, JSC Center Director Gerald D. Griffin selected Neil B. Hutchinson to be Program Manager of the Space Station and John W. Aaron as his deputy. Although both were veterans of the Apollo Program, neither had previous management experience equivalent to that entailed in the upcoming Phase B studies that would include daily interactions with other NASA centers, contractors, and international partners. They would have to develop those skills on the run. Two parallel efforts would be underway. Phase B Definition and Preliminary Design for the "initial station" and a multifaceted Advanced Development and Technology Program.

This latter program, similar to the supporting research and technology program NASA conducted during Apollo, was established to provide technology design options that would be developed to a

---

20. Funding in FY 1984 was provided by Office of Space Flight ($13.0 million), Office of Aeronautics and Space Technology ($6.0 million), Office of Space Science and Applications ($2.0 million), and Office of Space Tracking and Data Systems ($0.9 million). Data from Culbertson testimony to Subcommittee on Science, Technology and Space, April 3, 1985. The total spent during all of Phase A has been difficult to compile. For example, a MSFC document claims that $8 million was spent on Space Station studies from 1979-1983, a period that overlaps Phase A but does not differentiate on how it was spent.
21. Personal communication during interviews in 2003.

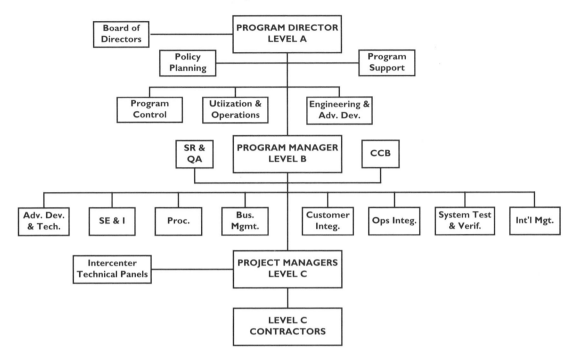

Figure 4. Program / Project Management Levels, from NASA TM-86652, Space Station Program Description Document, Book I – Introduction and Summary, Prepared By The Space Station Task Force, March 1984, Final Edition.

prototype level. Participating in this program would be the definition and preliminary design contractors, as well as others. Each would develop advanced technologies they believed necessary for the Space Station to meet its goals. NASA test beds, both on the ground and in space, would be made available to the contractors for test and evaluation and at the end of the definition phase selected technologies could then be incorporated as needed into the design phase. Some of the technologies not selected for the design phase would continue to be funded with the objective that they would be available downstream as alternatives to reduce program risk (cost and schedule) if the initial designs experienced problems.

During an interview for The Notre Dame Technical Review in October 1984, Hutchinson described the objectives of the Advanced Development and Technology program. In response to a question as to what technology would be the most difficult to develop, he stated, "I would put power generation at the top of the list. The cooling system will be another. If you are going to build something to last for twenty-five years you don't want it to have any moving parts, and thermal heat pipe technology, thermal busses as they are called, certainly lend themselves to that. I don't think that computing technology is going to be a significant problem, but I do believe that developing the technology to make the computing system we install in the station transparent to future technology, meaning that it will be upgradable without having to tear it apart, will be a tough job. Control of a very large structure in orbit is a significant engineering challenge. Developing and perfecting space construction methods is another one that will be a little tough."[22] As events unfolded, the technology challenges he chose to highlight turned out to be on the mark, some more stubborn to resolve than originally projected.

The decision to put Level B at JSC would be a constant irritant over the next two years as the other NASA centers strove to increase their level of participation. Centers with minimum involvement in the original work packages would attempt to justify a larger piece of the action. Similar problems occurred

---

22. An Interview with Neil B. Hutchinson, Manager of the Space Station Program, conducted by Robert Mahoney, Editor of the University of Notre Dame Technical Review, Winter 1984 Issue. The interview was conducted on October 2, 1984.

during the early days of the Apollo Program, the rivalry between JSC and MSFC being the most prominent. NASA centers do not take kindly to direction from another center designated as "lead" or, for that matter, direction from NASA Headquarters. The assignment of the systems engineering and integration (SE&I) functions to Level B rather than keeping them at an office in Washington, as was the case for Apollo, was another controversial decision that would be overturned. Eventually, major parts of JSC's responsibilities would be pulled back for a time to Washington.

## EARLY BUDGET CONTROVERSIES

With these decisions in place, the action shifted to the Administration and the Congress; how to agree on the funding required to proceed with the Phase B studies. This debate raged through the spring and summer of 1984. Beggs and the NASA Headquarters staff attempted to shape the outcome, both at the White House and in the halls of Congress. Eventually, with many compromises, Congress passed a FY 1985 Space Station appropriation of $150 million, much less than the $225 million NASA originally requested during the preliminary discussions with OMB which was compiling the Administration's overall FY 1985 budget request (The $150 million was augmented by $5.5 million with a supplemental bill approved on August 15, 1985).

In spite of these difficulties, these were heady days for the Space Station Task Force and all the NASA centers and contractors who were supporting the planning. Although the first budget was tightly constrained, the program was finally shown as a line item in NASA's budget. In Washington, this was always the first hurdle to overcome for a new program. And even though the budget was less than requested, senior NASA managers had experience on how to work around such constraints. The many members of Congress whose districts were involved in the program could always be depended on for support. Other Space station allies in Congress were numerous, strong, and united and also could be counted on for support in future budget debates.

The bitter wrangling between supporters and opponents that began with the FY 1985 budget would continue as each new fiscal year budget request was forwarded to the Congress. As a result, the program was never able to achieve the classic "bell curve" funding profile that is the hallmark of successful R&D programs. The flat yearly budgets the program was eventually forced to live with required NASA management to adopt a new way of doing business with its Centers and contractors. Program compromises, rather than good program management, were the only way the program would survive.[23]

In Book 6 – System Operations[24] of the Yellow Book series, the initial increment of the Space Station System was stated to include at least the following generic elements: a utility module, research and development labs, habitats, remote manipulating system(s), resupply modules, a multiple berthing adapter, shuttle docking and berthing capability, orbital maneuvering vehicle capability, accommodations for attached payloads, one or more orbital platforms including one in near-polar orbit, provisions for servicing payloads, and ground systems to support the flight elements. In March 1984, the list included the ability to support essentially all the "generic functions" described by the Space Station Task Force at the JSC briefing in January 1984. But the $8 billion dollar price tag that NASA had advertised for the "initial space station" did not include some of these elements, or represented only modest starts on critical components.

Perhaps the most surprising and troublesome number was the total projected to complete the definition phase, $614 million (Figure 5). In 1981, Donald P. Hearth, Langley Research Center Director, led a study team that examined why so many NASA programs had experienced cost overruns.[25] One of the conclusions reached by the team was that a major problem causing the overruns was that NASA did not devote enough resources to the programs during the definition phase and thus too many unforeseen problems arose leading to overruns. His team recommended that up to ten percent of the total program cost be expended during the definition phase. The straight math would have required a minimum of $800 million during the definition phase of an $8 billion program. The final program cost was so much

---

23. For a detailed discussion of the FY 1985 budget debate, see McCurdy, Chapter 23 Congress II, pps. 211-223.
24. TM-86652, Space Station Program Description Document, Book 6 – System Operations.
25. Issues in NASA Program and Project Management, NASA SP-6101 (04), edited by Francis T. Hoban, NASA, Spring 1981. Hearth Report recommendations contained in pps. 5-10.

higher than the $8 billion that the initial definition phase funding was far off the mark. The Hearth study also recommended that each program have a reserve of approximately ten percent of the total cost in the initial cost estimates.

Neither of these recommendations was followed. NASA was forced to skimp during the definition phase. Program cost growth in later years would confirm that NASA spent too little during the definition phase resulting in the inability to project program complexity and costs. At the start, Space Station books would carry a reserve that was too low to resolve problems when they arose. Perhaps the most interesting statement in the Hearth report was: "NASA, OMB, and the Congress should expect up to a 30 percent cost growth even if the project is well managed and there are no **major** (original emphasis) technical surprises." This last observation, based on extensive experience, was not what OMB and the Congress would buy into for the Space Station.

| MILLIONS OF 1984 DOLLARS | |
|---|---:|
| Logistics Module (2) | 820 |
| Multiple Berthing Adapter | 965 |
| Habitability Module | 1220 |
| Laboratory Module (2) | 1036 |
| Resources Module | 1855 |
| Platforms (2) | 730 |
| Teleoperator Maneuvering System | 190 |
| Servicing Capability | 230 |
| Operations Capability | 340 |
| Development | 7386 |
| Definition | 614 |
| Total | 8000 |

Figure 5. Space Station Program Cost Estimate, same source as Figure 2.

Left out of the cost estimation were such big ticket items as shuttle launch costs, operations, and payload test and integration. These three components were acknowledged to be very expensive. Shuttle support and modifications required for rendezvous and docking with Space Station elements, although an integral part of the program, were never included in the cost estimates. The Space Station complex would be the first NASA program where all the elements for a given mission would not be launched simultaneously. For example, when a Saturn V left the launch pad on an Apollo mission, all the pieces needed to complete the mission were on the vehicle and had been integrated and tested end-to-end as a complete system. The Space Station would be an assortment of parts delivered to KSC over a period of many years and eventually mated for the first time in Earth orbit. There would be no opportunity to carry out end-to-end testing on the ground. Interface specifications would have to be enforced and controlled at a level never before required. These costs, and many others, would be debated over the next fifteen years as the full scope of the program was defined. Opponents charged that the projected $8 billion was just a number invented by NASA to get its foot in the door, and to a large degree they were correct.

When the "Yellow Books" were published in March 1984, and with the assumption that the agency would receive the needed yearly appropriations, NASA committed to an overall schedule that would complete concept development in the first quarter of FY 1985. The next milestone, initial definition, would be completed by mid FY 1987 followed in turn by design, development, test and evaluation. First elements would be launched in CY 1991 and initial operational capability (IOC) would be achieved by the end of FY 1991 (Figure 6). This schedule discreetly showed two "authority to proceed" milestones at the end of the concept development phase and the initial definition phase. As most federal agencies and cabinet level departments are always proceeding based on the current year's appropriation, or funds carried over from the previous year, it would have been presumptuous on the part of NASA to have left these milestones out. If authority to proceed was given at the end of the definition phase in FY 1987, NASA believed it would have a green light to proceed toward placing a space station on orbit and manned in the early 1990's, with the potential to grow in capability in future years.

NASA published its Long Range Program Plan[26] in 1985. "Establish a permanent manned presence in space" was one of NASA's eight long-range goals that included developing a cost effective space transportation system, maintaining leadership in aeronautics research, and expanding knowledge of the

---

26. National Aeronautics and Space Administration, 1985 Long-Range Program Plan. Although NASA periodically updated its long-range plans, this document did not carry an SP number or specific date. The report chaired, I believe, by Stanley R. Sadin, Code RS, includes the interesting note: "This report is a working document. The more detailed report volumes and the full data base constitute an extensive set of documents available only to those who can demonstrate a need for them." The eleven part Summary in the author's files is 232 pages long.

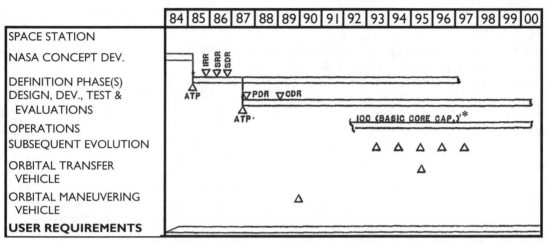

Figure 6. Space Station Overall Schedule, same source as Figure 2.

universe. The Plan unabashedly stated that all the goals were supported by the Administration's FY 1985 budget even though the Space Station budget request to Congress was only two-thirds of NASA's original request. Among other statements in the Plan was one that the first Space Station structure would be in space by 1992, a prediction that eventually proved to be overly optimistic. However, in 1985, all NASA and contractor briefings and brochures had captured the President's earlier words and would describe the Space Station as "The Next Logical Step." For manned space flight, it was the only game in town.

Budgets submitted by the Reagan Administration and those of subsequent administrations, modified by Congressional actions following their submission, would not support all of NASA's long-range goals. Compromises required to accommodate the changing fiscal climate impaired all the programs. But the goal most strongly effected was "a permanent manned presence in space," the Space Station. With the first budget submission, the die was cast that would lead NASA and Space Station supporters down a long, tortuous path of reduced capability, missed milestones, and cost overruns. In the following pages you will read how congressional micromanagement, NASA mismanagement, and vacillating support by all administrations, condemned the Space Station to a constantly changing and unpredictable future.

## — CHAPTER 1 —
## FIRST REALITY CHECK – PROBLEMS AND SOLUTIONS
### (APRIL 1984 - DECEMBER 1985)

In the spring of 1984, Space Station planning participants met at JSC in what was billed as a "skunk works" status review. The objectives of the review were to: (1) Identify the significant configuration requirements, (2) Establish the "Reference Configuration," and (3) Prepare the Phase B RFP. At this meeting the six original geometries were carefully reviewed except that the Raft and CDG 1 geometries had been combined and became the CDG-Planar concept. Six significant configuration requirements were identified: (1) A 75-150 kW bus power level available at IOC, (2) It had to be a large structure capable of servicing free-flyers, (3) It had to be able to simultaneously and continually view the Earth, anti-Earth, stars, and sun, (4) It had to be able to accommodate scientific airlocks for the above viewing, (5) It had to be compatible with tether operations, and (6) It had to be able to accommodate growth including adding power up to 300 kw.

As the review progressed, the Spinning Array (Spinner) and Streamline (new name Big T) geometries were eliminated. Three were chosen for further analysis: the Power Tower, CDG-Planar, and Triangular (now called Delta). The Power Tower and CDG-Planar had similar design characteristics with a central truss to which modules could be attached and large, wing-like solar arrays located at the ends of the truss. From the three geometries the Power Tower was chosen as the "Reference Configuration," for the Phase B studies (Figure 7). Its dimensions dwarfed the Soviet's *Mir*. It would have a 400 foot long central truss to which the pressurized modules would be attached. When fully assembled it would accommodate a crew of six. At the sun facing end (anti-Earth) a 200 foot long truss would hold the solar panels. Over the next seven years the "reference configuration" would undergo several name changes and grow in size and complexity as the program evolved responding to congressional and several administrations' direction.

With the Johnson Space Center leading the day-to-day space station management activities, NASA in the summer of 1984 prepared the several Request for Proposals that would be needed to begin Phase B. In addition to the four "lead" centers (JSC, GSFC, MSFC, and LeRC), project offices were established at the Kennedy Space Center (KSC), Langley Research Center (LaRC), Jet Propulsion Laboratory (JPL), Ames Research Center (ARC), and the National Space Technology Laboratory (NSTL, now named the John C. Stennis Space Center). Their primary roles would be to manage the many advanced development and technology contracts that would proceed at the same time as the definition studies.

Release of the RFPs awaited Congressional action on the FY 1985 budget that would authorize spending beginning in October 1984. The schedule for the initial definition phase was tight. It called for an Interface Requirements Review (IRR) at the end of the first quarter of FY 1985, followed by a System Requirements Review (SRR) in the first quarter of FY 1986 and a System Design Review (SDR) by the last quarter of that fiscal year. The result of the SDR would provide a picture of what the space station would look like. No more "geometries," no more concerns about showing what NASA really had in mind. All things being equal, when the pieces were assembled in orbit, it should appear very similar to the artist's rendition shown in Figure 7. After the initial definition phase, the Space Station Task Force had foreseen the need for continuing definition and design studies as new modules or capabilities might be added. But first things first – get the funds and complete the Phase B initial definition and design studies.

### MAN-TENDED OR PERMANENTLY MANNED?

Debate in the Congress on the FY 1985 appropriation centered on whether the space station would be highly automated and "man-tended," as some members advocated, or would be designed from the outset as a permanently manned facility, NASA's clear desire. Congressman Edward Boland (D-MA), chairman of the appropriations subcommittee with NASA oversight, insisted that Phase B include studies that would allow the Congress to assess the feasibility of at least starting the program as a man-

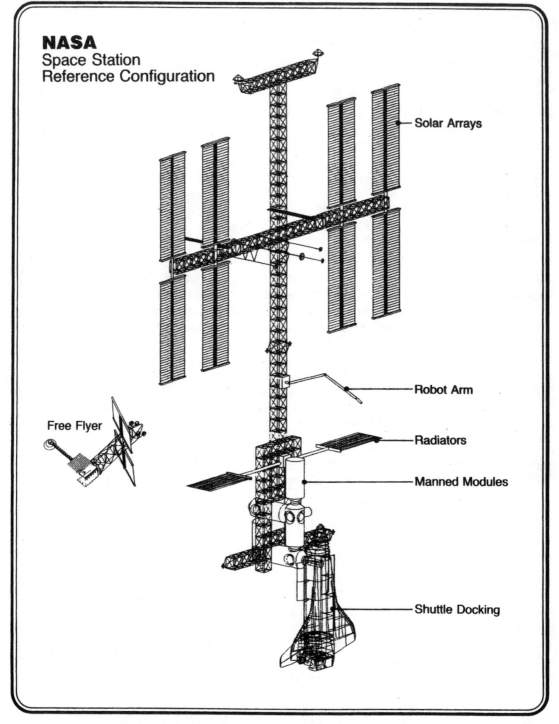

Figure 7. Power Tower "Reference Configuration" NASA 85-H-270.

tended operation. Boland's position was supported by several members of his subcommittee and other congressmen and senators. Over the objections of congressional supporters of the permanently manned space station, language was included in the NASA FY 1985 appropriations to study automation and robotics for use in the space station. NASA was to report back to the Congress no later than April 1985 on the results of the study.

Before the appropriations bill was passed, NASA had already been in discussions with the Senate Appropriations Committee to produce such a report and NASA had agreed to establish a congressionally sponsored study to report on automation for the space station. This agreement served to defuse a controversy that threatened to derail congressional support for the program at its outset. Eventually, Congress approved an appropriation of $150 million, the amount requested, and sent it on to the President for his signature. It was $69.5 million below NASA's original request to OMB.[1] It also included the requirement that NASA define how it would provide a man-tended approach that would be due to the Congress at the time a configuration was selected. Accommodating these additional studies further reduced the funds that would be available for the first order of business, the Phase B studies. On July 18, 1984, President Reagan quickly signed the HUD and Independent Agencies bill that included NASA's appropriation.[2] The program was about to get underway; significantly underfunded when considering the vast amount of work that lay ahead in Phase B.

Congressional approval of the FY 1985 budget did not silence the program's critics. The *New York Times*, among other members of the media, led a constant drum beat against the program from the time it had been announced by the President. Typical reactions from the naysayers can be summed up in excerpts from a *New York Times* November 1984 editorial. Titled "The Wrong Stuff," it explained: "the agency's next big goal is a manned space station . . . But what could a manned space station do that cannot be accomplished far more cheaply by the shuttle and by automated space platforms? NASA has no convincing answers, perhaps because the reason for pushing the space station is primarily bureaucratic. It's a big-ticket, make-work program to keep the agency busy after the shuttle."[3] These sentiments reflected the commonly expressed objections to the space station heard in Congress and from some in the scientific community.

### SPACE STATION PROGRAM OFFICE ESTABLISHED – PHASE B RFPs RELEASED

With a management framework in place at the Centers, in August 1984, Beggs created the Space Station Program Office in Washington (Level A) responsible for providing overall program direction and top-level requirements definition. Phil Culbertson, with the title of Associate Administrator for Space Station would be in charge, John Hodge would be his deputy. The Headquarters office would continue to be small, consisting of about 50 staff organized into four divisions (Business Management, Engineering, Operations, and Utilization), a policy and plans office, a program support office and a program scientist. As the Level A program office managers, Culbertson and Hodge could now look forward to the daily grind of maintaining control over a fractious empire of semi-sovereign states, the various NASA centers. Each center's constituency and allegiances, carefully developed and nurtured over the past twenty-five years, included their congressional representatives, local government officials, and businesses. Center directors were often more attuned to these interests than working as a close knit member of the NASA family. More on this subject as the story continues.

On September 26, 1984, a week before the beginning of the new fiscal year, NASA released four Work Package RFPs for Phase B definition and preliminary design. Responses were due by November 15th, an unusually short period for such complex proposals but reflecting the knowledge that the proposal teams were already up to speed based on their earlier association with the Phase A studies. Deviating from standard NASA practice, the Work Package RFPs contained few specifications. In this way, it was expected that the contractors would take the opportunity to provide NASA with their best ideas on how to design the various end products called for in the RFPs. Essentially all the aerospace companies had already aligned themselves into teams under the aegis of a small number of "prime" contractors with some subcontractors appearing on more than one team. In order to assure competition, NASA announced it would select two competing contractors for each Work Package. The total value of the eight contracts would be approximately $52 million, a small sum compared to the future, anticipated contracts that would be awarded in the next phases. The Source Selection Board would be run under

---

1. For a description of the FY 1985 budget debates, see McCurdy, Chapter 23 and Afterword.
2. Public Law 98-371, An Act Making Appropriations for the Department of Housing and Urban Development, and for Sundry Independent Agencies (98 Stat. 1225), 98th Congress, 2d session, July 18, 1984.
3. Quote included in: Space Station Program History: 1981-1987, Dr. Terence T. Finn, Deputy Director, Policy Division, Office of Space Station, February 12, 1988.

the direction of Neil Hutchinson's office and barring any problems, he hoped to have the contractors onboard by April 1985. Many RFPs for the Advanced Development and Technology Program were also released early in FY 1985 and would continue to be released throughout CY 1985.

## AUTOMATION AND ROBOTICS STUDIES

In anticipation of the requirement to produce a report for a robotic or man-tended space station, language that would be included in the NASA FY 1985 Congressional appropriation, NASA organized in the spring of 1984 an in-house team, the Advanced Technology Advisory Committee (ATAC). Contracts were then signed with a design team of five aerospace companies. ATAC's ensuing report would be used as a guide for the Phase B contractors as their work progressed. The team selected by NASA Headquarters Space Station office, was comprised of Boeing, to examine operator-systems interfaces; General Electric, to study space manufacturing; Hughes Aircraft, to study subsystems and mission ground support; Martin Marietta, for the design of autonomous systems and assembly; and TRW Systems, to examine the problem of satellite servicing. The initial reports would be due in April 1985.

In addition to the above, SRI International, under NASA contract, formed a team to provide an overall technology plan and the California Space Institute established an Automation and Robotics Panel (ARP) to provide an independent study. The ARP's mandate was very broad. It was asked to provide guidance on the application of advanced automation and robotics (A&R) applicable to space station operations at IOC and recommend how the space station could be designed to accept advances in automation and robotics technology as the program matured.

First to report with seventeen, detailed, wide-ranging, recommendations was the California Space Institute.[4] The seventeen recommendations were divided into five major categories: (1) Design the IOC Space Station to accommodate major evolution and growth in its use of Automation and Robotics, (2) Implement advances in Automation and Robotics throughout the Space Station during IOC, (3) Apply Automation and Robotics funds to advance the technology base, meeting Space Station needs, (4) Establish management structures that ensure rapid, ongoing advancement of Automation and Robotics, and (5) Promote interactions among and benefits to national communities. The Panel suggested that NASA pursue a strong applications research program and that the program be funded at thirteen percent of the total Space Station costs; a minimum acceptable level would be seven percent. At JSC and other NASA offices, the recommendations were received with mixed enthusiasm.[5] If the program office agreed to conduct all the studies, it would mean that the funds, already stretched thin, would be further diluted and delivery of the Phase B results would slip.

One month later, ATAC submitted its report.[6] Included in the report were references to the other seven studies as well as providing ATAC's own thirteen recommendations. In summary, ATAC stated that "Because automation and robotics are so important, all Space Station Program elements (the "core" station, orbiting platforms, orbital maneuvering vehicles, and ground facilities) will make substantial use of these technologies as part of the basic program." Although the Committee did not go as far as the ARP and suggest a significant level of funding, their report made it clear that it believed that automation and robotics should play a major role in the program. "As a result of its study, the Advanced Technology Advisory Committee believes that a key element of the 'right' technology for the Space Station era is extensive use of advanced general-purpose automation and robotics. This could include many systems and devices (such as computer vision, expert systems, and dexterous manipulators) that have been made possible by recent advances in artificial intelligence, robotics, computer science, and microelectronics."

---

4. An Independent Study of Automation and Robotics for The National Space Program by the Automation and Robotics Panel, Administered by California Space Institute, University of California, NASA Grant NAGW629, Cal Space Report CSI/85-01, 25 February, 1985.
5. Based on a interview (3-12-04) with Dr. David R. Criswell, Senior Technical Advisor and Program Director for the California Space Institute during the study. He also recalled conversations with the staff of Senator Jake Garn (R-UT) during which it was indicated that the FY '85 budget would require NASA to conduct automation and robotics studies because of the need to keep up with the recently announced Japanese programs in these technologies.

The eight contractor and ATAC studies covered a range of subjects, but with the exception of teleoperated systems, similar to the proposed Canadian mobile servicing system, they did not include high priority applications for the permanently manned space station. If the recommendations were followed, automation and robotic systems would only be utilized in the period leading up to a permanently manned station. A permanently manned station would be designed to take advantage of astronaut capabilities to build and service the Space Station. Nevertheless, the studies reinforced the Congressional mandate requiring NASA to study a man-tended space station. As a result the studies would continue for several years. And a man-tended configuration, supported by automation and robotic systems, would continue to be listed as an option. As the program progressed, most of the work would be shelved. It is interesting to note that all the NASA centers, with the exception of the Goddard Space Flight Center (GSFC), NSTL, and LeRC were represented in these early A&R studies yet GSFC was soon given the lead in developing automation and robotic applications for the Space Station.

## PHASE B SCHEDULE

Appearing before the Subcommittee on Science, Technology and Space on April 3, 1985, to explain the FY 1986 budget request, Culbertson reviewed the progress the program had made.[7] In anticipation that all the Work Packages would soon be signed, he laid out the Phase B schedule. "The Space Station Program will begin a 21-month period during which engineering information will grow in detail as our options are focused to a single design . . . In mid FY 1986, as an outcome of the Systems Requirement Review, we will select a Space Station baseline configuration . . . preliminary design . . . will commence and extend into the first quarter of FY 1987." During his testimony, he explained how the FY 1986 budget request of $230 million would be allocated. The two largest segments would be spent on the system definition contracts ($74 million), and advanced development ($82 million).

He completed his statement reviewing the status of international activities and stated that Memoranda of Understanding (MOU) had been negotiated to coordinate activities during Phase B. He pointed out that a key ingredient for the international partners was a long-term requirement to utilize Space Station capabilities. "The fact that this program will span several decades dictates that a true partnership must incorporate on all sides a commitment to utilize the Station. It is envisioned that the ownership and responsibility for sustaining engineering of internationally provided elements will rest primarily with the international partner. The financial and technical responsibility for the international components of both Phase B and Phase C/D rests with the participating countries." At the end of Phase B, MOUs for Phases C and D would require a new round of negotiations with the three partners to iron out issues that could not be completely defined until Phase B was completed. This second round of MOUs was not signed until 1988, well after Phases C/D were underway. However, it was never a true partnership of equals. As the political climate changed, NASA made unilateral program modifications without consulting with its longtime partners, and at times the partners modified the terms to reflect their requirements.

## BUDGET BATTLES BEGIN

In a typical year, senior government bureaucrats will be closely involved with three different budgets. The budget that allows for current year spending, the next fiscal year's budget that is being explained and defended before the many committees and subcommittees of the Congress during seemingly endless appearances, and the fiscal year's budget request for two years down the road that is being reviewed by the OMB. Guidelines for submitting this latter budget might be disseminated by OMB as early as Christmas thus requiring long hours of internal debate and processing during and after the holiday season. As the process winds its way through the various levels of the bureaucracy, during which the participants are sniffing the changing winds of the Congress during the ongoing hearings,

---

6. Advancing Automation and Robotics Technology for the Space Station and for the U.S. Economy, Advanced Technology Advisory Committee, NASA Technical Memorandum 87566, submitted to the United States Congress, April 1, 1985. Two separate Executive Overviews exist, one dated March 1985, the date of the Technical Report, the second April 1, 1985. The differences between the two versions are minor, the earlier version omitted a Preface, and Introduction. The recommendations are the same in both versions.
7. Statement by Philip E. Culbertson, Associate Administrator for Space Station, before the Committee on Commerce, Science and Transportation, Subcommittee on Science, Technology and Space, April 3, 1985.

guidelines might be modified. By late spring or early summer an agency's budget two years hence is taking final shape. In rare instances, it may be subject to last minute appeals that may go as high as the President if an agency believes the final numbers are too low. Final changes are made by the end of the calendar year in order for the President to discuss his new programs in the state of the union address. The monstrous budget is delivered soon after to the Congress.

During his April 1985 testimony to the Subcommittee on Science, Technology and Space, Culbertson did not mention the fact that the $230 million was $50 million below the agency's original "scrubbed down" request to OMB.[8] Such information is considered "privileged" in the Executive Branch and agency heads are not permitted to disclose original request numbers in open sessions. Only one agency, DOE (and its predecessors the Atomic Energy Commission and ERDA) has a special Congressional dispensation called the "Holifield Tables,"[9] that permits it to discuss the request to OMB. Agency requests might be "leaked" to staffers but would never be a part of the record. NASA's friends in Congress who had the "leaked" numbers might try to get those testifying to say "yes we need and requested more" but those testifying were required to say it was sufficient and soldier on. The consequence to the Space Station program of staying within lower budgets was that schedule and content changes would have to be made.

To make matters worse Congress, armed with the newly passed Gramm-Rudman-Hollings Deficit Reduction Act of 1985, began to take the easy route to budget reductions by cutting those programs classified "discretionary," the category that covered all NASA programs. During the summer and fall of 1985, the House and Senate argued back and forth on what the Space Station appropriation should be and finally passed an FY 1986 appropriation of $205 million, $75 million below the NASA request to OMB in the spring of 1984. Thus began the chronic underfunding of the Space Station. The Reagan Administration soon assigned the space station a low priority and the next administrations, except for an occasional short-lived elevation, never raised its priority. Ultimately, OMB and Congress micromanaged the program through the budget process.

## WORK PACKAGE CONTRACTORS SELECTED

Two weeks after Culbertson's appearance before the Subcommittee, the Work Package contracts were signed. Administrator Beggs was the source selection official who made the final selections. In keeping with Beggs' desire to maintain competition during this phase of the studies and assure that there would be a pool of contractors able to respond to future RFPs, two prime contractors were selected for each Work Package.

Work Package #1, under the management of MSFC, would define the "common module," the pressurized module in which the crew would live and conduct research. For the common module, the work package also required the contractors to define its internal systems such as power and data management, and how the module would be outfitted as living and working quarters. Also included in the WP #1 contract was the design of the environmental control and life support system and the designs of the orbital maneuvering and orbital transfer vehicles. Boeing Aerospace Company, Seattle, Washington, and Martin Marietta Aerospace, Denver, Colorado, were selected as the prime contractors.

Work Package #2 was released by JSC. Contractors were asked to define and design the overall station architecture, assembly and structure. This included the integration of the utility, thermal and attitude control systems, space shuttle interfaces, communication and tracking, the data management system,

---

8. Space Station FY 1986 Budget Review Presentation to the Administrator, May 7, 1984. The $280 million was considered a minimum request based on the anticipation that the FY 1987 request would be on the order of $600 million to get Phases C and D started.
9. In 1957, during the early days of the Cold War, the Joint Committee on Atomic Energy, concerned that the administration then in power might not be requesting enough funds in the annual budgets submitted to the Congress for the Atomic Energy Commission (AEC), under the leadership of Senator Chet Holifield (D-CA), required that internal AEC budget requests to the Bureau of the Budget be provided to the Congress, the "Holifield Tables." This permitted the Congress to have the details if it wanted to second guess the president's request for AEC. This mandate carried over to the AEC's successors, ERDA and DOE, the only agencies known to the author that are required to provide the Congress with this transparency in the budget process.

and EVA requirements. Because JSC was the home of the astronauts and was most familiar with their needs and desires, it would also have responsibility for certain aspects of module outfitting, overlapping responsibilities assigned in Work Package #1. The prime contractors selected were McDonnell Douglas Astronautics Company, Huntington Beach, California, and Rockwell International Space Station Systems Division, Downey, California.

Work Package #3, to define and design the unmanned platforms, was released by GSFC. This work included how the platforms would be equipped and serviced, and identify potential payloads to be attached. The two primes selected were General Electric Company, Space Systems Division, Philadelphia, Pennsylvania; and RCA Astro Electronics, Princeton, New Jersey.

Work Package #4, for the design of the power generation system was the responsibility of LeRC. Not in the running originally for a Work Package, Andrew Stofan, LeRC Center Director, had lobbied hard for this assignment. Lewis had been a major player in developing various types of power conversion systems for terrestrial energy use, including solar energy. When terrestrial energy research was eliminated by Beggs, it left a cadre of skilled engineers at LeRC with little to do and facing layoffs. It was a good fit. Rockwell International, Rocketdyne Division, Canoga Park, California; and TRW Federal Systems Division, Redondo Beach, California, were selected as the prime contractors. (Figures 8 & 9 have a complete listing of all the primes and subcontractors.)

## SPACE STATION PROGRAM
## WORK PACKAGE STRUCTURE
### PRIME CONTRACTOR TEAMS FOR DEFINITION AND PRELIMINARY DESIGN

| WORK PACKAGE 1 – MARSHALL | | WORK PACKAGE 2 – JOHNSON | |
|---|---|---|---|
| BOEING | MARTIN MARIETTA | ROCKWELL | McDONELL DOUGLAS |
| United Technologies Corp. | Hamilton Standard | Grumman Aerospace Corporation | RCA |
| Aerojet Tech Systems Co. | McDonnell Douglas Technical Services Co. | Harris Corporation | Honeywell. Inc, |
| Rocket Research Co. | Wyle Laboratories | Sperry Flight Systems | International Business Machines Corporation |
| Rockwell International | Hughes Aircraft Co. | Intermetrigs Inc. | |
| Teledyne Brown Engineering | Hercules, Inc. | TRW, Incorporated | |
| TRW-Space And Tech. Group | Honeywell, Inc. | | |
| Telephonics Corp. | | | |
| Thermacore, Inc. | | | |
| Westinghouse Electric Corp. | | | |
| Camus, Inc., | | | |
| Perkin-Elmer | | | |
| Umpqua Research Co. | | | |
| Fairchild Republic Co. | | | |
| Garrett Corporation | | | |
| General Electric Co. | | | |
| Life Systems, Inc. | | | |
| Lockheed Missiles & Space Co. | | | |
| Hughes Aircraft Co. | | | |
| Oao Corporation | | | |
| Ltv-Vought Co. | | | |
| Space Communications Co. | | | |
| Sundstrand Energy Systems | | | |

Figure 8. Work Packages 1 & 2 Contractor teams.

## SPACE STATION PROGRAM
## WORK PACKAGE STRUCTURE
### PRIME CONTRACTOR TEAMS FOR DEFINITION AND PRELIMINARY DESIGN

| WORK PACKAGE 3 – GODDARD | | WORK PACKAGE 4 – LEWIS | |
|---|---|---|---|
| GENERAL ELECTRIC | RCA | ROCKETDYNE | TRW |
| Federal Systems Division – TRW Space & Technology Group<br>Teledyne Brown Engineering<br>Grumman Aerospace Corporation<br>Perkin Elmer<br>Spar Aerospace Limited<br>Integrated Systems Analysis, Inc.<br>Essex Corporation | Lockheed Missiles and Space Company, Inc.<br>Ball Aerospace Systems Division<br>Computer Sciences Corporation | Ford Aerospace & Communications Corporation<br>Garrett Pneumatic Systems Division<br>Harris Corporation<br>Sundstrand Energy Systems | General Electric Company<br>General Dynamics – Convair Division<br>Grumman Aerospace Corp.<br>Life Systems, Inc.<br>Mechanical Technology, Inc.<br>Perkin Elmer<br>United Technologies Power Systems (UTC) |

Figure 9. Work Packages 3 & 4 Contractor teams.

In March 1985, a few weeks before the Phase B contracts were awarded, Hutchinson, the Level B day-to-day manager, established the Space Station Control Board (SSCB).[10] The Board's primary functions would be to control technical and design requirements, scheduling, and allocation of resources while the Phase B studies were underway. Twenty-one members were assigned to the Board with representatives from NASA Headquarters, Hutchinson's office, the four "Work Package" offices, LaRC, KSC, and JPL. The Board would report to Hutchinson and his deputy, John Aaron. As noted earlier, each Work Package contractor was given wide latitude during its studies to design their elements in anticipation that they would arrive at innovative solutions for their assigned work. The Control Board would then be responsible for sorting through potentially disparate designs and select those elements that would work best together to define the "reference configuration."

Also at this time Hutchinson established a two level decision making structure that reported to the SSCB. The next level below the SSCB consisted of four boards and panels: the Systems Integration Board (SIB), the Operations Panel, the Mission Integration Panel, and the Management Information Systems Control Board. The next level was named Technical Integration Panels (TIPs). Each of these latter panels was charged with developing options for each element of the Space Station and then report their recommendations up through the chain to the SSCB.

While the work package contractors would be studying specific space station systems and subsystems, the NASA in-house teams mentioned above would be studying how to integrate elements that were still in the process of definition. Ultimately the SIB, chaired by JSC's Al Louviere, became first among equals and was the clearing house that processed and accepted or rejected all the change requests. If differences would arise between the SIB and the TIPs, they could be appealed to the SSCB. The SIB also had the responsibility to pull all these inputs together so that a top-level Space Station Program Definition and Requirements Document could be issued to replace the "Yellow Books." Sound

---

10. Charles N. Crews (Executive Secretary, Level B Space Station Control Board (SSCB)) to Distribution, "Minutes for 25 March 1985, Level B SSCB," 12 April 1985. The twenty-one members were the Program Manager, Neil B. Hutchinson; the Deputy Program Manager, John W. Aaron; the Manager of the Space Station Program Management Office Documentation and Configuration Management Office, Charles N. Crews; managers of the Space Station Program Office's Systems Engineering and Integration Office, Data Management Systems and Operations Office, Customer Integration Office, Program Management Office, Program Management Office, and International and External Affairs Office; (Headquarters) Space Station Program Scientist; Johnson Space Center directors of safety, reliability, and quality assurance, research and engineering, and space operations; managers of Space Station Project Offices at Marshall, Johnson, Goddard, and Lewis; the manager of the Space Station and Advanced Projects Office at Kennedy; managers of Space Station Offices at the Jet Propulsion Laboratory and Langley; chief of the Space Station Commercialization Office at Ames; and a representative from the Johnson Space Center National STS Program Office.

complicated? Not really. Such management hierarchies were a standard approach at NASA for controlling complex programs.

## Space Station Operations Review

Book 6 of the "Yellow Books," Systems Operations, had been written by the Space Station Operations Working Group in 1983 and 1984 without the benefit of a specific space station configuration in mind. Now, with the contractors all working with the "reference configuration," Culbertson and Hodge felt the need to relook at Space Station operations unconstrained by the information that NASA had included in the Work Package RFPs. In April 1985, they invited twenty-three experienced industry managers and consultants, including many former senior NASA managers, to participate in a Panel, chaired by Charles W. Mathews, to review Space Station operations.[11]

Following a series of NASA briefings to bring the participants up to date, the Panel made a number of recommendations. Their first concern was that a new operations baseline was needed to replace the existing baseline established by the Operations Working Group in 1984. They recommended that the new baseline be placed at the top of the Space Station program planning hierarchy and drive the design. Furthermore, they recommended that customer relations should be placed at the top of the operations actions and noted that it did not appear that there were strong, dedicated customer offices at all levels in the organization. This latter recommendation repeated a concern provided to the Space Station Task Force in 1984 by the Task Force on Scientific Uses of the Space Station (TFSUSS); the need to develop a science oriented management structure. Unfortunately, these recommendations were never fully implemented.

The Mathews panel questioned the proposed crew size and felt that a six person crew would be insufficient and "grossly overcommitted" in terms of the functions they would be asked to perform. The Panel felt that for crew safety a simple return vehicle should be attached to the space station at all times and that depending on the shuttle fleet for rescue was questionable. And finally, they saw no evidence that NASA had set up a formal maintenance and logistics support group and were concerned that if funding were to be constrained in the future these important functions would be short changed.

In response to the Panel's criticism on the lack of a strong customer office, in April, NASA formally appointed Dr. David C. Black as the Space Station Program Chief Scientist. Black came to his new duties on a leave of absence from the NASA Ames Research Center, however, he was still carried on the rolls as an Ames employee. NASA already had a Commercial Programs Office whose agency-wide mission was to focus on commercial projects that would be a part of any program with a mandate to incorporate commercial considerations in design and operations. However, the Panel did not consider the office to be very effective. As you read on, you will find that many of the concerns expressed by the Panel in 1985 would surface again, and again.

## International Partners Sign Up

With the definition and design phase underway, NASA pushed to have the MOUs with the international partners signed. As mentioned earlier, negotiations had dragged on since June 1984. To some degree NASA was charting new ground with these MOUs as the partners end products, when finally incorporated, would have major impacts on the program. Coordinating the negotiations with the State Department, the last MOU was finally signed in June 1985.[12]

Skipping ahead a few months, as part of a briefing at NASA Headquarters in August 1985, John Hodge described what the international partners "are considering."[13] Canada's contribution would be the most critical. It agreed to study construction and servicing requirements and design a mobile servicing system (MSS) that would be attached to the space station's external structure. It would be designed to

---

11. Participating panelists and a summation of the results of the seminar can be found in Space Station Operations Seminar, Columbia Lakes Conference Center, West Columbia, TX, April 22-25, 1985, Seminar Proceedings, June 14, 1985, Technical & Administrative Services Corporation (TADCORPS), 600 Maryland Ave, S.W., Washington, D.C. 20023, (NASA Contract No. NASW-4040).
12. The MOU with Canada was signed on April 16, 1985, with Japan on May 9, 1985, and with ESA on June 3, 1985. Dates from Space Station Presentation to NASA Headquarters, given by John D. Hodge on August 9, 1985.

manipulate the large, heavy modules and structural elements brought to the Space Station as shuttle payloads. Without a fully functioning MSS to move all the elements around and assist in their mating, assembly on orbit would be impossible. Canada also agreed to study new solar array systems for the free flying platforms and remote sensing facilities.

The European Space Agency (ESA) would study the design of a pressurized laboratory module, utilities module, and a servicing vehicle. It would also study free-flying, man-tended platforms for various applications. Japan proposed to design a multipurpose experiment module (JEM), a experiment logistics module, attached experiment pallets, a small external manipulator system that would be attached to the JEM, and a scientific airlock. Over the next few years, some of these studies would be deleted or cut back, but the core contributions, the Canadian MSS, ESA's laboratory module and man-tended free-flyer, and the Japanese experiment module would survive.

Hodge also discussed some of the more difficult items that remained to be decided with the international partners in the months ahead. They included: protection of intellectual property rights, pricing principles, operational costs and the possible use of barter agreements to offset costs. A high priority was the participation of European astronauts as space shuttle crews that would be trained at JSC to install and service their elements. It was hoped that these issues would be resolved during the next round of negotiations.

## NATIONAL COMMISSION ON SPACE

Sold as "the next logical step" the Space Station, when operational, would be just that, a step, but to where? For those who viewed space as the new frontier, being confined to a low Earth orbit space station was only a partial fulfillment of their vision; been there, done that. In Reagan's next State of the Union address on February 6, 1985 he stated: "Our second American revolution will push on to new possibilities not only on Earth but in the next frontier of space. Despite budget restraints, we will seek record funding for research and development." Space enthusiasts in Congress had already pushed through an authorization[14] to define "the next frontier." Thus the President was on firm ground to make such a bold proposal. In March 1985, the President appointed fourteen members (one more added later) to the National Commission on Space. Congress charged the Commission to "formulate a bold agenda to carry America's civilian space enterprise into the 21st century."[15] Reagan chose former NASA administrator Thomas O. Paine to chair the Commission. Paine had been administrator during the years, 1969-1970, when the first Apollo landings had taken place. The other fourteen members had equally impressive resumes coming from both the private sector and government and included former astronaut Neil A. Armstrong and Nobel Laureate Luis W. Alvarez.

The Commission immediately began to hold a series of public forums and workshops around the country to provide background for their deliberations. On May 15, 1985, Culbertson gave the Commission a Space Station briefing and brought them up to date on the status of the program.[16] All the program's bells and whistles were shown in full color artist drawings. But just one month after the start of the Phase B studies, Culbertson acknowledged a small slip in the schedule. Instead of the IOC beginning in early 1992, he showed the Commission a new milestone chart with IOC in 1993 reflecting the smaller than requested budget and later than anticipated contract signings for the Phase B studies. One of the adjustments was the lengthening of Phase B from eighteen months to twenty one months. However, the $8 billion dollar figure, now called the cost to achieve "initial capability," was still the baseline number needed to fulfill Reagan's mandate to have the Space Station operational "within a decade."

13. Space Station Presentation To NASA Headquarters, John D. Hodge, Deputy Associate Administrator, Office of Space Station, August 9, 1985.
14. Authorization for the Commission was included in the NASA Authorization Act of 1985, P.L. 98-361, Title II, July 16, 1984. Reagan then issued Executive Order 12490 on October 12, 1984 setting up the Commission but did not appoint the members until March 29, 1985.
15. Pioneering the Space Frontier, The Report of the National Commission on Space, Bantam Books, May 1986. A complete listing of the Commission's members and nonvoting members as well as the support staff, can be found in the report.
16. Space Station Presentation to the President's National Commission On Space, May 15, 1985, Philip E. Culbertson, Associate Administrator, Office of Space Station. A series of forty-two vugraphs used in the presentation, many full color artists renderings of various aspects of the Space Station as then proposed.

An interesting sidelight of the briefing to the Commission was a discussion of the DOD relationship. Prior to Reagan's decision to approve the space station, and while it was being hotly debated inside the Administration, NASA had attempted to obtain an agreement from DOD to utilize the space station and was turned down. In spite of this lack of support, a USAF Space Command officer served on the Space Station Task Force, and Culbertson revealed that an officer was also serving in the program office in Houston. Now that it was an approved program, DOD apparently had a change of heart and was examining "the potential of a permanently manned space station." He went on to say that "NASA believes DOD might well utilize space station's laboratory capabilities in a research mode." Such interest on the part of DOD was not surprising considering that DOD had already tried, and failed, to convince a previous administration to support their Manned Orbiting Laboratory (MOL) program.

Culbertson completed the briefing by discussing the potential evolution of the program. Four future options were described: (1) A large space station in low-Earth orbit; (2) A full geosynchronous capability; (3) A return to the Moon and lunar bases; and (4) Manned Mars exploration. Various complimentary paths were shown originating at the initial space station and then branching out to achieve each or all of these options.

When the Commission's final report was released in May 1986, it included a review of NASA's ongoing efforts to provide a permanent Space Station. Among the many ambitious goals recommended by the Commission was one to transform the Space Station into a Spaceport in low Earth orbit. Saying that the Commission recommended ambitious goals is an understatement, especially when viewed from a more modern perspective. Among other recommendations was the need to expand NASA's budget as a percentage of GNP. In the Commission's final report, Pioneering the Space Frontier, the four options described by Culbertson were endorsed. In spite of the Commission's vigorous endorsements, the "record funding" promised by President Reagan in his State of the Union address never materialized.

## Reference Update Review

In preparation for the Phase B Interface Requirements Review (IRR) and the Systems Requirements Review (SRR), scheduled to occur in early 1986, NASA conducted a Reference Update Review (RUR-1) during July 1985. The purpose of the RUR was to reduce some of the Space Station design options that were being studied by the Phase B contractors. Recall, each contractor was given some latitude in the designs they chose to study in order to permit them to propose creative solutions to some of the unique problems they had been asked to solve. The RUR was the point in time when NASA focused on the key options being studied so that the two reviews scheduled a few months later could more easily make the critical selection of the new "baseline configuration" that would replace the "reference configuration" for all further work.

All the NASA centers with a piece of the action as well as the contractor teams and the international partners participated in the RUR. A major problem had surfaced prior to the RUR. The "work package" contractors, as they progressed with the Phase B studies, were indicating that large cost increases over earlier estimates should be anticipated for the elements included in each work package. The first cost estimation red flag had been raised. Over the eight day period during which the RUR was conducted, five subjects received the most attention: the overall configuration and related problem of the module attachment pattern, selecting the atmospheric pressure that would be maintained within the habitation modules, how to establish criteria whereby evolutionary growth could be accommodated, and how to organize a work breakdown structure that would reflect the complicated interactions of the many pieces that would eventually be brought together on orbit.

This last subject, how to organize the work breakdown structure, was an important issue at the RUR. Its design and detail would organize the program and would be an essential management tool. Work breakdown structures evolved from DOD projects that required a large number of contractors and subcontractors to work together to successfully complete a project. Individual pieces of a project would be identified down to the smallest element with the responsible entity identified along with cost and schedule milestones. These in turn would be formed into a larger, hierarchical matrix encompassing all the pieces with the final product emerging at the head of the matrix. A well-planned work breakdown structure allowed the project manager to closely monitor the progress of each piece and if an element was identified as being in trouble, appropriate resources could be brought to bear. Depending on the complexity of a project, five or more levels might be established in the work breakdown structure. The Space Station was just such a complex program. To assure success, all the thousands of pieces would have to be closely tracked, especially in view of the tight cost under which NASA had promised to deliver a Space Station, within a decade. Developing the final work breakdown structure for the Space Station would take several years before it was completed to management's satisfaction.

A design option that consumed considerable debate at the RUR was the question of what should be the working pressure in the habitation modules. Two pressures were being considered, 14.7 and 10.2 psia. Each had advantages and disadvantages. The experimenters and other potential users preferred 14.7, normal Earth pressure. Their major need for this pressure was that many would be comparing the results of their work on the Space Station to experiments conducted on Earth and the fewer variables involved, the better. However, operating at this pressure would then entail either developing high pressure space suits to be used during required EVAs or long acclimation periods in an airlock to allow the astronauts to transition to available space suits designed to operate at lower pressure. The reverse adjustment would be needed to return to the ambient working pressure in the habitation modules. Neither of these outcomes was attractive as high pressure suits would be difficult and costly to develop. Consuming hours of astronaut time to adapt to the lower pressure suits and then back to cabin pressure was also considered an unacceptable obstacle. Astronaut time while on orbit was already recognized as a precious commodity. RUR #1 adjourned with this and many other issues still unresolved.

At the end of his August 1985 briefing to Headquarters management, Hodge outlined nine challenges that remained for the program: (1) Design for "permanence," maintainability and evolution, (2) Build to cost and schedule, (3) Manage systems engineering and integration in-house, (4) Establish an efficient technical management and information system, (5) Incorporate new technologies, balancing cost, schedule and technical risk, (6) Understand operational costs, and incorporate life cycle cost planning, (7) Orchestrate the international dimension, (8) Maintaining customer focus when time, money and engineering begin to pinch, and (9) Assembly and checkout on orbit. Hodge's willingness to discuss these challenges before an audience of NASA's senior management, well after the Phase B studies were underway, was a candid admission of the many problems the program faced. His emphasis on the many different types of cost problems he anticipated is particularly telling.[17] Contractor warnings on cost growth had become a major concern that would have to be addressed.

Bringing systems engineering and integration in-house would soon take place. However, the program would continue to struggle to find answers to his other challenges up to first element launch, thirteen years later. One concern, "maintaining customer focus when time, money and engineering begin to pinch," was never resolved. Many of the compromises that were made over the years as the result of schedule and money problems were at the expense of providing customer capabilities. Hodge correctly forecast this problem.

In August Culbertson decided to find ways to reduce or defer the Phase C/D development costs and provide options to accommodate the expected results from the Phase B studies. He wrote to Neil Hutchinson[18] asking him to "initiate, in an effort involving both Levels B and C, a study of the potential impact on configuration, utilization, operations, program and schedule risks and both development and operating costs: of a number of potential variations to the reference configuration or contract direction, as defined in the Phase B RFP." He laid out in detail two categories of actions: "the first are those which lead to a reduction of direct and wrap around labor costs, both those incurred by prime contractors and by the government; the second are a number of potential content deferrals or reductions in system capability or complexity." In other words, Culbertson was warning that based on forecast budget constraints NASA may have to rethink the whole program. Eighteen months after receiving Reagan's approval, Culbertson was confronting the need to return to the drawing board to redesign the program's architecture.

---

17. This conclusion is based on discussions with John Hodge in April 2004.
18. NASA memo, from S: Philip E. Culbertson, to Neil B. Hutchinson, Subject: Space Station Program Cost Estimates, dated 14 August 1985.

## POTENTIAL USERS REVIEW DESIGN

August 1985 was a busy month on several other fronts. The SESAC Task Force on the Scientific Uses of Space Station (TFSUSS) whose chairman Peter Banks had testified before Congress earlier in the year, held its second summer study.[19] For one week team meetings were held in the disciplines of Astronomy and Astrophysics, Solar-Terrestrial Processes, Solar System Exploration, Earth Observations, Microgravity, Physics and Chemistry in Space, and Life Sciences. Among the major objectives of the study were those to develop user requirements, and to explore issues related to international participation. In addition, four cross-discipline teams were formed to review the overall Space Station configuration, the utilization of the free-flying platforms, science operations, and communications and information systems that would be available to science experiments and experimenters.

As would be expected as a result of such a study attended by over 175 scientists, engineers, and contractors, including some from overseas, the discussions and recommendations were wide ranging. One of the important recommendations was: 'The essential need for a truly long-term, manned capability." And in a similar vein, the Task Force stated that "it is essential that NASA look ahead to the activities that are anticipated over the 25 to 30 year life span of the core facility and its associated elements." With respect to cooperation with the international partners, the TFSUSS had been in close contact with ESA scientists and had reached agreement on several key issues that would be explored in the "near future." Foremost was the selection process for experiments that would use the Space Station. Discussions would be held on having a "peer-evaluated" process to assure meritorious projects. Discussions would also be held to resolve collaboration, resource allocation, and a suitable international science management structure, including funding.

The Report's Conclusions and Recommendations ended with this warning: "Finally, throughout the post-IOC planning process, the intrinsic growth limits must be defined carefully in terms of the supporting transportation and communication systems. The limits set by the STS delivery of equipment and personnel to the core station, the lack of frequent access to polar orbiting platforms, the limitations of the TDRS communication system, and other such problems will act in concert to set boundaries of reasonable growth." As you read through all the recommendations and concerns, the scientific community, as represented by the disciplines and researchers in attendance, was supportive of the Space Station, but feared it might not be able to live up to its full potential if capability in addition to that already planned was not provided.

On August 20, 1985, the Space Station Office signed an MOU with Space Industries Inc. (SII) to exchange information on their respective programs during the Phase B studies. SII had been formed in 1982 by former senior NASA engineer and JSC director of engineering and development Maxime "Max" A. Faget. Max had signed an earlier MOU with NASA whereby SII would define and develop, with private funds, an industrial space facility (ISF). The ISF would be a habitable, modular platform projected to be ready for launch on a shuttle by 1989. It would function primarily as a materials processing facility but could provide other functions including use as a substitute for some projected shuttle space station test bed flights. It was being designed to operate unmanned for extended periods after shuttle deployment. The new MOU had the potential to go one step further. If mutually desirable, the ISF docking system would be adapted for mating with the Space Station thus providing another laboratory module at IOC.

Before Hutchinson responded to Culbertson's August letter, a Reference Update Review (RUR) #2 for the Phase B studies was held in Washington in November. RUR #2, produced some surprising decisions. Starting with the Phase B reference configuration, the Power Tower, the attendees proposed several major changes that would eventually be adopted in the new baseline configuration. The most significant modification was to go from the single truss Power Tower to a dual truss design. The primary reason for this change was that it allowed for greater flexibility in module attachment and growth. Other design considerations agreed to included a standardized module length of 43 feet with modules arranged in a figure eight pattern, maintaining the safe haven concept versus a crew return

---

19. SESAC Task Force on Scientific Uses of Space Station, Space Station Summer Study Report, March 1986. Although the Task Force met in August 1985, the report was not released until March 1986.

vehicle recommended earlier by the Mathews Panel, and providing a microgravity level of $10^{-5}$ g's for users. Final adoption of these changes would await the System Requirements Review (SRR) in March.

In a vugraph used for a later presentation, Hodge indicated that these changes resulted in an overall "more capable" Space Station but did not explain how or if the changes would increase or lower cost. Two months after RUR#2 was completed, Neil Hutchinson made the decision to accommodate the users, a 14.7 psia internal atmosphere would be adopted. He would cross the bridge to developing a high pressure EVA suit later.[20] Final decisions on the other areas discussed at the RUR were deferred to a later time. Culbertson took advantage of the RUR to hold a follow-on meeting to discuss the major issues that concerned him. In addition to the senior space station managers who would already be in Washington for the review, he invited the eight "work package" company managers. In his invitation to the meeting he listed three issues for discussion. Establishing "a sound configuration for a cost which the government can afford, putting in place an effective government / contractor management structure for design, development, test and evaluation (DDT&E), and winning approval for the development program."[21]

The discussions that took place could not have foreseen the difficulties NASA would soon face. The train threatened to come off the tracks when OMB announced shortly after the meeting in December that NASA's FY 1987 request would probably be pared back to $100 million, just enough to complete the definition phase. Culbertson and Hodge's worse fears had come to pass, and to make matters even more difficult, Administrator Beggs, the Space Station champion, might not be around to run interference at the White House and reverse the OMB locomotive.

## BEGGS RESIGNS – MANAGEMENT TURMOIL

Within a year after the go-ahead for the Space Station, NASA Administrator Beggs was beset by accusations of misconduct during his tenure at General Dynamics Corporation, prior to returning to NASA as Administrator. As he fought to clear his record, he came to the conclusion that he could not accomplish this and at the same time manage NASA during a period of great change. In December 1985, he took a leave of absence but kept an office at NASA Headquarters. Two months later he would resign and, eventually, successfully contest the charges against him. His departure set off a chain reaction over the next six months throughout NASA's top management, both at NASA Headquarters and at several centers.

Having notified the Administration of his decision to take a leave of absence, the White House immediately cast about to fill the position of Deputy Administrator, vacant for over one year following the resignation of Hans Mark in September 1984. At the end of November 1985, over Beggs' objections, the White House appointed William R. Graham, a nuclear weapons consultant to the Department of Defense, as Beggs' new deputy. Graham's appointment was quickly approved by the Senate.

Before taking his leave of absence, Beggs promoted Culbertson to the position of General Manager. Robert C. Seamans was the second NASA manager to fulfill this function (an informal title, Seaman's actual title was Associate Administrator) while serving under NASA's first Administrator T. Kieth Glennan and second Administrator, James E. Webb. With Webb's concurrence, Seamans crafted the position to centralize under his control the management of the day-to-day internal functions of a new, expanding agency leaving Webb to be Mr. Outside.[22] When Seamans was elevated to Deputy Administrator in 1965, he continued carrying out these duties under his new title and there was no need to fill the position of general manager, nor was it filled by succeeding administrators. Beggs, harboring concerns about how Graham would manage NASA, decided to reinstate the position. Culbertson had served in a variety of senior positions at NASA since the early days of Apollo and was intimately familiar with the agency's workings.

---

20. How Hutchinson arrived at his decision is covered in a series of memos from Charles N. Crews to distribution of minutes of the Level B Space Station Control Board for August 1, August 29, September 12, October 2, and October 17, 1985.
21. NASA memo, from Philip E. Culbertson, to Distribution, Subject: Space Station Managers Meeting, dated: November 8, 1985.
22. Aiming at Targets, Robert C. Seamans, Jr., NASA SP-4106, The NASA History Series, 1996, p. 80.

His most recent position, providing management oversight of Space Station planning, the most complex NASA program, made his elevation to General Manager a logical decision. With Beggs' departure on his leave of absence, Graham moved from Deputy Administrator to Acting Administrator on December 4, 1985. Once again the position of Deputy Administrator became vacant in effect making Culbertson NASA's second in command. (During his short tenure, Graham never had a deputy.) Two weeks on the job as Beggs deputy, and his previous government experience, had not prepared Graham for such a high profile demanding position. Overnight he was managing an organization of over 21,000 civil servants in some 20 field centers and facilities. Graham could not have chosen a worse time to take over the helm of a troubled agency. He was immediately confronted with the problems coming to a head with the FY 1987 budget and, most importantly, the projected funding shortfalls for the Space Station. The budget would be put to bed within a month. Even more demanding would be his responsibility two months later to guide NASA through the trauma of the *Challenger* accident.

A few days after his appointment as Acting Administrator, Graham met with Culbertson to discuss how he wanted Culbertson to perform as General Manager; the major areas he would control and those that he wanted Culbertson to manage. Appropriately, he would be the contact with other agencies and coordinate the interactions with the White House and Congress. Culbertson would provide the day-to-day oversight of all the agency programs. This split in responsibilities was similar to the arrangement Webb had agreed to with Seamans some twenty years earlier. On this basis, NASA's two top managers plunged into the troubled waters. Graham immediately ran into difficulties on Capital Hill as he attempted to represent NASA in the runup to the FY 1987 budget hearings.

Gerald Griffin, JSC Director, compounded the top management upheaval when he announced in December that he would resign and take a position in the private sector. Graham was faced with replacing Griffin, an Apollo veteran, at NASA's most visible center. JSC, home of the astronauts, managed the Space Shuttle program and critical pieces of the Space Station. He would be forced to select the director by choosing someone with whom he had little direct, working relations. Five days before the *Challenger* accident (January 26, 1986), he chose Jesse W. Moore. Moore, a NASA and JPL veteran was the Associate Administrator for Space Flight before taking the job at JSC and was well versed in Washington politics. As JSC center director, he would be exposed to a new brand of politics at a crucial time.

Meanwhile, Culbertson focused on the FY 1987 Space Station budget problems. Back in the spring, he had tried to convince Beggs to add over $1 billion in additional funding to the out-year budget's projections, sums much higher than the amounts discussed earlier with OMB. Beggs turned him down.[23] As a result, and fearing future tight budgets, he was forced to provide the direction to Hutchinson described earlier. Now he was in charge. He informed OMB that NASA would appeal the $200 million that OMB proposed to include in the President's FY 1987 budget request, $100 million more than OMB's first pass back. If the President accepted OMB's $200 million it would still be $400 million below NASA's final "scrubbed down" request to OMB that had been predicated on starting Phases C and D at the beginning of the new fiscal year. If the President approved only $200 million, the program would be delayed by one or two years.

Culbertson was given 12 minutes to make a presentation to Reagan. He would be followed by a 12 minute rebuttal from OMB. This turned out to be a fortunate sequence. As he gave his pitch, Culbertson placed a new Space Station model based on the Power Tower design in the middle of the table and saw that the President was interested. When he finished, he pushed the model over to the President and while OMB gave their counter arguments the President became engrossed examining the model and not listening to OMB Director James Miller's arguments. At the end of the presentations, Reagan agreed to restore some of the funds and, as Culbertson and the OMB staff left the meeting, a few unkind remarks were exchanged accusing Culbertson of using the ploy of showing the President the model and distracting him while Miller was speaking.[24] Through multiple changes in

---

23. FY 1987 Budget Spring Review Presentation to the Administrator, Office of Space Station, 15 May 1985.
24. This account, as well as those describing Culbertson's demotion later in the chapter, are based on conversations with Culbertson during April 2004, and James Miller in July 2004.

administrations and OMB leadership, lower level OMB staff that carried over from administration to administration never supported the program and continued to raise roadblocks in the years ahead.

Although Culbertson had won a partial victory, the final FY 1987 submission, $410 million, was still far below the amount he asked for in the spring, and $190 million below NASA's final request to OMB. If the administration's budget was approved by the Congress, it would mean that at the very least some elements would have to be deferred. At this point, less than two years after receiving the President's approval, the Space Station program was having serious problems. Agreement on the "baseline configuration," required to proceed to the Phase C/D studies, was still in a state of flux and decisions on many other critical design details had been deferred until the spring of 1986. The administration's new out-year budget forecasts would not allow the program to proceed along the path NASA had identified to the Congress just nine months earlier. Work Package contractors had already raised a red flag that NASA had underestimated the cost to deliver the elements contained in each work package to achieve IOC. NASA's Space Station managers would soon be impacted, as would all NASA programs, by the *Challenger* accident at the beginning of 1986. The fallout and recovery from the accident would relegate Space Station problems to a secondary concern.

# — Chapter 2 —

## The Baseline Configuration – Everyone on the Same Page?

### (January 1986 - April 1987)

In spite of the management turmoil at NASA Headquarters and the changes that would take place at JSC with Griffin's resignation, Space Station planning continued on its scheduled course. With Culbertson's promotion to General Manager, John Hodge was elevated to the vacant position of Space Station Associate Administrator, but only in an "acting" capacity. Although Culbertson had blunted OMB's efforts to completely cripple the program with the FY 1987 request to Congress, budget concerns, both real and projected, became the overriding Space Station management problem.

Instead of the sacrosanct $8 billion number (in 1984 dollars) that had been advertised as the cost to provide the initial operational configuration (IOC), the contractors were now estimating $12 to $14 billion to arrive at that point. These estimates were "scrubbed" at Headquarters and JSC and then reduced; however, additional costs were added, by Level B management and the project offices, to take into account costs not included in the contractor's estimates. These additional costs were associated with NASA management, integration of the international components, potential contractor award fees, and program reserve. These additions, had not been included before in the program cost forecast and brought the total projected cost at IOC to over $17 billion.[1] Although the $17 billion was not accepted by Hodge and his staff as a final number, word of the higher cost estimates soon leaked out signaling the start of the cost estimate disputes with OMB and Congress that would continue without letup.

These new numbers forced Hodge to call a meeting of Level A and B managers. It was held in Houston, four days before the *Challenger* accident. The senior JSC attendee was John Aaron, Deputy Space Station Program Director, along with seventeen other Headquarters and JSC managers. Further reducing or "scrubbing" costs would have to continue and Hodge asked that all levels of management establish a list of actions that could be deferred while still maintaining the release of the Phase C/D RFPs at the beginning of FY 1987. If this schedule could not be maintained, a permanently manned capability would not be possible by January 1994, the announced target date. He also suggested that in order to provide user capabilities at IOC the international partners be asked to commit to providing specific facilities as soon as possible and agree to make access to these facilities available to all the partners. In this way U.S. user facilities that might be deferred because of budget constraints would be included later as part of NASA's plan for evolution and growth.[2]

### The Challenger Accident

On January 28, 1986, the space shuttle *Challenger* exploded shortly after liftoff resulting in the loss of the crew. President Reagan immediately announced the appointment of a review board to determine the cause. In the aftermath of the accident, in addition to the management turnover already underway, other major changes struck NASA's top management. One of the victims was Culbertson. He had been at KSC to observe the launch and when the magnitude of the accident became clear, as senior manager present (Graham did not attend), he immediately began to make arrangements to accommodate the families of the crew, who were at KSC to observe the launch.

Vice President George H.W. Bush arrived that afternoon. Culbertson took him to meet the families and then proceeded with preparations for memorial services at JSC. After making initial arrangements to fly senior NASA managers from Washington to Houston, he returned to Washington to complete the planning. At this point a controversy arose as to who should be included in the NASA contingent

---

1. Projected costs can be found in Level A/B Cost / Content Review File, NHO SSHP-HDC held by the NASA History Office.
2. NASA memo, from SP: Robert Freitag, to Distribution, Subject: Minutes of Level A/B Strategy Meeting, January 23-24, 1986, dated 5 February, 1986.

traveling to JSC. Graham refused to include Beggs who Culbertson had placed on the list. As a result, Culbertson elected not to travel on the NASA charter but with Beggs on a plane provided by DOD for other dignitaries. Eventually he attended the ceremonies at JSC with Graham and other NASA Headquarters managers. Three days after returning to Washington from Houston, Graham called him into his office. He told Culbertson he could no longer trust him and asked him to resign. Culbertson refused and Graham took the only course open to him, demoting Culbertson to manage a small, future programs office. (In Civil Service terminology a horizontal transfer, same grade level, few responsibilities.) With the staff struggling to understand their roles as a result of the many senior management changes that were taking place, the *Challenger* accident tipped the scales and morale within the Agency plummeted.

## MANAGEMENT CHANGES AND DESIGN REVIEWS

Shunted aside, Culbertson's influence on the Space Station was still being felt. In February Hodge and Hutchinson convened a follow-up meeting to the January Level A/B strategy meeting in Houston. Held in Washington, this time managers from Level C were also in attendance to discuss their first look at how they might defer actions to stay near the old $8 billion target. It was agreed to follow the strategy outlined by Culbertson in 1985 and call the deferrals "evolution and future growth" in order to avoid the appearance that the program was being downsized. Agreement was reached on which elements would be deferred. Instead of four U.S. supplied modules at IOC, the U.S. would provide laboratory and habitation / station operations modules. The other two modules would be laboratories provided by ESA and Japan. Other deferments included reducing OMV capabilities, robotic servicing, and new space suit development. The latter deferment meant that during initial assembly, EVA's would be conducted using the already developed shuttle EVA space suit. With these changes, the projected out-year costs would be reduced but would still be above the $8 billion target; the new number was $9.3 billion.[3]

Additional senior management changes were announced in February. Astronaut Richard H. Truly was transferred to Headquarters to be the new Associate Administrator Office of Space Flight. Three days after Truly agreed to take the job, he asked his close friend Thomas L. Moser, Director of Engineering at JSC, to come to Washington as his deputy. Moser reluctantly accepted the position and moved his family to the Washington suburbs.[4] These two management moves would soon prove to be critical as NASA attempted to recover from the *Challenger* accident, and was beset by questions as to its ability to manage complex programs like the shuttle.

Space Station Phase B interface requirement and system requirement reviews (IRR / SRR) were conducted in January and March 1986 at JSC. These reviews, with attendance including all the "work package" contractors, coordinated the four work package studies. To assure that all the contractors were working to the same set of assumptions, further adjustments were made in the specifications contained in the original Phase B RFPs. With these adjustments the results of the Phase B studies would be on target to support the preparation of the Phase C/D RFPs scheduled for release in October.

Apprehensive about the growing budget squabbles in the Congress, Hodge began to rethink the program's management structure and "work package" assignments. He proposed some radical changes that would shift some of the responsibilities from JSC to MSFC and Washington. Neil Hutchinson, apparently disappointed with his proposed reduced role, resigned and took a position in the private sector. With Hutchinson's departure John Aaron, his deputy, became the Acting Manager of the Space Station Program Office at JSC.

The system requirements review (SRR) in March took into account the latest results from the Phase B studies and marked the point at which the basic characteristics of the Space Station were decided. The dual-keel baseline configuration recommended at the RUR #2 meeting was adopted for the Phase C/D studies. The physical structure now resembled a large rectangle of four metal trusses connected at the four corners. The dimensions of this framework were 310 feet by 150 feet. Midway in the long

---

3. NASA Memo, from S: John Hodge, to Distribution, Subject: Convergence Process, dated 18 February 1986.
4. Based on interviews with Moser in June 2004.

dimension, was a 150 foot fifth truss to which the various modules would be attached (Figure 10). The dimensions for the U.S. modules were tweaked from RUR #2 and were standardized at 44.5 feet long with a diameter of 13.8 feet. The ESA and Japanese modules, with slight differences, would have essentially the same dimensions. One of the advantages of this decision was that all the equipment and experiment racks could be standardized for both the U.S. and international modules. It was decided that all the modules would operate with internal pressures of 14.7 psi. Internal pressures would be reduced to 10.2 psi 24 hours prior to EVAs in order to reduce the pre breathing time needed to accommodate the space suit pressure of 4.3 psi and lessen the chance of inducing the bends when reentering.[5]

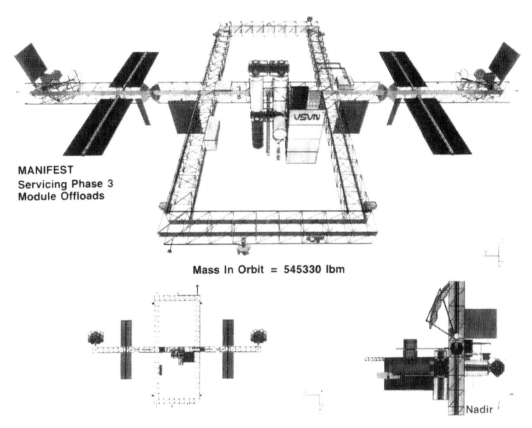

Figure 10. Space Station dual-keel configuration after assembly flight 17.

With the SRR dual-keel decision came other design changes. The most significant was the decision to modify the power generation system. A total of 75 kW would be provided when fully operational at IOC, 50 kW allocated to users and 25 kW for Station housekeeping. However, it was decided to provide the power with a hybrid system that included both standard photovoltaic arrays (25 kW), and 50 kW with solar dynamic heat engines. A spin-off from the terrestrial energy R&D NASA conducted in the 70s and early 80s, a solar dynamic system consists of a compound parabolic mirror to concentrate the sun's energy and generate high temperatures (up to 2000 degrees F) and use that heat to drive a turbine and generate electric power. The advantage of such a system is that compared to a photovoltaic array generating the same amount of electricity, a solar dynamic system is more efficient and can be one-fourth the size of a photovoltaic array. There were several disadvantages. A solar dynamic system could not be packaged for transport on a shuttle as compactly as a photovoltaic array. Also, the technology for the solar dynamic system was still evolving, especially the small Brayton cycle turbine that would utilize the high temperatures generated by the mirrors. Although it looked good in concept,

---

5. The Space Station, A Description of the Configuration Established at the Systems Requirements Review (SRR), Office of Space Station, NASA Headquarters, Washington, D.C. 20546, June 1986.

technological challenges would eventually force the program to abandon the approach and rely solely on photovoltaic systems. But in the summer of 1986, that was selected as the way to proceed.

Because of the many EVAs projected to service the Space Station and the co-orbiting free-flyers, it was decided to maintain development of a space suit especially designed for the Space Station. It would be deferred as mentioned earlier, and the shuttle suit would be used to complete initial assembly. But once initial assembly was completed, an EVA suit that maintained an internal pressure of at least 8.4 psi was needed to complete assembly and carry out other EVA tasks. The driver for this decision was the already described desire to avoid lengthy pre-breathing time required to go from the high internal Space Station pressure to a lower suit pressure. To operate at this higher pressure, a "hard" suit was already in the early stages of development. It provided easier articulation at the joints thus reducing fatigue and permitted an astronaut to carry out tasks that would be difficult or impossible in the "soft" shuttle suit. The same discussions and tradeoffs were conducted during the Apollo Program. At one point, the Houston Manned Spacecraft Center was directed to speed up a "hard suit" for Apollo and post-Apollo EVAs, but development was discontinued when missions scheduled to follow *Apollo 17* were canceled.

### JAMES FLETCHER RETURNS AS ADMINISTRATOR

In March 1986, the Reagan Administration determined that more experienced leadership was needed at NASA. The Presidential Commission on the Space Shuttle *Challenger* Accident (referred to hereafter as the Rogers Commission after its chairman) was scheduled to report its findings in May. It was evident from the public hearings held by the Commission that it would be highly critical of NASA management. The White House turned to Dr. James C. Fletcher who had been NASA Administrator at the end of the Apollo program (1971-77) and had returned to academia at the University of Pittsburgh. He was asked to turn the agency around, stem the outflow of experienced managers, and recruit new blood to replace the aging workforce. With some reluctance, he agreed to return.

While his nomination was being reviewed by the Senate, on April 2, 1986 Hodge provided a background briefing to Fletcher on the Space Station Program.[6] The briefing was interesting in what it covered and what was omitted. Acknowledging concerns arising from the *Challenger* accident, Hodge indicated that the program was dependent on using the shuttle. Available expendable launch vehicles (ELVs) could not do the complete job.[7] In addition, flights that were being planned to use the shuttle in order to validate some of the program's advanced technology were now in doubt. Hodge told Fletcher that he still expected the Phase C/D contracts to be released in October and signed by May 1987. The critical design review (CDR) would take place in mid-1989 after which fabrication would begin with the first element launched in 1992. Man-tended operations would begin in 1993 and IOC would be in early 1994. This schedule would require shuttle flights for assembly and logistics in 1992 and 1993. To justify his optimistic outlook, he quoted from President Reagan's February 1986 State of the Union address after the *Challenger* accident in which he said, "this nation remains fully committed to America's space program. We're going forward to build our Space Station." Hodge believed that since Fletcher was the Administration's choice to succeed Graham he would have some clout in the White House to overcome OMB's lack of support or outright opposition.

The program architecture he described to Fletcher remained essentially unchanged from the elements used as the starting point for the Phase B studies. The "Base" station still included laboratory and habitation modules, attached payloads, a satellite berthing and assembly capability, the OMV, and unmanned polar and co-orbiting platforms. The major difference was the decision to select a dual keel

---

6. The discussion that follows is based on "Background Material on the Space Station Program for Dr. James C. Fletcher," John D. Hodge, Acting Associate Administrator, Office of Space Station, April 2, 1986.
7. To students of history, it should not be surprising how events reoccur over time. After the *Columbia* accident in 2003, NASA reexamined the use of ELVs to complete the construction of the International Space Station (ISS). As reported in *Space News*, May 10, 2004, William Readdy, Associate Administrator for Human Space Flight, told a Senate panel that using ELVs to deliver to orbit the rest of the ISS elements would delay completion for four or five years and add considerably to the cost. Senator Sam Brownback (R-Kan), Chairman of the Senate Commerce, Science, Technology and Space Subcommittee, indicated he didn't like that answer and would pursue the inquiry in further hearings.

structure. He stated that the international partners were onboard with these changes including the early launch of the ESA and Japanese laboratory modules to make up for two U.S. modules that would not be available at IOC but deferred to a later date.

With regard to the international partners, he reviewed the actions being undertaken by each of the partners and described their contributions in detail with the most time spent reviewing Japan's interests. Although Japan was a full partner, there was an undercurrent of competition and fear that Japan would eventually outstrip the U.S. in space technologies and the use of space. He also outlined some of the concerns of the other partners, as well as the U.S. worries. Foremost among the international partners concerns was whether or not they would have meaningful roles to play, and in view of the continuous budget battles whether there was a guarantee of program continuity. In this regard, they questioned if the announced schedule would be kept. They also continued to worry that they might not be allowed access to some U.S. elements in view of the possibility that DOD might be a "user" of some facilities. This last issue continually raised alarms with the international partners as well as in Congress and with potential U.S. industrial and academic users. DOD had changed its mind several times on whether or not it would use the Space Station and this issue would not be resolved until 1988.

On the U.S. side, Hodge said the major concern was the increase in management complexity in terms of integration and allocation of functional capabilities for key international elements not completely controlled by NASA. With the changes that resulted from the strategy meeting in February, some of the international contributions were now on the critical path to successful IOC. For these concerns, Hodge assured Fletcher they were all being addressed and that "Canada, European Space Agency and Japan participate in Space Station Program as genuine partners."

With a final series of charts, Hodge outlined the vision of a continuously evolving and growing program. The Space Station would be a permanent facility and the initial design would allow "scarring" (NASA terminology that means the ability to add or attach new elements to the basic structure) without incurring great additional cost or added complexity. In this way Space Station capability would grow as needs arose. As envisioned in the earliest studies, this could include support of a lunar base, commercial manufacturing, and servicing geosynchronous satellites. In turn, this would mean the ability to provide additional capacity such as more power and to be able to accommodate larger crews. He cautioned that the cost to fulfill this vision was not completely understood but this would be clarified as the program matured. However, the major concern that Hodge failed to bring to Fletcher's attention was the escalating cost estimates for the overall program. The intent of the briefing was clear. Hodge wanted to assure Fletcher that despite some uncertainties NASA could maintain a robust, growing program. As Hodge remembers, Fletcher did not ask many questions.[8] He was primarily listening as this was just one of many program briefings scheduled before he was confirmed on 12 May.

## THE SPACE SHUTTLE – A POTENTIAL SPACE STATION SINGLE POINT FAILURE

When the Space Station Task Force first formed in 1982, the initial studies relied entirely on using the space shuttle to carry major elements to Earth orbit for assembly and servicing and to carry the crews. At that time, NASA was still in the throes of justifying the high cost of the shuttle fleet's development and operations. Management argued that these costs would be easily amortized and reduced over a period of many years predicated on the need and ability to fly thirty or more flights per year.[9] A Space Station destination requiring multiple flights fit nicely with NASA's long range plans to fully utilize the shuttle's capabilities. During early planning, the Orbital Maneuvering Vehicle was an alternative to carry small payloads from Earth to the Space Station. ESA was also considering a new launch

---

8. From an interview with John Hodge, May 18, 2004.
9. Economic Analysis of the Space Shuttle System, Prepared for National Aeronautics and Space Administration, Washington, D.C. 20546, January 31, 1972. The analysis showed a potential of 403-681 flights over a twelve year period. In a letter to Senator Walter F. Mondale, dated April 25, 1972, Administrator Fletcher refers to the Mathematica study and writes that even at a reduced flight rate, the shuttle would reach a break-even point at 360 flights and thus would "represent a good investment."

capability that would service the Space Station using the Ariane rocket but, again, only for small payloads or, perhaps, crew changes.

The Space Transportation System (STS) (aka the shuttle) had its maiden flight in April 1981 with Commander John Young, NASA's most experienced astronaut, at the controls and Robert Crippen his copilot. Before the first flight, in May 1980, NASA had published a guide for government and non-government utilization of the shuttle.[10] The payload carrying capability to low-Earth orbit for a 28.5 degree inclination mission was advertised as 65,000 pounds and for a 57 degree inclination mission, 56,000 pounds. Cargo bay maximum payload envelope was quoted as 15 feet in diameter by 60 feet long. A remote manipulator system (RMS) with a 50 foot manipulator arm, located in the cargo bay and operated from the crew compartment, would be available to assist in payload activities such as the launch and recovery of large satellites.

These same numbers would constrain the design of the initial space station elements, however, it was anticipated that shuttle performance would be upgraded in later years to permit carrying heavier payloads, especially to higher inclination orbits. By November 1982, JSC was circulating a long range schedule based on four shuttles with launch dates for assigned payloads extending out to September 1987.[11]

The *Challenger* accident on January 28, 1986, the 25th shuttle launch, exposed for all to see serious shortcomings in NASA management's chain of command and the way it conducted its business. More importantly, from the perspective of the Space Station, it portended potential severe impacts. The shuttle fleet was grounded for two years and eight months until *Discovery* lifted off on September 28, 1988. The investigation and remediation of the problems uncovered took much longer than anticipated. Some observers thought that NASA was overcautious in returning to flight. Many thought there would be only about a one year hiatus, but caution won out over a more aggressive approach to return to flight.

The Presidential Commission on the Space Shuttle *Challenger* Accident, chaired by William P. Rogers, former Secretary of State in the Nixon administration, consisted of twelve additional very experienced members including former astronaut and Vice Chairman Neil A. Armstrong, renowned test pilot Charles E. Yeager, and Nobel Laureate Richard P. Feynman. It began its hearings and deliberations in February 1986 two weeks after the accident. Operating under the terms of a Presidential Executive Order,[12] the Commission's mandate was to; "(1) Review the circumstances surrounding the accident to establish the probable cause or causes of the accident; and (2) Develop recommendations for corrective or other action based upon the Commission's findings and determinations." Carrying out its mandate, the Commission held a series of closed and public hearings at many locations across the country that extended into April 1986. The Commission report was published in June 1986.[13]

A reviewer of the Commission's recommendations could come to the conclusion that they all influenced, to some degree, how the shuttle would be utilized for Space Station assembly and maintenance, but none were specific to the Space Station. Most dealt with perceived management or pre launch process shortcomings that contributed to the *Challenger* accident. Sadly, several of the recommendations closely resemble those made by the *Columbia* Accident Investigation Board (CAIB), seventeen years later. NASA's safety oversight was found wanting. The Rogers' Commission recommended that NASA establish an Office of Safety, Reliability and Quality Assurance, and that was done. Limited options for launch abort and crew escape were criticized. One reason given by NASA management during the hearings as to why better options were not available – program cost. Maintenance safeguards and reporting of performance trends were considered inadequate and it was recommended that the practice of cannibalizing one shuttle to quickly provide parts for another launch be stopped. The Commission believed that relying on the shuttle as the agency's prime launch capability "should be avoided." A few recommendations were directly related to shuttle design deficiencies.

---

10. Space Transportation System, Reimbursement Guide, National Aeronautics and Space Administration, JSC-11802, May 1980.
11. STS Flight Assignment Baseline, National Aeronautics and Space Administration, JSC-13000-8, November 1982. Interestingly, the distribution list for this document, although containing 97 NASA Headquarters staff, did not include any senior managers assigned to the Space Station Task Force.
12. Presidential Executive Order 12546, signed by President Reagan, February 3, 1986.
13. Report of the Presidential Commission on the Space Shuttle *Challenger* Accident, June 6, 1986, Washington, D.C.

After receiving and reviewing the Commission's report, the President asked that NASA report back in 30 days and describe how and when the Commission's recommendations would be implemented. NASA, of course, had been closely following the Commission's deliberations and a number of actions were already underway. On July 14th NASA sent the President an overview of its plan including the milestones by which it would measure its progress.[14]

In the aftermath of the accident, NASA faced the reality of still relying solely on the shuttle for the success of the Space Station. Arguments, pro and con, were waged as to the wisdom of building a replacement shuttle to maintain a fleet of four shuttles, considered by NASA to be the minimum number to service all the anticipated needs. Several studies were conducted between 1985 and 1987 to understand the issues involved in developing alternatives, however, they never went beyond preliminary analyses and none were ever implemented.

With a little foresight, and now with 20/20 hindsight gained from the *Columbia* accident in 2003, Congress and the Reagan Administration should have anticipated flight interruptions of similar length if another shuttle accident occurred and taken more seriously the need to have alternative ways to build and service the Space Station. A robust, heavy lift ELV program started in 1987 would have served many other purposes besides supporting the Space Station. Instead, the Administration soon would take a different approach (NSDD-42 discussed in Chapter 3) that would delay and complicate the development of a heavy lift ELV.

## PROGRAM CONTINUES TO CHANGE

With Fletcher waiting in the wings until his confirmation (although he had already made the obligatory reconfirmation appearances before the Senate), the task of representing NASA in April at the FY 1987 Congressional budget hearings fell to Graham. Two major Space Station issues were predicted to be points of contention at the hearings, the $410 million budget and management changes that would effect the program as a result of the *Challenger* accident. It was well known that NASA was very disappointed with the Administration's Space Station budget request but it would have to put on a happy face and defend the President's submission. This would be done by explaining the stepped-up role that the international partners had agreed to, thus reducing the need for NASA's near-term funding. Overall Space Station capabilities would remain the same at IOC and U.S. elements would just be deferred. Another major issue would be the expected management and procedural changes that would be recommended when the Rogers Commission reported. Graham would pass the buck to Fletcher to describe at some later date how NASA would respond, as the Rogers Commission had not yet submitted its report.

Since early February Hodge also had been contemplating making major changes in Space Station management. In March, he made a presentation with recommendations to the Space Station Program Management Council consisting of Center Directors from the several NASA field centers involved in the program.[15] He pointed out that overlap issues had surfaced between the four work package managers. It was his belief that the Level B lead center located at JSC, put in place by Beggs, would not function well during the Phase C/D studies and hardware development, specifically the overlap in responsibilities between JSC and MSFC. Hodge was concerned that with Jerry Griffin's departure as JSC's Director, a senior manager with whom he had a long-standing working relationship, these overlaps would become more and more difficult to resolve. He did not foresee the same problems with the other two work package managers at LeRC and GSFC.

His proposed solution was to split the program into five parts with two main elements. The "pressurized environment" consisting of the habitation and laboratory modules, interconnecting structures, airlocks and associated subsystems such as the ECLSS would be assigned to MSFC. The second main element, "structure and architecture," included the truss, attitude control, data management, communications and tracking, and associated subsystems, and would be managed by JSC. The other parts, such as the power

---

14. Report to the President, Actions to Implement the Recommendations of the Presidential Commission on the Space Shuttle *Challenger* Accident, July 14, 1986, Washington, D.C.
15. John D. Hodge, Work Package Reassessment, Briefing to the Space Station Program Management Council, 19 March 1986.

system and free-flyers would remain at LeRC and GSFC. His proposals were not met with universal acceptance. JSC was especially unhappy as it felt that some of the proposed MSFC responsibilities, including the design and integration of the ECLSS, should remain in Houston.

Progress on another issue simmering in the background was contained in a report to Congress in mid-April. NASA and its contractors submitted their second, mandated report on the impact of designing the Space Station as a man-tended facility.[16] The analyses showed that a man-tended Space Station would provide useful early capabilities but it would not result in "substantial" cost reductions (some in the program believed it would add cost) and would not substitute for a permanently manned facility. NASA proposed to phase man-tended features into the permanently manned Space Station over a 3-5 year period after initial deployment of the "basic station." This compromise was accepted by Congress and the man-tended option became a major addition to the schedule during the assembly sequence that would lead to full capability.

## PHILLIPS REPORT – ANDREW STOFAN SELECTED AS SPACE STATION MANAGER

Prior to his confirmation, in anticipation of the Rogers Commission findings, Fletcher requested in April that the National Academy of Public Administration undertake a study of NASA management. Although the study, under the direction of U.S. Air Force General Samuel C. Phillips (hereafter referred to as the Phillips' report), would not formally submit its report until the end of December, Phillips provided an early briefing to senior NASA management at the end of June on his committee's findings relevant to the Space Station. His briefing included several recommendations for restructuring Space Station management. One would significantly change the direction of the program: "Large multi-center programs should be managed by a strong program director at Headquarters supported by a competent program office in the Washington area." Not a surprising recommendation; that was the way NASA successfully managed Apollo during Phillips' tenure as Apollo Program Director. Changes to Space Station management conforming to the Phillips' report recommendations were promptly approved by Fletcher.

Fletcher announced that the Space Station Level B (the "competent program office") would be moved from Houston to the Washington area, location TBD. Its new title would be Level A Prime (A') (Figure 11). The second part of Phillips' recommendation, "a strong program director," was implemented when Fletcher quickly selected LeRC's Director, Andrew J. Stofan for the position.[17] Stofan's selection was a bitter disappointment to Hodge who thought he deserved to be named the full time Associate Administrator Space Station Program, the position he had been filling on an acting basis for the last seven months.

Immediately following his decision to reorganize Space Station management, Fletcher sent a memo to the center directors involved in the program explaining his reasons for the changes and what would be done.[18] When word of the pending changes became known, turmoil and protests erupted; first at JSC and soon after from the Texas congressional delegation. Rumors abounded at JSC that thousands of jobs would be lost or transferred to Washington and other centers.

## ASTRONAUT GORDON FULLERTON CRITICAL OF DESIGN

Further complicating the management changes, astronaut C. Gordon Fullerton, provided an analysis to JSC Level B management that was published in July by *Aviation Week & Space Technology*. It described his concerns with the new Space Station design.[19] Fullerton called into question many aspects of the "baseline configuration," especially those that required the astronauts to assemble and maintain in space such a large structure during EVAs.

---

16. The NASA plan to implement the Advanced Technology Advisory Committee (ATAC) recommendations was submitted to the Senate Commerce Committee on April, 18, 1986.
17. NASA Headquarters Release 86-84, NASA Announces Space Station Associate Administrator, New Program Structure, dated 30 June 1986.
18. NASA Directive, from A: James C. Fletcher, to Distribution: "Space Station Management Directive," dated 27 June 1986.

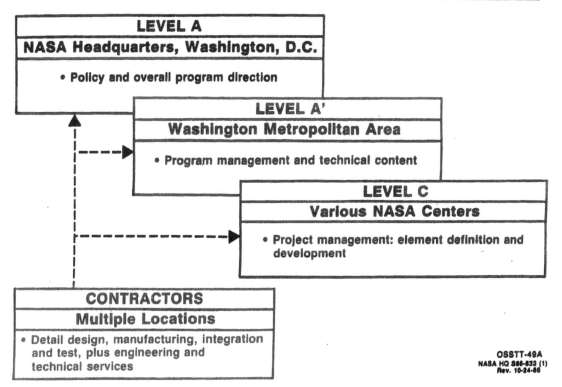

Figure 11. For the management hierarchy shown in this figure, beginning in 1986 Stofan headed Level A, the Level A manager was Tom Moser. Level C managers are named in note 23.

After flying as commander on shuttle flight 51F in August 1985, Fullerton had been assigned by the astronaut office to follow Space Station developments, a typical type of assignment for an astronaut when programs were approved. Astronaut input of this kind was never ignored. Keep in mind, astronaut safety was always uppermost in the minds of JSC engineers, especially in the aftermath of *Challenger*. EVAs were risky no matter how well planned and practiced. A requirement to perform multiple EVAs just compounded the concerns. If design changes would reduce such risks, they would be made.

In the *Aviation Week* article, correspondent Craig Couvalt listed the key issues that concerned the astronaut corps and the JSC engineers. First was the worry that with one less shuttle the fleet would not be large enough to support the Space Station as well as the rest of a crowded flight manifest. Initial projections for the Power Tower design required 8-10 shuttle flights to achieve IOC. The dual keel design increased the number to 19 or more flights.

Another Fullerton concern was that the dual keel design, with modules suspended in the middle of the trusses, was too complex to be assembled by astronauts performing EVAs from the space shuttle. At the same time, Fullerton felt that the design minimized the astronaut's EVA role once the major elements were in place because there would be poor access for maintenance of equipment used outside the laboratories and living quarters. With the habitation and laboratory modules framed by the main trusses, visibility from inside would be obstructed. He also raised concerns that the new design did not include provisions for a crew escape vehicle, relying solely on the shuttle to return the crews in an

---

19. Launch Capacity, EVA Concerns Force Space Station Redesign, Byline Craig Couvalt, *Aviation Week & Space Technology*, July 21, 1986. In a conversation with Couvalt on 1 June 2004, he could not recall who leaked Fullerton's internal memo, however, the article includes several quotes from Fullerton that he assumes he obtained after the memo became known.

emergency. When a shuttle was not docked, there would be no emergency return. Regardless of the type of emergency, the crew would have to remain in the safe haven and wait for the next shuttle to arrive which could take up to one month or more.

Many aspects of the habitation module designs were also criticized, including inadequate separation of work, exercise, and eating areas from the sleep quarters. The ability to use the Canadian mobile servicing center during assembly of major elements was also questioned. Finally, there was a growing concern at JSC that NASA was pressing ahead with a complex design and misrepresenting how difficult it would be to achieve program schedules. Reading between the lines, twenty years later, Fullerton was telling senior management the program was too complex and proceeding on too fast a track based on a design that would be difficult to assemble. Couvalt quoted Fullerton as saying he may have overstated his concerns for "shock effect."

Whether or not he overstated the situation, Fullerton's "shock effect" got everyone's attention. John Aaron announced that changes were required. Design changes and downsizing studies began at the four Phase B contractors. Meanwhile, following the Phillips' report recommendations, Hodge had issued instructions to change the management roles for work packages one and two, essentially the same changes that he had proposed earlier in February.[20] And all of this was happening during Stofan's first days in Washington before he could get his feet on the ground. Stofan found himself in the middle of an internal management and Space Station design tug of war. To add to his problems, members of the House Appropriations Subcommittee for HUD and Independent Agencies that had oversight of NASA's budget were crying foul because they had not been consulted about all the changes. Washington politics would be a rude reawakening for a NASA center director who had become accustomed to being king of the hill in the hinterlands of Ohio. Probably the greatest problem he inherited was the $8 billion dollar cost figure that NASA had concocted and defended since the earliest days of the program. By this time, the new $9 billion plus estimate was well known outside the agency.

In addition to Congressional concerns over management changes, at a hearing before the House Space Science and Applications Subcommittee on July 21, Congressman Bill Nelson (D-FL) dug into the concerns raised by Fullerton on the Space Station "baseline configuration." Nelson asked why these concerns had not surfaced earlier and was told by Hodge (he had not yet announced his retirement after being passed over for Associate Administrator) that it was not clear why JSC had waited so long to raise these issues. And then, surprisingly, Hodge said that no funds were being spent to look at alternative designs even though it was well known that the JSC contractors had begun such studies and they would soon be formalized by Stofan.[21] This apparent lack of communication between JSC and NASA Headquarters at such a crucial time, while the design was undergoing major modifications, points up the continuing problems NASA Headquarters experienced when dealing with JSC and vindicated Phillips' recommendation to move Level B to Washington.

### FLETCHER CALLS TIME OUT – CETF FORMED

To defuse the criticism that his reorganization directive created, Fletcher agreed to hold off carrying out any further changes for 90 days and in the meantime try to mend his Congressional fences. This delay prevented Stofan from moving ahead to implement the management changes he had in mind; changes that would have to be in place before the Phase C/D RFPs were released. Instead, he began to hold a series of briefings to get up to date on the program. Although LeRC was one of the four Level B work package centers, with the responsibility to develop the power systems, Stofan had not been closely following all the complexities of the evolving design and the growing unrest in the Congress.

The delay, however, provided time to investigate the design concerns highlighted by Fullerton and others. In August Stofan convened a Critical Evaluation Task Force (CETF).[22] He decided to select a neutral site for the Task Force deliberations and it convened in August at the Langley Research Center (LaRC) in Hampton, Virginia. W. Ray Hook, LaRC Space Station Office manager, chaired the

---

20. NASA Memo, from S: John Hodge, to John Aaron, Subject: Work Package Disposition, dated 30 June 1986.
21. The background information for this discussion is based on a July 28, 1986 *Aviation Week & Space Technology* report headlined: "NASA Managers Divided on Station," datelined Washington but without a correspondent's name.

meeting. Its charter was to: "Perform a configuration look-back to develop options and, to the extent possible, validate the baseline configuration." While reviewing the baseline configuration, the CETF would attempt to develop technical options that would address the identified concerns. In addition to the Task Force, Stofan created an Executive Technical Committee that would meet in Washington to provide guidance and review the CETF findings. Chaired by Stofan, its membership included staff from Headquarters, the five Level C project offices (these five managers also served on the CETF), representatives from their respective engineering divisions, plus the astronaut Flight Crew Office. By this time, Clark Covington had been named to fill the position (downgraded to Level C with the establishment of Level A' in Washington) vacated by Hutchinson. Both he and John Aaron, who resumed his role as deputy, were included on the Executive Technical Committee.[23]

CETF membership included a cast of hundreds from the five NASA centers, the Phase B contractors, and users. The international partners participated at various times when specific topics of interest to them were reviewed. Five design elements were examined: (1) The assembly process, (2) Safety, (3) Operations, (4) Transportation capacity, and (5) Early productivity.[24] Over the next several weeks the various subgroups met and addressed the many concerns that resulted from the new "baseline configuration."

Perhaps the most significant modification was in the assembly sequence that added elements called resource nodes. The need for such nodes (small, separate, pressurized chambers) had been studied in past years but no decision had been made to include them in the basic design. Now they were considered to be essential elements. There would be four nodes. Two attached to the ends of the US laboratory and habitation / operations modules, and two between the U.S. modules and the ESA and Japanese modules (Figure 12). The resource nodes would ease and reduce the first assembly activities and subsequent EVAs by providing expanded volume and storage space, and ports for airlocks and shuttle docking. By adding the resource nodes, the assembly sequence and overall design would address many of Fullerton's concerns on how difficult he thought it would be for the astronauts to carry out assembly. Overall, the recommended changes resulted in reducing the number of shuttle flights to arrive at the man-tended, early science capability, from seven to five, and reduced the number of flights to achieve the permanently manned configuration (PMC) from twelve to eight[25] (Figure13). As a result of the CETF changes, the dual-keel configuration would be delayed for some time and both the man-tended and permanent manned configuration would be on orbit before the four main trusses would be assembled.

When Ray Hook reported to Fletcher in mid-September that the CETF found the Baseline Configuration to be "basically sound," in fact many major changes had been recommended and eventually would be incorporated in the design. Hook summarized the advantages of the proposed CETF modifications for Fletcher. They would permit earlier science returns, reduce assembly risks and the number of EVAs, improve safety, add pressurized volume, improve the ability to maintain the Space Station, and accommodate the new STS launch capacity constraints that were in the process of being implemented as a result of the *Challenger* accident. CETF also calculated a new number for the amount of EVA hours needed to reach PMC, 120 man hours. This was a considerable reduction over previous estimates of 416 man hours. Recommendations to provide a crew emergency return vehicle (CERV) were deferred and the time that crews could remain in the safe haven was extended from 28 days to 45 days. Fletcher was not completely satisfied with all the recommendations and asked that additional analyses be done to understand what additional costs should be included, the potential role of expendable launch vehicles, and how the program office should be organized.

In need of an experienced Deputy to share the management load, Fletcher tapped Dale D. Myers. Myers had been Associate Administrator for Manned Space Flight at the end of the Apollo Program,

---

22. NASA News Release: 86-116, NASA Announces Plan for Space Station Review, dated: 20 August 1986.
23. The Level C managers were: JSC: Clark Covington, MSFC: Luther Powell, LeRC: Ronald Thomas, KSC: Charles Giesler, and GSFC: Ronald Browning.
24. This information and that in the following paragraphs was extracted from: Space Station Configuration Chronology (1983-1988), Tom Moser, Program Director, SSPO, not dated, held in the Author's files.
25. The nineteen flights mentioned were to achieve complete assembly. To avoid confusion as you read on, the C in PMC and IOC, were used interchangeably to mean "configuration" or "capability. Later, to be politically correct, PMC became PHC, permanent human capability.

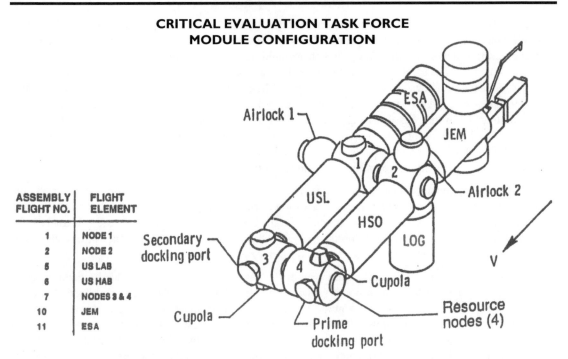

Figure 12. The proposed module configuration resulting from the CETF review showing the locations of the resource nodes relative to the other modules. (Moser)

## CRITICAL EVALUATION TASK FORCE
## BASELINE MODIFICATIONS

| MAJOR CONCERN | BASELINE | CETF REVISED BASELINE |
|---|---|---|
| Excessive EVA | | |
| - During Assembly | 48 MH/FLT | 24 MH/FLT |
| - Total to PMC | 416 MH | 120 MH |
| - Post PMC EVA Maint. | 800 MH/YR | 240 MH/YR |
| Restrictive Pressurized Volume (US) | 15,200 FT3 | 19,320 FT3 |
| Early Science | | |
| - Man Tended Capability | 7 ASS'Y FLTS | 5 ASS'Y FLTS |
| - Attached Payloads | 10 ASS'Y FLTS | 3 ASS'Y FLTS |
| Maintenance | | |
| - Equipment Moved From Outside | — | 2,730 LB |
| - External Maintained Items | 300 | 110 |
| Transportation Limitations | | |
| - Man Tended Capability | 7 FLTS | 5 FLTS |
| - Permanently Manned Capability | 12 FLTS | 8 FLTS |
| Safety | | |
| - Safe Haven Duration | 28 DAYS | 45 DAYS |
| - Eva Risk | | Much Reduced |
| Phasing Of Mobile RMS | Absolutely Needed After FLT-2 | Mobile Arm Phased Beginning with FLT MB-3, Fixed Arm Added at FLT MB-4 |

Figure 13. Modifications to the "baseline configuration" addressing the major concerns voiced by astronaut Gordon Fullerton and others. (Moser)

and more recently, DOE's first Under Secretary. This was a sound decision because Fletcher had agreed during his confirmation hearings to support a replacement shuttle for *Challenger* in order to maintain flight schedules. Prior to joining NASA in 1970, Myers had managed the Apollo Command and Service Module development at North American Rockwell, now the shuttle builder. At the beginning of October, before Myers was confirmed, Graham resigned and took a White House position as Director Office of Science and Technology Policy.

As a result of the CETF studies, Stofan announced at the end of September realignments in the four work packages. Preliminary design of resource nodes would be added to WP #1. WP #2 at JSC would be responsible to define how the resource nodes and airlocks would be outfitted. This meant that JSC would oversee additional contracts and reduce the number of jobs that might be lost as a result of the program restructuring. Acceptance of resource nodes as new elements (estimated to cost $200 million each) reduced some of the anxiety in the Texas congressional delegation. GSFC would now have the responsibility to release an RFP to define the design for the flight telerobotic servicer (FTS). WP # 4 contracts at LeRC were not changed. *Defense Daily* reported that the Texas congressional delegation was satisfied with the new assignments and that the "Texas delegation stands ready to push forward."[26]

In spite of the accommodations made to please the Texas delegation, key supporters for the Space Station, the budget guillotine continued to hover overhead. The resource nodes would add at least $800 million to the overall costs. Those who were closely following the program, in and out of Congress, knew that NASA's new $9 billion estimate to achieve IOC (not yet officially endorsed) was as unrealistic as the original $8 billion estimate.

Both the configuration changes and the new management direction in which Space Station was headed raised new concerns in the user community that was rapidly reflected in Congress. In September the House Committee on Science, Space, and Technology requested NASA to recommend "alternative approaches to science operations management."[27] In response, NASA put a study team together staffed primarily from the Headquarters Space Station and OSSA offices. Skipping ahead a year, the team's report was submitted in August 1987.[28] The study examined several management alternatives and came to the conclusion that each science discipline preferred a different approach. Each favored development of science payloads as the sole responsibility of the users. Ignoring this desire, the study recommended that development of flight experiments continue to be maintained within NASA but that management interfaces be simplified. Other recommendations dealt with defining the science infrastructure for users including the need to coordinate international requests to use Space Station facilities.

In order to maintain schedules, NASA's initial request to OMB for FY 1987 was over $1 billion in new funds. Later in the budget cycle, by juggling start dates for some major pieces, the request was reduced to $600 million. Congress finally appropriated $420 million for FY 1987, $10 million above the administration's final request. But the full effect of receiving $420 million versus NASA's higher requests were still unknown as well as the impact on the FY 1988 request that was being prepared predicated on receiving $600 million in FY 1987. Budget estimating had become a great guessing game with little resemblance to conducting sound program management. In September Stofan asked John Aaron to review all the different estimates and assumptions that defined the program costs in order to meet the announced IOC date and then develop new cost options.[29] A Cost Containment Team consisting of staff from Headquarters and the Work Package managers was immediately formed, reporting to Aaron.

## STOFAN ORGANIZES WASHINGTON OFFICES

Although still constrained by Fletcher's 90 day "time out," Stofan continued to look ahead to how he would structure his office and Level A Prime. In September he established the Space Station System

---

26. *Defense Daily*, September 30, 1986, p. 156.
27. House of Representatives Committee on Science, Space and Technology, Report 99-829, September 16, 1986.
28. Space Station Science Operations Management Concepts Study, August 1987. The report did not carry the usual SP number, and it is not clear if or when it was transmitted to the House committee that made the request.
29. NASA Memo, from S: John Hodge, to Distribution, Subject: Planned Procedures to Establish Space Station Cost Containment, dated 12 September 1986.

Engineering and Integration Task Force. The objective of the Task Force was to recommend what the role of Level A Prime should be and how it should be organized to perform its functions. He chose Lawrence J. Ross who had served as his Director for Space Flight Systems at LeRC and had been elevated to center Deputy Director after Stofan's departure.

Ross formed the Task Force with members from Headquarters, the four "work package" centers plus KSC, LaRC, and JPL. Of note was the participation of Richard Kohrs from JSC, of whom more will be heard from later. The Task Force began with the understanding that General Phillips' recommendations, that a "competent program office in the Washington area" should report to a "strong" program director, was the model Stofan wished to install. Apollo and Shuttle programs were reviewed to understand what the staffing levels were for offices performing similar functions. The Task Force's conclusion was that 200 to 400 civil servants would be needed augmented with 1,000 or more support contractor personnel. The larger numbers would be closer to the Apollo experience. Because the new Level A Prime office, with the primary function of providing system engineering and integration would start from a zero base, the Task Force recommended that staff with the needed skills and experience be recruited from NASA centers plus adding some new hires. They cautioned that problems would be encountered trying to relocate so many people from the NASA centers, most with strong ties to their current locations and a general aversion to working in Washington.[30]

In September Franklin D. Martin transferred from GSFC where he was the Director Earth and Space Sciences Division to become Stofan's deputy. In October Moser, after some arm twisting by Stofan, moved from his position as Truly's deputy, and a more senior position, to become the Director Space Station Program Office (Level A') and began to assemble his staff. Moser recalls telling Fletcher that if he took the job he believed that he would need a staff of over 380 to manage the oversight of such a complex program.[31] (By December, as he became more familiar with the job, he had raised his estimate to 478.) As predicted, convincing longtime civil servants at NASA centers to move to Washington was difficult. In order to meet the higher cost of living, it would require that those moving to Washington would have to be promoted and offered grade levels of GS-15 or SES and the number of such positions available was tightly controlled. It would take Moser many months to fill most of his key positions and many lower level positions were never filled.

With concerns continuing to mount in the Congress as to the direction the Space Station program was heading, at the end of October 1986 Congressman Bill Nelson, Chairman of the House Subcommittee on Space Science and Applications, wrote to Daniel J. Fink, Chairman of the NASA Advisory Council. He questioned the ability of the new Space Station configuration proposed by the CETF "to accommodate high quality attached payloads at an early time." Based on this concern, he asked Fink to provide an "independent review."[32] Fink responded reminding Nelson that the TFSUSS, under the direction of Dr. Peter Banks, had been providing the type of review he requested. Although the Task Force was scheduled to finish its work, Fink wrote that he would extend its life so that it could participate in the review of the draft RFPs and the other issues that Nelson identified related to the science payloads. Nelson also wrote "that NASA should establish a formal Space Station Advisory Committee under the Council." Fink and Stofan eventually would agree to Nelson's latter request and the Space Station Advisory Committee was established at the end of 1987.

Meanwhile, release of the Phase C/D RFPs, until very recently scheduled for the start of FY 1987 (October 1, 1986), was delayed while the Aaron budget exercises were underway. A new release date of early February 1987 became the target. If this new release date held, it meant that contracts would not be signed until the end of summer, a slip of at least five months.

For the remainder of 1986 and into January 1987 the Cost Containment Team, now led by Moser and Aaron, would devote their energy to finding a cost number that NASA could live with and that Fletcher

---

30. L.J. Ross, Space Station System Engineering and Integration Task Force Report to Mr. Andrew J. Stofan, 14 October 1986.
31. Based on interview with Moser, 6-30-04.
32. Letter to Dr. Daniel J. Fink, President D.J. Fink Associates, signed by Bill Nelson, Chairman Subcommittee on Space Science and Applications, dated October 27, 1986. Fink's response on NASA letterhead in the author's files is undated, but usually Congressional inquiries were answered in a week, especially if the inquiry came from the chairman of a NASA oversight committee.

could defend in the final battles with OMB and at the FY 1988 Congressional budget hearings. It was known that Congress, supported by a large contractor lobby, would hold out for the originally promised Space Station capabilities; OMB would hold NASA to the original $8 billion cost estimate. The two positions were mutually incompatible. Costs of various options ranging from $16.5 billion to $17.4 billion were being tossed around like ships in a storm. Each option required design changes, deletions or additions in capability, and schedule changes that would defer IOC. The $8 billion estimate was completely off the table.

Three days before Christmas, Stofan and the Cost Containment Team briefed Fletcher and Myers on their recommendations. By making a few modifications to the CETF revised baseline, they proposed a $13.2 billion (FY 1984 dollars) program. This number included a reserve of $2.5 billion, or almost 20 percent of the total program cost, a major increase from the amount included earlier. Fletcher considered the total cost to be still too high and the cost exercises continued. The next day, NASA announced that Fletcher had approved a new Space Station organization.[33] In addition to identifying a number of individuals who would assume senior management roles, Moser was given a second hat and would serve as Stofan's Deputy Associate Administrator for Development. Earlier in the month, Stofan had held a retreat for all his senior managers to set the course for the battles that lay ahead. At the retreat it was decided that Moser needed additional management clout, especially in regard to controlling program costs, and the new position was created and agreed to by Fletcher.[34]

## PHASED PROGRAM TASK FORCE

A few months after the CETF completed its work, Stofan and Moser began a new round of studies conducted by the newly appointed Phased Program Task Force. Its objective differed from the CETF; reduce the cost of the program without significantly reducing Space Station capabilities and without precluding the ability to add capability later in the program; a reprise of the problem that Culbertson and Hodge attempted to solve one year earlier. Now, however, finding a solution was more urgent. It was clear that many members of Congress, both in the House and Senate, were becoming more difficult to please. Even strong Space Station supporters, in the aftermath of *Challenger* and the announced program changes, were becoming harder to convince that NASA was on the right track.

To accomplish their objective, the Task Force decided that the easiest route would be to divide the program into two parts: Block 1 and Block 2 (referred to later as Phase I and Phase II). Block 1 would include those elements leading to PMC (based on the CETF configuration) after eleven assembly flights. In addition there would be two polar-orbiting, unmanned platforms, one supplied by ESA and one by the U.S.[35] The initial crews would be four to six after the two U.S. modules were attached. A new EVA man hour estimate to reach PMC, 156 hours spread over eight missions, was calculated by the Task Force. This was 36 hours more than projected during the CETF study. Crew size would increase to eight after three additional assembly flights carrying logistics and the JEM and ESA modules. A complete Block I Space Station composed of all the revised baseline configuration elements would be operating at the end of sixteen flights (Figures 14 & 15). Observational science would be limited, but a substantial program of materials processing could be conducted, as well as life science investigations. Automated technology payloads also could be accommodated.

Block 2 would be sold as a growth capability based on Block 1 and leading to the full dual-keel configuration with a satellite servicing bay, and include a co-orbiting platform. Block 2 would add various new modules and attached payloads and provide additional power for users by incorporating solar dynamic power systems. When the Block 2 Space Station was completed, it would have the full capabilities envisioned earlier in the program.

A review of the draft Phase C/D RFPs by a user group consisting of NASA, academe, and industry representatives was conducted at the end of 1986. Clearly, Block 1 would not be as attractive to

---

33. NASA News Release 86-180, Space Station Organization Approved, dated December 23, 1986.
34. Ibid. 31.
35. Details of the CETF and Phased Program Task Force findings are taken, primarily, from: Space Station Configuration Chronology (1983-1988), Tom Moser, Program Director, SSPO, no date but probably put together in February 1988, the last date shown at the bottom of the charts.

## REVISED BASELINE
## CONFIGURATION ASSEMBLY SEQUENCE

| | | | |
|---|---|---|---|
| MB | MANNED BASE | | |
| P | PLATFORM | | |
| L | LOGISTICS | | |
| FTS | FLIGHT TELEROBOTIC SERVICER | | |
| JEM | JAPANESE EXPERIMENT MODULE | | |
| ELM | JAPANESE LOGISTICS MODULE | | |
| EF | JAPANESE EXPOSED FACILITY | | |
| ESA | EUROPEAN SPACE AGENCY | | |
| MSS | MOBILE SERVICING SYSTEM | | |
| PMC | PERMANENTLY MANNED CAPABILITY | | |
| TCS | THERMAL CONTROL SYSTEM | | |
| WTR | WESTERN TEST RANGE | | |

| Flight Number | | | |
|---|---|---|---|
| 1 | MB 1 | 1/2 PV, NODE, TRUSS, FTS | |
| 2 | MS 2 | 1/2 PV, NODE, TRUSS | |
| 3 | MB 3 | TCS, AIRLOCK, PAYLOADS, MSS | |
| 4 | MB 4 | AIRLOCK, PAYLOADS | |
| 5 | P-1 | U.S. POLAR PLATFORM (WTR) | |
| 6 | MB-5 | U.S. LAB MODULE | ← MAN TENDED |
| 7 | L-1 | LAB MODULE OUTFITTING | |
| 8 | MB 6 | U.S. HAB MODULE | |
| 9 | P-2 | ESA POLAR PLATFORM (WTR) | |
| 10 | MB 7 | NODES, CUPOLAS | |
| 11 | MB 8 | CREW (4), LOGISTICS | ← PERMANENTLY MANNED CONFIGURATION (PMC) |
| 12 | MB 9 | JEM, EF#1 | |
| 13 | L-2 | LOGISTICS, CREW, PAYLOAD | |
| 14 | MB 10 | ESA MODULE | |
| 15 | L 3 | CREW, LOGISTICS | |
| 16 | MB 11 | EF#2, ELM | ← COMPLETE REVISED BASELINE CONFIGURATION |

Figure 14. Revised Baseline Configuration Assembly Sequence. OSSTT-910J. (Moser)

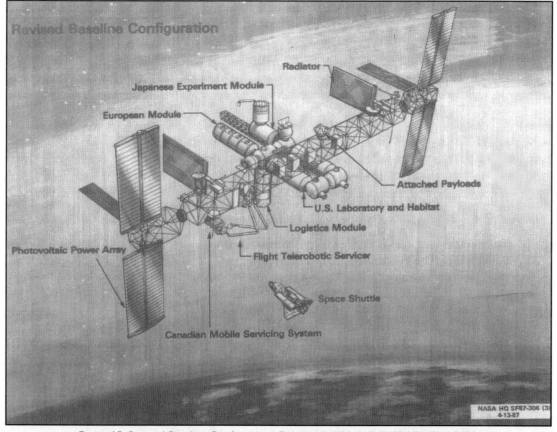

Figure 15. Revised Baseline Configuration Drawing NASA HQ SY87-276 (3) 4-3-87.

potential users as the originally proposed design. Observational science requiring very large pointing payloads, would be restricted. Only a limited number of materials processing payloads could be accommodated especially if they imposed a requirement for large amounts of electric power. Life science and animal experiments would likewise be restricted because of a limited amount of available crew time. For all investigations there would have to be a substantial reliance on automation and ground control monitoring, to reduce the need for data processing and crew time. In spite of these restrictions, the "user community" gave its blessing to the RFPs. As 1986 was drawing to a close, the Phase B definition and preliminary design, underway for the past twenty-one months, was completed. The CETF "baseline configuration" was accepted by all the parties participating in the review as the starting point for the Phase C/D RFPs.

After considering several other options and further discussions, Fletcher decided to go forward with a two part program in the hopes that both Congress and OMB would agree to the phasing. Together, Block 1 and Block 2 would cost $14.5 billion, considerably more than the $8 billion number the program had been promoting from its inception. Having earlier rejected the $13.2 billion cost Fletcher now agreed to an even larger number. However, there were new factors that caused the elevated cost. The Cost Containment Team had recommended that the total include a reserve of $3.1 billion. As a general rule of thumb, the amount of reserve needed in a program is highly dependent on the program's maturity. The lower the maturity the higher the reserve needs to be. Fletcher, using past experience as a guide, upped the recommended reserve to $3.8 billion, his recognition that the program was a long way from maturity. Other changes proposed by the Cost Containment Team added $600 million.

To accommodate Space Station cost escalation and growth in all programs, NASA's request to OMB for FY 1988 was $10.569 billion including $1 billion for Space Station. Fletcher and Myers elected to brief OMB the first week of February 1987 on the new Space Station cost commitment of $14.5 billion. In their presentation to OMB they argued that if allowed to pursue a phased program it would be possible to project flat budgets of approximately $2 billion per year in the out-years and still provide a fully capable Space Station. They thought this would appeal to OMB but their arguments were rejected. Still smarting over the previous year's budget dispute, OMB maintained its position that the program should be deferred.[36] The FY 1988 budget submitted to Congress reduced NASA's total request by $1.088 billion and the Space Station portion was reduced to $767 million.

There could be no further pretense. The Space Station program was being driven by just one consideration – cost. NASA would manipulate the numbers, schedules and content trying to keep the program alive while continuing to provide some semblance of a useful facility. Beggs' earlier declaration that you could buy a Space Station "by the yard" had come full circle. But now the question had become: "What kind of a Space Station was the country buying?

Complicating the congressional debates on NASA's FY 1988 budget, an old issue boiled to the surface again. What would be DOD's role in the Space Station? During discussions with DOD and OMB in conjunction with the FY 1988 request, DOD wanted to expand their potential usage of Space Station to include "national security purposes" of undefined character. This proposed change in language leaked out and raised red flags with both Congress and the international partners. With the MOUs they signed in 1986, all had agreed that the Space Station would be used for "peaceful purposes in accordance with international law." It was understood that "peaceful purposes" included some types of military activity. This was clarified to mean that weapon systems would not be carried but that certain types of research could be conducted. However, this clarification did not completely satisfy Congressional critics. During a hearing before the House Committee on Science, Space and Technology on April 5, 1987, Fletcher tried to cool the debate. He pointed out that there was agreement that major weapon systems such as laser beams would not be carried on the Space Station but, as an example, "research on semiconductors would be fair game."[37] After additional hearings on the subject, the issue remained

---

36. These conclusions are based on: Memorandum For the President, From: James C. Miller III, Director, Subject: Revised Cost Estimates for the Space Station, dated February 10, 1987, and statements in Miller's book, Fix the Budget, Urgings of an "Abominable No-Man," Hoover Institution Press, Stanford, 1994.

37. Administrator James Fletcher responses to questions before the House Committee on Science, Space and Technology, 100th Congress, 1st Session, February 5, 1987.

unresolved awaiting further action by the Administration and agreement on the language in the new round of MOUs being negotiated with the international partners.

Behind the scenes another important decision was being made. One of the Phillips' Study recommendations to Fletcher in June 1986 included how to locate the new Space Station Program Office. He suggested that it should be "more than 15 minutes and less than 45 minutes" from NASA Headquarters. Moser had taken the recommendation to heart and had been looking in the Maryland and Virginia suburbs for a good location. Forty five minutes would be better than fifteen if for no other reason than it would discourage frequent visits from Headquarters staff. He had already moved his office three times in three months and was getting desperate. Finally, in April, through his JPL support contract, offices were leased in Reston Virginia, forty five minutes from downtown Washington, if the traffic was light.[38]

Meanwhile, Congress was becoming more agitated as leaks of the new costs were circulated. The Congressional Budget Office (CBO) suggested canceling the program in order to take a bite out of the growing deficits.[39] Their analysis showed that as much as 80% of all shuttle launches would be needed for Space Station assembly and servicing in the mid-1990s, an unacceptably high percentage in view of other payload commitments. During hearings before the Senate Commerce Committee, Subcommittee on Science, Technology and Space defending the use of the shuttle, Fletcher stated that in regards to manned space flight "I think we may have lost the competitive edge to the Soviets." He also expressed concern that the U.S. was falling behind Europe and Japan in the field of microgravity research, a rather surprising suggestion in view of the partnership that had developed for such research on the Space Station. The threat of foreign competition (Soviet) had justified Apollo. Perhaps he thought that this appeal would strike a chord again and save the Space Station. Two of the leading critics at the hearing in February were Senators Albert Gore (D-TN), and John Kerry (D-MA) who argued that the *Challenger* shuttle replacement should be canceled.[40]

## CONFLICTS CONTINUE WITH OMB AND CONGRESS

Because of concerns about the sudden cost growth and probable disputes with Congress in the upcoming hearings, OMB would not give NASA permission to release the Phase C/D RFPs. OMB Director James C. Miller III explored his options to have an independent analysis made of NASA's new cost commitment. Under a withering headline: "NASA's Engineers Plan More Plumbing," the *New York Times* chimed in with its analysis. "The agency's main future project is a space station, assembled from a scaffold and modules flown up on 32 shuttle trips. The station serves no great goal, just a multitude of minor missions to muster support from all possible users . . . Many of its missions could be met in other ways, but supplying hardware is what keeps NASA busy. Like the shuttle, the space station is not an end but a means, infrastructure, built for when a President someday decides what to do with it."[41] The *Times* article reflected the thinking and arguments put forward by critics in and out of Congress.

To appease OMB, and have some control over the cost analysis, Fletcher turned to an old standby. When NASA faced opposition to its programs, the National Academy of Sciences, National Research

---

38. Based on interview with Moser, 6-7-04.
39. For those unfamiliar with the workings of the Congress, the Balanced Budget and Emergency Deficit Control Act of 1985 also mentioned in Chapter 2, called the Gramm-Rudman Act, took effect in the last half of FY 1986. As a result, the Congress and Executive Branch departments and agencies agreed to reduce spending over the next six years at which point the budget was projected to be in balance. The deficit goals went from $171.9 billion in FY 1986 to zero in FY 1991. The total reduction targeted for FY 1988 was $108 billion. NASA was an attractive target because it fell into the category of "discretionary" spending as opposed to expenditures on programs such as Social Security and interest on the debt that must be funded. Threats to reduce NASA's appropriation were taken very seriously by NASA managers.
40. This background based on a story, "Space Station Faces Possible Two-Year Deployment Delay," filed by Theresa M. Foley in the February 9, 1987 issue of *Aviation Week & Space Technology*.
41. The *New York Times*, February 12, 1987. The *Washington Post*, not to be outdone in criticizing a Republican Administration's program, on February 26, 1987 printed a story by reporter Kathy Sawyer, "Space Station in Deep Trouble" that echoed some of the same concerns.

Council (NRC) could be counted on to provide authoritative analyses that might slap NASA wrists but would usually find a solution or make recommendations acceptable to all parties. Responding to Fletcher's request, NRC chose Dr. Robert Seamans to chair the review. Jumping ahead a few months, the review was completed in September 1987. Among its many findings, it endorsed the revised Baseline Configuration but stated that the nation's long-term goals in space should be clarified before committing to go forward with Block 2. A prime concern was how NASA arrived at its cost estimates. In particular it told NASA to pay close attention to how it estimated and controlled operating costs (GAO was also studying this issue at this time). NRC warned that deployment would be difficult and risky recommending that NASA upgrade the shuttle and develop ELVs to provide logistics support to reduce the number of shuttle flights. Concern was also expressed about orbital decay of the complex during early assembly if for some reason reboost were delayed. Lack of spare hardware in case equipment was lost during deployment was an additional concern. Finally, it concluded that the Space Station program represented a management challenge for NASA. If the Space Station was to be successful, the commitment to the program must be "national in character."[42]

This last conclusion by the NRC emphasized the dilemma that NASA faced; softening support in the administration that first proposed the program and outright hostility from some members of a Democratic controlled Congress. How much influence a lame-duck president could exert on the Congress was open to question when it was well known that strong critics of the program existed within Reagan's own administration.

For the next two months, from the first week in February until the beginning of April Fletcher, Myers and Stofan, supported by their staffs, trudged back and forth to multiple Congressional hearings. But Congress was not buying how the $767 million would be spent. A series of hearings in the House and Senate culminated on April 7 with a hearing before the House Appropriations Committee, Subcommittee on HUD and Independent Agencies, chaired by Edward P. Boland (D-MA). At the top of Boland's concerns was the amount of electric power that would be available for Block 1. He was particularly concerned that if the ESA and Japan modules were assembled at the beginning of Block I there would not be enough power for U.S. users. He made it clear that these modules should not be a part of the initial assembly. In negotiations with OMB to lower costs, NASA had agreed that only 50 kW would be provided at IOC. Boland considered this to be too little and insisted that it be increased to 75 kW. Then, in a striking display of Congressional arm twisting, he said, "Before we sign off on releasing the RFPs . . . we need to be confident that those basic tenets are understood."[43]

Congressional oversight and micromanagement to this level, approving RFPs, was highly unusual, to say the least. It now appeared that in order to release the RFPs, NASA had to obtain both OMB and Congressional consent. If this continued, schedules would never be met. With the pressure on to release the RFPs and maintain any semblance of schedule, NASA agreed to modify the design.

## STS Modifications and Study of a Mixed Launch Fleet

In addition to the management changes that the Rogers Commission recommended in 1986 to assure the safe operation of the shuttle, NASA implemented modifications to the STS that effected shuttle performance when used for Space Station missions. Correction of the O-ring joint design, whose failure was considered to be the cause of the accident, was relatively straight forward and that redesign did not effect shuttle performance except for tighter, temperature constraints at launch. Changes were also made to beef up the nose wheel, main landing gear and brakes to provide larger safety margins for launch-aborts-to-landing when full payloads would still be onboard. Consider that commercial aircraft that encounter an emergency soon after takeoff normally jettison fuel to reduce landing weight. Such a procedure was not possible for shuttles. Once demated from the large expendable tank and SRBs, shuttles had to land with their full launch payloads. These changes added some weight to the basic orbiter.

---

42. Report of the Committee on the Space Station of the National Research Council, September 1987, Washington, D.C.: National Academy Press. The Committee issued an interim report in June that focused primarily on the Space Station cost estimates and the difficulty of arriving at a good cost estimate.
43. Hearings before the U.S. House of Representatives, Committee on Appropriations, Subcommittee on HUD and Independent Agencies, 1988 Appropriations, 100th Congress, 1st Session, 7 April 1987.

All shuttles were not created equal. One, *Columbia*, weighed more than the other remaining two. In addition, depending on the mission, allowable payload weights would vary. As described earlier, the maximum that shuttle payloads could weigh prior to any modifications for changes made after the *Challenger* accident was 65,000 pounds when launching to a low-Earth orbit and a 28.5 degree inclination. This was the inclination chosen for the Space Station, however the orbital altitude that the Space Station was designed to maintain was 220 nautical miles, slightly higher than that at which maximum payload could be delivered. To reach this higher altitude reduced the payload launch weight. With the new design changes and other launch restrictions, short duration Space Station missions could count on less than 46,000 pounds for payloads. For flights that would entail longer times on orbit when EVAs were conducted, total payloads would be reduced approximately 3,500 pounds more to account for 1900 pounds of EVA expendables and equipment and 1500 pounds for the docking module. In addition, two thousand pounds were usually held in reserve to be available in case of late payload weight growth. Thus, as a result of the *Challenger* accident and other factors, to achieve Space Station PMC would require two more flights than the CETF planned (10 versus 8), or the overall Station configuration would have to be simplified.[44]

Prompted by the *Challenger* accident, additional studies were done to examine the need for a mixed fleet of launch vehicles including a heavy lift launch capability that could augment or substitute for the STS. The National Commission on Space in May 1986 projected in their report that the shuttle fleet "will become obsolescent by the turn of the century." They recommended three major needs "would have to be met in the next 15 years: cargo transport to low Earth orbit; passenger transport to and from low Earth orbit; and round-trip transfer beyond low Earth orbit." The Commission did not get into the detail of what a mixed launch fleet would consist of that would serve all those needs.

This detail was included in a study[45] conducted by NASA Headquarters and completed at the end of 1986. Briefly referred to earlier in the chapter, the study task force addressed three main issues: (1) the appropriate mix of shuttle and ELVs, (2) policies and practices that NASA could use in planning launch services, and (3) how to promote the commercialization of launch services. The study proposed a mixed fleet launch strategy for the periods 1988 to 1995 and 1995 to 2010. The latter time period, covering Space Station assembly, would use a mix of ELVs some of which were still in the testing or development stages. A heavy-lift ELV would be used whenever possible to off-load large payloads, such as science spacecraft, that were scheduled for launch on the shuttle. In this way, heavy Space Station or other payloads that required astronaut interaction or EVA could get priority on the shuttle. The second study, requested by NASA, was conducted by the NASA Advisory Council and was completed in March 1987.[46] Its recommendations closely tracked the earlier study but added a sense of urgency in view of the impact of the *Challenger* accident and the effect another accident might have. An interesting conclusion to the studies was the realization that using the shuttle for logistics resulted in a large penalty. The "up" payload was always larger than the "down" payload by as much as 28,000 pounds. Not a surprising revelation when you think of the missions, but it brought home the inefficiencies of using such an expensive launch vehicle for mundane resupply and argued for using ELVs for these flights.

More than one year after the accident, the shuttle fleet was still grounded and further delays were expected. The consensus of both reports was clear; future space programs needed a heavy-lift, expendable launch vehicle. The Advisory Council also supported the need for a fifth shuttle and recommended investigating how to modify the STS to have the capability to launch heavier payloads. Studies were already underway to double shuttle stay times on orbit by increasing the amount of expendables they could carry and by improving the efficiency of their usage for what would become known as the extended duration orbiter (EDO). (STS had a name change at this time, NSTS, National Space Transportation System. But both names were used interchangeably in later years. The titles included the orbiter, the external tank and the solid rocket boosters.)

---

44. The numbers provided here were extracted from Space Station Configuration Chronology (1983-1988), Tom Moser, Program Director, SSPO, not dated but probably put together in February, 1988, the last date shown at the bottom of one of the charts.
45. Report of the NASA Mixed Fleet Study Team, NASA Headquarters, December 30, 1986.
46. Report of the Task Force on Issues of a Mixed Fleet, NASA Advisory Council, March 16, 1987.

Following the recommendations of the above studies, the Space Station Program Office began another transportation study in the spring of 1987. Participants were brought in from the four work package centers and LaRC. Over the next six months the study reviewed how to reduce the number of shuttle flights required for assembly, decrease risks during on-orbit assembly and module outfitting, and reduce the effect of losing a Space Station element due to either a launch accident or failure of the element to operate correctly once delivered on orbit.

Four options were examined that could lead to reducing the number of shuttle flights: (1) utilizing the new NSTS that incorporated the modifications resulting from the *Challenger* accident recommendations, (2) using an enhanced NSTS that would have advanced, higher performing, solid rocket motors (ASRMs), (3) developing the shuttle-C that would have overall better performance, and (4) using a combination of existing ELVs and shuttles.

New numbers were being developed for the post-*Challenger* NSTS, configured with ASRMs and operating the space shuttle main engines (SSMEs) at 104%. They indicated that the launch weight could increase and that the payload could grow from 46,000 pounds to approximately 56,000 pounds by the mid 1990s. The latter weight, more than 15% higher than the original specifications, meant that the shuttle could carry more equipment and reduce the number of assembly and logistics flights. Most importantly, the changes would allow the landing weight of the shuttle to increase from 211,000 to 230,000 pounds, improving landing safety margins.

A modified shuttle, Shuttle-C, had been a gleam in the eyes of shuttle management and the Space Station staff for several years. It would require a new NSTS design configured with standard solid rocket boosters and external tank, but the orbiter itself would be greatly modified and operate with improved avionics. The good news was that the Shuttle-C SSMEs would run at 100%, the preferred power setting, versus 104%. With these changes Shuttle-C could carry 98,000 pounds. All other things being equal, Shuttle-C had the potential to reduce by almost half the total number of flights needed for assembly. However, the added cost of development made it very unlikely that it would be available in the near term.

The analysis of unmanned, expendable launch vehicles included five boosters that either existed or were close to being operational. They were Titans II, III, and IV, Delta 3920, and Atlas / Centaur. Titan payload weights increased from 1,500 to 26,000 to 30,000 pounds for each upgrade. Delta 3920 listed its payload capacity as 7,000 pounds and the Atlas / Centaur could carry 11,500 pounds. Shroud sizes varied for each booster which meant that if they were used for Space Station assembly, payloads would have to be tailored for each weight and to fit within each shroud. The Titans III and IV provided the most advantages but each had serious disadvantages such as availability (they were being procured for military payloads) and the inability to restart the second stage for rendezvous maneuvering. Even if modified, the expendable launchers would only be useful to deliver payloads if the OMV were also available to retrieve the payloads and carry them to the Space Station. Under the best of circumstances it would be a number of years before the OMV was ready for flight. An unmanned, shuttle derived vehicle (SDV) was given a cursory look. It would have the largest payload capacity, estimated to be 100,000 pounds, however, it would also be the most costly to develop. The study also examined resupply and reboost requirements for the Space Station as they would effect the number of required flights.

When the Transportation Study completed its work, it made a number of recommendations. Foremost, the program should stay with the baseline NSTS, but improved with the expected enhanced capabilities to launch and land higher weights. Using the enhanced NSTS, but with increased safety margins, ten or eleven assembly flights would still be required to achieve PMC depending on the decision to include two or four nodes. However, a total of twenty flights would be needed to complete Block I assembly, an addition of four flights above the last assembly schedule. With each analysis the assembly sequence and number of flights changed depending on the assumptions used. The study settled on a flight rate of five per year versus the eight flights that had been thought necessary to accommodate all anticipated users. This rate would still provide a little margin for unplanned contingencies. Four person crews would occupy the Space Station at PMC and crew rotation would occur every 45 days. To reduce the number of crew rotation flights, it was proposed that crew stay time gradually increase to 180 days.

When Block I assembly was completed, crews could grow to eight. The final recommendation was to continue to study options to increase NSTS margins and develop Shuttle-C.[47]

These recommendations were not out of line for the conditions of the day, but if they were followed the Space Station would continue to be completely dependent on the shuttle for success. Budget realities undoubtedly shaped the recommendations, and as a result NASA lost the opportunity to actively pursue a heavy lift ELV alternative that would be useful to the Space Station and other applications. Some Senators, such as Donald W. Riegle Jr. (D-MI), were sympathetic to the need for ELVs but NASA could not muster enough support to convince OMB and Congress that it needed alternative launch vehicles. At that time, DOD had forty-three expendable launch vehicles on order compared to NASA's one. DOD was on record requesting that it be the lead agency to develop a new heavy lift ELV. Several large payloads that DOD had committed to launch on the shuttle were now in jeopardy and needing a heavy lift ELV if shuttle flights were reduced or restricted. In this instance, and others, the inability of NASA to obtain needed funding early in the Space Station program set the stage for subsequent problems that the program would be forced to face. Eventually, an alternative launch vehicle would be needed and it would be provided by an unexpected source, especially if you were looking into your 1987 crystal ball.

Two other options might possibly help reduce the number of scheduled logistic and crew change-out flights. ESA and Japan had begun parallel studies using the Ariane-5 and H-II boosters, respectively. It was uncertain if they would be man-rated, however, they were both being designed for logistic missions requiring rendezvous and docking. When they would be available was still unknown. Before the changes were made to reduce the U.S. requirement to supply lab and habitation modules by moving up the launch of the ESA and Japanese lab modules, ESA had been planning to use the Ariane to deliver their module. Now it would have to be delivered by the shuttle. The Japanese Experiment Module (JEM) had always been scheduled for launch on a shuttle flight, the H-II rocket was not being designed to carry the JEM. Thus, as Space Station planning went forward, the only sure carrier available for assembly and crew rotation was the shuttle. NASA, that prided itself in having redundancy in its systems and options in case of emergencies, was now wedded to a single, somewhat fragile system, the NSTS, to successfully complete the Space Station.

---

47. Data for the Transportation Study derived from Space Station Configuration Chronology (1983-1988), Tom Moser, Program Director, Space Station Program Office, no date.

## — Chapter 3 —
## Replacing Challenger — More Studies — More Paper
### (April 1987 - January 1989)

Without a heavy lift ELV that could support the Space Station, replacing *Challenger* became a high priority. During Senate confirmation hearings, Fletcher was grilled on the Administration's position for the replacement. Concerns voiced were not confined to its need for the Space Station but also on the impact the needed additional funding would have on other NASA programs. Those in favor of the new shuttle understood that Space Station assembly would be drawn out and difficult if the shuttle fleet was confined to three orbiters. They also knew that large communications and science payloads already under development would be effected. Those opposed to building a fifth shuttle had multiple reasons including using it as a pretext to advocate canceling the Space Station. More overtly, opponents made the argument that within a constrained budget NASA would have to shift funds or cancel programs to find the necessary resources, something they were unwilling to approve.

Negotiations between NASA and OMB to provide the funding for the new shuttle were difficult. Estimates of the cost by NASA's Office of Space Flight fluctuated widely but were on the order of $1-$2 billion. Some spare pieces left over from the fabrication of the first four shuttles were in storage at Rockwell. How easy or difficult it would be to use them was a matter of conjecture. Some thought it would greatly reduce the cost, others saw problems. Adding to OMB's discomfort, NASA was asking for new funding to modify the existing NSTS design to improve its performance as previously discussed. When NASA's FY 1987 appropriation was finally approved, Congress added over $3 billion to the President's budget request to fully fund the replacement shuttle. Some in Congress rebelled at this large increase but they were in a distinct minority. GAO had warned in July 1987 that building a new shuttle plus modifying the existing fleet in response to the *Challenger* accident would not come cheap. The bill had just been delivered with the cost distributed within several NASA budget line items.

Backing up in our story as discussed earlier, the NASA Management Study chaired by General Phillips in 1986, commissioned immediately following the *Challenger* accident, provided NASA with many important observations and recommendations. In December 1986 Phillips made an in-depth presentation (104 charts) of his study's recommendations to senior NASA management. The presentation undoubtedly reflected his wisdom and character and when he spoke people listened. Wisely, several recommendations provided earlier in June 1986 concerning the Space Station program (covered in the previous chapter), had already been implemented.

Two themes in the Phillip's report's many recommendations were repeated in the briefing. The shuttle recovery program "must not be placed at risk." Needed resources must be made available. The report endorsed the Rogers Commission recommendations on the need to revise the shuttle management structure and improve communications. It warned that the long hiatus in space flight activities could raise problems at return to flight. Launch teams would be out of practice and would need careful supervision. The Phillips study provided specific suggestions on managing the "extensive interrelationships" between Space Station and Space Shuttle programs. One of the recommendations was to create a new organizational level, a "super AA," to whom both the Associate Administrators for Space Flight and Space Station would report. The reasoning for this recommendation was that once assembly flights began the Space Station and space shuttle would actually operate as one program.[1] It would take three years for NASA to come around to this view and, in a small difference with the recommendation, the Associate Administrator of Space Flight would wear both hats. Former astronaut William B. Lenoir would be elevated from Associate Administrator for Space Station to Associate Administrator Office of Space Flight with the Space Station Office reporting to him in his new and expanded role.

---

1. Presentation to the Administrator National Aeronautics and Space Administration, Report of the NASA Management Study Group, December 16, 1986, S.C. Phillips.

Another voice on the subject of how NASA should recover from the *Challenger* accident was heard in March 1987. The American Institute of Aeronautics and Astronautics (AIAA) had conducted its own, detailed analysis of NASA programs.[2] Their study concluded that with the drive to eliminate the federal budget deficit, the U.S. civil space program faced a "crisis." One chart included in the AIAA Assessment showed a comparison of federal agency outlays from 1965 to 1985. Of all the agencies, several showing growth of 354% to 1771%, only NASA's outlays were negative by 54%. Among their many recommendations, those that would affect the Space Station included replacing the *Challenger* as quickly as possible, substantially increasing budgets for microgravity research, strengthening international space station cooperation and, in a bow to Congress, "The Space Station should remain on schedule for deployment in 1994, but its design should be sufficiently flexible to accommodate user demands commensurate with budget, schedule, and operational constraints."

One other AIAA recommendation, applicable to Space Station, made the case that multi-year appropriations should be approved for projects that require long-term commitments. This struck at the heart of the problem afflicting the Space Station. It was a plea often raised by government agencies responsible for conducting large programs that extended over many years before completion. Requiring yearly budget approval from Congress tied the hands of managers trying to run efficient programs. They were never sure if the next year's funds for which commitments existed would be forthcoming forcing them to make program decisions that were not always in the best interests of running a sound program.

Studies conducted by the AIAA, whose membership represented almost all the aerospace industry, were done by knowledgeable participants and usually carried some weight. However, in this case, the Congress turned a deaf ear to the "crisis" and the argument for multi-year funding. Thus Congress continued to influence minor to major details of the Space Station for years to come through control of the purse strings.

## GAO Disputes Runout Cost Estimates – Progress Made

The budget debates of 1986 carried over to 1987. At the request of Robert A. Roe (D-NJ), Chairman of the House Committee on Science, Space, and Technology, the Government Accounting Office (GAO) began a review in January 1987 of the cost methodology used by NASA to arrive at its new Space Station runout costs. The GAO report, delivered in July, provided another new and still higher cost estimate. GAO looked beyond NASA's $14.5 billion estimate that had been submitted to OMB. In their analysis GAO costed a number of items not included in the $14.5 billion but would be required to complete the program. Analyzing projected annual operating costs from fiscal years 1987 to 1998, GAO concluded that the program would add $5.3 billion to the $14.5 billion. NASA had only included in the $14.5 billion an estimate of operating costs after FY 1998 at about $1 billion per year. Then, GAO examined what they termed "excluded costs" that they estimated "could add billions of dollars." The "excluded costs" included such items as the crew emergency return vehicle and additional modifications to the shuttle fleet.[3]

The simple arithmetic if you agreed with the GAO analysis, and GAO said they ran the numbers past NASA, would give a total of over $20 billion, the first time a $20 billion estimate reared its ugly head. In NASA's defense, it had never said these costs were included in the $8 billion and $14.5 billion estimates. Most of these excluded costs were spread around in other parts of NASA's budget or were still uncosted. NASA chose not to identify these additional costs at this time, budget examiners would have to ferret them out on their own.

Finally, accommodating all the meddling by Congress and OMB, two Phase C/D RFPs were released at the end of April 1987. The first RFP was to develop a phased Space Station (a revised Block I) with a requirement to provide 75 kW of photovoltaic electric power (per the Boland agreement). The second RFP, called the enhanced configuration, would add solar dynamic systems to bring total power up to

---

2. U.S. Civil Space Program: An AIAA Assessment, Approved by the AIAA Board of Directors, Edited by Jerry Grey, Stanley Rosen, Johan Benson, March 1987.
3. GAO Fact Sheet for the Chairman, Committee on Science, Space, and Technology, House of Representatives, Space Station, NASA's 1987 Cost Estimate, July 1987, GAO/NSIAD-87-180FS.

125 kW. Offerers were requested to submit proposals for each option.[4] Because of the late release of the RFPs, it was expected that contracts would not be signed until the fall. If all actions proceeded according to this new schedule, first element launch would occur in March 1994 and consist of one-half of a photovoltaic array, a node, a segment of the truss, and part of the reaction control system and tank. Other elements would follow. After a year of budget concessions and design compromises, the program appeared to be settling down.

Although the budget disputes delayed the release of the Phase C/D RFPs, progress was made in other critical areas. In May Boeing Computer Services was awarded a contract to provide a technical management information system (TMIS) that would maintain, distribute, and archive in one system data generated by the three NASA management levels and contractors. TMIS was a critical management tool that in addition to tracking all the technical and organizational interfaces, including those with the international partners, would be used to support management decision making. It sounded like the answer to any managers prayer but in the years ahead it would not perform as well as expected. Also in May an RFP was released for a Phase B study of the flight telerobotic servicer (FTS), a device that would support the astronauts EVAs during assembly and servicing and complement the Canadian RMS when it became available. Two contracts would be awarded to Grumman and Martin Marietta in November for the FTS Phase B definition studies.

In June, Lockheed Missiles and Space Company was selected to develop the software support environment (SSE) that would provide a common foundation for all future program software development, integration, and maintenance. Its primary goal was to minimize costs and risks during software development and determine if off-the-shelf software could be used. And finally, Grumman Aerospace was selected in July to provide broad program support for the Level A' office. Although still in the process of transferring personnel from the centers and hiring new staff, by August Moser had decided how he wanted to organize his office and new organization charts were published for both his and Stofan's offices (Figures 16 & 17).

### Ride Report – Mission to Planet Earth

After serving on the Rodgers Commission, astronaut Sally K. Ride, the first American women in space, was asked by Administrator Fletcher in the spring of 1987 to chair a task group "to define potential U.S. space initiatives, and to evaluate them in light of the current space program and the nation's desire to regain and retain space leadership." Recovering from the *Challenger* accident would not be easy. Fletcher conceded that NASA had fallen from the high esteem in which it was held when he first led the agency. Until the Space Station was launched, the Soviet's *Mir* space station had no competition. NASA would have to demonstrate that it had a bright future if for no other reason than to respond to the many critics who were using the *Challenger* accident and the escalating cost of the Space Station to make the case that NASA was already overcommitted and budgets should be reduced. The Ride report also would attempt to bolster NASA's image and morale by identifying potentially exciting and affordable long-term goals. The Task Group participants, composed almost entirely of NASA personnel from the various centers and headquarters, started their study by asking two questions: "Where do we want to be at the turn of the century?" and, "What do we have to do now to get there?"

When the interim report was published in August 1987,[5] the answers to the questions sounded familiar: "four bold initiatives were selected for definition, study, and evaluation." They were: (1) Mission to Planet Earth, (2) Exploration of the Solar System, (3) Outpost on the Moon, and (4) Humans to Mars. All the initiatives would be dependent on utilizing an evolving Space Station beginning with the Phase I configuration.

Mission to Planet Earth would be a very ambitious program that would add many more projects to the Space Station than already planned including four sun-synchronous polar platforms and five geostationary platforms. It was proposed that this initiative would include the international partners already committed to the Space Station as well as other U.S. government agencies such as NSF and NOAA.

---

4. NASA News Release: 87-65, NASA Issues Requests for Proposals for Space Station Development, dated 24 April 1987.
5. Leadership and America's Future in Space, A Report to the Administrator By Dr. Sally K. Ride, August 1987.

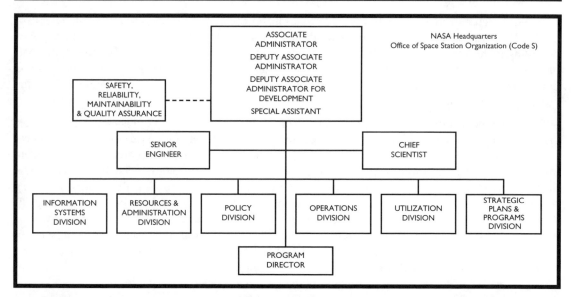

Figure 16. NASA Headquarters (Level I) Office of Space Station Organization chart. (1987 OPS report)

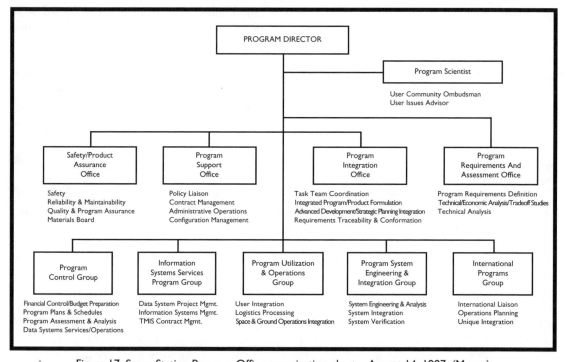

Figure 17. Space Station Program Office organization chart – August 14, 1987. (Moser)

For the Outpost on the Moon, the Space Station would serve as an "operational hub in low-Earth orbit" for crews and equipment before being transported to the Moon. This initiative proposed frequent trips to the Moon after the year 2000. Sending Humans to Mars also envisioned a major role for the Space Station. First it would be used for life sciences research to validate long-duration human space flight. Then it would serve as the jumping off and return point for manned round-trip "sprints" to Mars of approximately one year duration. The first of these "sprints" was proposed to take place before 2010. The Space Station would also be used to receive the automated return of Mars samples and avoid the possibility of contaminating Earth with unknown pathogens.

In addition to the four initiatives, the Ride Report delved into the issues of transportation and needed technology. It supported the conclusions of the "Report of the Task Force on Issues of a Mixed Fleet"[6]

---

6. Report of the NASA Mixed Fleet Study Team, NASA Headquarters, December 30, 1986.

that said that NASA urgently needed a heavy-lift expendable launch vehicle. On advancing space technology, the Report took the position that the present course that NASA was taking was a "status quo caretaker path with no potential growth."

Assessing whether or not such a bold program could be realistically embraced, the Report rationalized that in the absence of fiscal and resource constraints, "America ought to be doing this." Providing support for this national duty but at the same time urging some caution, the Report quoted from the Rogers Commission, "The attitude that enabled the agency to put men on the Moon and to build the space shuttle will not allow it to pass up an exciting challenge – even though accepting the challenge may drain resources from the more mundane (but necessary) aspects of the program." In other words, some NASA programs might have to suffer in order to support these new goals. The problem of course is that one person's "mundane" is another's top priority. Space priorities are set, ultimately, by the Congress and the White House responding to very diverse constituencies. Congress and the Administration would partially endorse and fund the initiatives recommended for Mission to Planet Earth and Exploration of the Solar System. However, the funds needed to start development for Outpost on the Moon and Humans to Mars were not approved and would have to wait for President George H.W. Bush to make a proposal in 1989. Although Congress had provided some new funds for the Ride task group initiatives, as the beginning of FY 1988 drew near, Congress had not been as kind to the Space Station request.

Another study was undertaken in the same time frame as the Ride task group by the NASA Advisory Council.[7] This study focused on international relations and overseas competition and growing competence in space activities. Highlights of the report pointed out that the Soviet Union would be contending for world leadership and was organizing a major marketing program of Soviet space services. Specifically, they were targeting western nations "through discussions with individual scientists and through new government agreements." Glasnost had just appeared and how nations would react to a more open USSR was still unknown. While conducting the study, members of the NAC Task Force traveled to the Soviet Union and returned impressed by what they had learned. In a similar vein, the Task Force pointed out the advances being made in Europe and Canada and stated that although these countries were partnering on the Space Station the "Europeans have been restive under the limitations of a subordinate relationship." Japanese programs were given close scrutiny. They reported that Japanese space capabilities were increasing rapidly and that Japan was seeking to avoid dependency on other nations for critical technologies. Chinese and Indian capabilities were also discussed in the report noting the progress both countries were making.

As a result of these observations, NAC came to several conclusions: "U.S. monopoly of access to space or dominance in space activity will no longer have substance . . . leadership in specific areas and technologies will in many instances be shared and in other areas will be lost." In the view of the Task Force, continued and increased cooperation would become more important. The major deterrent to cooperation in space with foreign nations had been the fear that it would result in the transfer of sensitive technology. NAC urged that this issue be reexamined and that a new policy be adopted to accommodate these fears. This rather dour outlook reinforced what some members of Congress had been forecasting. As the new fiscal year got underway, more problems than solutions seemed to be surfacing for the Space Station program and all of NASA. A bogeyman threatening U.S. space leadership, as was the case for Apollo, always helped at budget time to convince the Congress to pass bigger budgets.

Coincidentally, the release of the NAC report was immediately followed by former NASA Administrator, Thomas Paine, publishing a report of his trip to Moscow to celebrate the 30th anniversary of Sputnik.[8] The Soviets put on a real show describing the programs they would conduct

---

7. NASA Advisory Council, Task Force on International Relations in Space, International Space Policy for the 1990s and Beyond, October 12, 1987.
8. "Observations on a Trip to Moscow to Attend the International Space Future Forum Celebrating the 30th Anniversary of Sputnik," Thomas O. Paine, Director – The Planetary Society, Former Administrator – NASA, Former Chairman – National Commission on Space, October 19, 1987. The full report is in the author's files with attached articles on the Forum from Aviation Week & Space Technology dated October 12, 1987. Excerpts of Dr. Paine's report were printed in the American Astronautical Society Space Times, vol. 27, no. 1, January-February 1988.

in Earth orbit and missions to the planets and deep space. Examples: A series of satellites to Mars including landers and sample return, and opening the first robotic lunar surface laboratory. Paine was impressed, writing that the Soviets were confident and open with their plans and forging ahead on many scientific and technological fronts and that cooperation with European and Japanese space capabilities was increasingly important. He asked the question: "Does this portend a Spenglerian technological decline of the west?" He answered his question by writing: "Of course not." But went on to warn that "NASA's problem is confined to a tiny triangle bounded by the White House, Capital Hill, and the Pentagon." A fair observation and still on the mark. The motivation for his concerns, the Soviet grand plans, were mostly bluster and would soon be relegated to the "dust bin of history" as the sitting President might have declared.

## Mood in Congress Deteriorates

In spite of the big boost in NASA's overall FY 1987 budget to fund a new orbiter, from a Space Station perspective, fiscal year 1988 began on the same low note as FY 1987 had ended. During the Authorization and Appropriation hearings on the FY 1988 budget Congress asked NASA to supply six new Space Station analyses covering such disparate subjects as personnel levels and satellite servicing. It seemed as soon as one study was completed Congress would ask for two more. Uncertainties and problems abounded and support from the Administration was waning as evidenced by the FY 1988 budget battles with OMB resulting in the Administration requesting almost $300 million less for Space Station than NASA requested. With the Phase C/D RFPs on the street Congressional hearings on NASA's FY 1988 budget continued. During initial hearings by the House Authorization Subcommittee on Space Science and Applications at the end of March 1987, Congressman Nelson cautioned NASA that his committee might not forward a budget that would allow NASA to fund all of its programs at the levels requested. He also warned that funding needed by the Space Station to continue on its projected schedule would not be approved if it meant a reduction in funds needed for other programs, in particular ongoing science programs. True to his word, the Space Station request for $767 million was immediately under fire. Fiercely attacked in the House, the Space Station appropriation barely survived. Things were more friendly in the Senate where a floor vote maintained Space Station funding by a wide margin, 84 to 12. But after the conference committee met and the smoke cleared, the FY 1988 appropriation had been reduced to $393 million.

Phase C/D proposals had been submitted and were being evaluated when the final House and Senate votes were tallied. Adding together the Administration's fiscal years 1987 and 1988 requests to Congress, NASA had been short changed $351 million in the final Congressional appropriations. More importantly, OMB in FY 1987 and FY 1988 had reduced NASA's internal requests by $478 million. When combined, these cuts of over $800 million required NASA to look at options to further reduce costs and reevaluate how to modify the Phase C/D schedules and deliverables. The appropriations would not allow the originally scheduled work to be done.

## Space Station Operations Task Force

In October 1987 the Space Station Operations Task Force (SSOTF) released its report.[9] It examined, in detail, Space Station operations. The Task Force had toiled for eight months and included an enormous number of participants from NASA, industry and academe, as well as representatives from the international partners, consultants and support contractors. Two cochairmen, Peter Lyman from JPL and Carl Shelley from JSC oversaw four panels. The goal was to produce a framework that met program objectives of safe and user-friendly operations and would support the participation of domestic users and the international partners. It focused on the mature operations phase of the Space Station (post IOC) when both manned and unmanned elements would be operating. In addition, the SSOTF was asked to assure that the framework it would recommend would accommodate the development phase and evolutionary growth. Altogether, a difficult job since the Task Force was required to factor in operations costs and required support systems, and at this time very little was known of the experiments and manufacturing processes that might be conducted on the Station.

---

9. NASA, Space Station Operations Task Force, Executive Summary and Summary Reports, accepted by Andrew J. Stofan, Associate Administrator for Space Station, October 1987.

Briefly summing up this massive effort, the SSOTF provided 30 recommendations divided into five broad categories: Program Operations Management; Space Operations and Support; User Integration and Accommodation; Logistics Operations Support; and Systems Development. The recommendations covered such issues as operations costs and pricing policy, crew training, the several different information systems that would be needed to track and coordinate operations, and roles and responsibilities for such a complex undertaking. For this latter point, the SSOTF devised a matrix to identify where engineering responsibilities would reside at NASA and the international partners, a critical issue alluded to when Hodge first briefed Fletcher. This was a major contribution that helped clear the air between Level II and Level III as it defined the responsibility for design, payload integration, maintenance, and sustaining engineering for all the systems and payloads. Time would tell if the recommendations would be workable.

As late as mid-October 1987, Dale Myers was still telling the House Subcommittee on Space Science and Applications, that NASA was on track to award the contracts in November.[10] In his testimony Myers reviewed the progress made in the last year and noted that the NRC had endorsed the Space Station baseline configuration. Addressing some of the NRC's concerns, he told the Subcommittee that he disagreed that deployment would be "difficult and risky" and stated that NASA was confident that the Station could be deployed safely. NASA had looked into the NRC concern on orbital decay and had added one flight to the assembly sequence to avoid any problems. This meant that twenty assembly and logistics flights would be required to complete Phase I. In regard to spare hardware, he agreed that NASA should continue to assess the issue but that a preliminary look estimated that having spare equipment could add another $250 million to the program. Finally, on the issue of cost estimation and the impact of operating costs, he promised that NASA would send the Subcommittee a recently completed analysis.

As promised, NASA soon forwarded their operations cost management study.[11] It explained in some detail how the Space Station Office would control operations costs. It identified the key operations cost drivers: transportation, information systems, sustaining engineering, and logistics. For each it suggested how costs would be monitored. To control information systems costs, for example, the program office would use TMIS and SSE, support contractor programs described earlier. Controlling logistics costs would pose, perhaps, the most difficult challenge as it required managing user equipment and the provision of consumables and spares, including returning much of it to Earth after each flight. Because of the complexity of the bookkeeping, bidders on the Phase C/D RFPs were asked to describe how they would perform logistics support analyses for the hardware they would provide.

It is not clear whether or not this report convinced the Congressional oversight committees that NASA had mastered the problem of cost estimation. What is clear is that nowhere in the report did NASA provide an estimate of what to expect in the way of operating costs. Until this time, only GAO had made a preliminary estimate of these costs.

A companion report to the operations cost management report, Space Station Development Plan,[12] was also forwarded to Congress in November 1987. The Plan brought the Congressional committees up to date on the Space Station program after all the permutations that had taken place in the last year and reflected the new budget numbers. It included the development strategy and management approach, procurement approach, safety considerations, and a new schedule.

The Plan also discussed how the program was addressing the issue of emergency return in the event that a shuttle was not docked at the Space Station. A definition study would be initiated in January 1988 for a crew emergency return vehicle (CERV). In concept, the CERV could be launched on either the shuttle or an ELV and remain berthed at the Space Station for extended periods of time. A brief review of how the Space Station would work with the international partners was also discussed. The

---

10. Statement by Dale D. Myers, Deputy Administrator, before the Subcommittee on Space Science and Applications, Committee on Science, Space and Technology, October 14, 1987.
11. Space Station Operations Cost Management, Submitted to the Committee on Science, Space and Technology, U.S. House of Representatives, Office of Space Station, NASA Headquarters, October 1987.
12. NASA, Space Station Development Plan, Submitted to the Committee on Science, Space and Technology, U.S. House of Representatives, November 1987.

schedule provided assumed Phase C/D contract awards in November 1987. That would start the clock toward scheduling a first element launch at the end of first quarter 1994, man-tended capability (MTC) by the end of first quarter 1995, and permanently manned capability (PMC) in early 1996. A reference assembly sequence and manifest was appended that showed seven flights to achieve MTC, eleven flights to PMC, and twenty one flights, including two Ariane missions, to finally complete Phase 1. The bad news was that the number of flights had grown by one in just a few months. The good news was that ESA would provide two of the logistic missions. In spite of the large reduction in anticipated funds, NASA jockeyed the schedule to maintain first element launch in 1994, a date that would prove to be unattainable.

## NASA Courts the Private Sector

Private sector involvement in the Space Station was the subject of a NASA workshop held in Nashville Tennessee in November 1987. From the earliest days of the Space Station Task Force, NASA had courted the private sector to interest it in developing experiments and manufacturing processes that could take advantage of the unique, microgravity environment the Space Station would offer. With the Phase C/D contracts soon to be signed that would lead to metal bending in a few years, NASA was anxious to get more companies involved and be in a strong position to show critics that useful research would take place at an early time. A few cost sharing grants with private companies had resulted from the early industry contacts, and the Small Business Innovation Research (SBIR) program had met with some success attracting industry partners.

The Nashville workshop was the first follow-up meeting after NASA released its Commercial Space Policy in August 1986. The Policy contained guidelines on how the private sector could participate in Space Station development. Participation could be in two forms, as a user of Space Station capability, or as a provider of hardware services. Demonstrating how seriously NASA was depending on private sector involvement, responders to the Phase C/D RFPs had to include a plan in their proposals to show how they would assist NASA in securing private sector participants that would use or provide services to the Space Station.

At the Workshop, NASA discussed the guidelines for commercial enterprises. Attendees were encouraged to submit specific experiments or describe services that they would deliver. In return, NASA would provide incentives, where appropriate. Most importantly, NASA said it would honor and protect proprietary rights for information contained in the proposals received and the data collected during the course of the work. In response to NASA's earlier overtures, several projects were already underway. Private ventures were studying protein crystal growth, and the design of an electrophoresis experiment to determine if there were benefits to manufacturing drugs in a very low gravity environment.

## Phase C/D Studies Begin – Cost Reductions Studied

Dale Myers' prediction that the Phase C/D contracts would be signed in November was only off by a few weeks. In mid-December four letter contracts for detailed design and development were awarded to Boeing, General Electric, McDonnell Douglas, and Rocketdyne. Boeing was the prime contractor for Work Package 1 managed by MSFC and would be responsible for designing the modules, life support system, station propulsion, and developing a concurrent logistics plan.

The Work Package 2 prime contractor McDonnell Douglas, managed at JSC, was responsible for designing manned and other selected systems, station architecture, assembly, and shuttle interfaces. Work Package 3 for the design and fabrication of unmanned platforms was awarded to General Electric and would be managed by GSFC. Rocketdyne won the competition to design and build the power systems for Work Package 4 managed at LeRC. Each prime, with the exception of General Electric, was supported by most of the subcontractors with whom they had worked during Phase B. General Electric had slimmed its team down to one major subcontractor. In effect, although there was some new teaming, the two contractors selected for each of the four Phase B work packages had been whittled down to one for each Phase C/D Work Package.

While not listed as work package centers, LaRC, KSC, and JPL also were given important Space Station responsibilities. LaRC's role was perhaps the most important during the Phase C design work.

It would provide system engineering and integration (SE&I) support to Level II and assist Level I in defining Space Station evolution. It would also be the focal point to oversee technology development, and the lead center representing users of the Space Station conducting technological experiments. KSC's role would increase in importance as Space Station elements were delivered for final test and integration prior to their assembly flights. A small JPL staff would work at the Reston office to define and assess top-level program requirements.

How the contracts would be adjusted to respond to the reduced FY 1988 budget was a problem to be dealt with at a later date. The good news was that Stofan was able to carry-over $100 million of unspent FY 1987 monies to add to the pot. Given the late contract signings for Phase C/D, a new schedule was announced. Preliminary Design Review (PDR) would be held in January 1989 and Critical Design Review (CDR) in August 1990. First element launch still was projected for the first quarter of 1994 followed by man-tended capability in the first quarter of 1995. The permanently manned Station would slip and be on orbit by the fourth quarter of 1995 and the revised baseline configuration, Phase I, would be complete by the fourth quarter of 1996.

Faced with the reality that Space Station budgets would not be at the level requested, in November Stofan began yet another analysis of how to reduce costs. (Previous cost exercises were conducted in December 1985, December 1986, and January-February 1987.) The analysis looked in detail at each of the thirteen major Space Station systems based on the information generated during the Phase B studies.[13] Cost drivers for each system were carefully examined and evaluated based on the impact of some "what ifs": changes in phasing, requirements, and ground rules. Tom Moser once again led the effort realizing that he faced a difficult task of balancing political, fiscal, and technical hurdles.

From the thirteen systems, six were singled out as candidates for cost savings: ECLSS, EVA, power, the mobile transporter, communications and tracking, and propulsion. Savings for each system were estimated. For example if the ECLSS went from a closed loop to an open loop system, savings of $226 million might be realized. For power, if the solar dynamic system was dropped, or deferred, it would save $26 million. Total savings that might be realized if modifications were made to all six system was estimated to be $626 million. However, each modification might incur penalties in other areas such as added weight and additional resupply as the original system concepts were optimized to reduce weight and logistic requirements. The power cost savings although relatively small, if adopted, would have the largest impact. At IOC the Station might not have enough power to accommodate all the users and supply Station housekeeping.

Moser's briefing to Myers on the results of the study covered two full days, December 18 and 19, 1987 as the pros and cons of each modification were discussed. His summary chart kept the hope alive that eventually the Space Station would fulfill its early vision. The last column was titled "Grow Back Later." In this column he listed ECLSS retrofit to a closed loop system, development of the hard suit and other EVA improvements, more power, and added propulsion and communication that would bring the design back to the last approved configuration. The cost to make these additions was estimated at $979 million.[14] In the end, no immediate actions were taken. NASA would have these options in its back pocket to use as circumstances dictated. Eventually, some would be incorporated as modifications to the Phase C/D contracts and the program plan.

Concern about how Congress was manipulating the Space Station budget prompted another configuration study in January 1988. Moser decided he needed a fallback solution that would permit him to accommodate users earlier than currently planned, and demonstrate NASA's good intentions to work with the users. This exercise was called Early Man-Tended Capability (EMTC). The goal was to devise a milestone in the 1994-95 timeframe that would allow some man-tended work to be carried out and still be compatible with the program continuing to evolve to Phase I. The plan proposed would start with experiments on the Spacelab carried on shuttles during 1990 and 1992. This would be followed with the

---

13. Report on the Review of Space Station Systems Reduced Cost Option, Prepared by the Space Station Program, Presented to the Deputy Administrator, December 18-19, 1987.
14. Based on interview with Moser, July 6, 2004. The thirteen systems were: Environmental Control and Life Support, Propulsion, EVA Activities, Mobile Transporter, Communications and Tracking, Electrical Power, Thermal Control, Data Management, Guidance Navigation and Control, Fluids, Utility Distribution, and Mechanical.

launch and utilization of the Industrial Space Facility between 1992 and 1994, and early man-tended operations from 1994 to 1996 based on a modified Space Station configuration. A neat evolution, and doable, but with major penalties to the program. It would require total program replanning and because it would divert some funds and use shuttle flights scheduled for assembly, there would be a delay of two to four years in achieving the full man-tended capability. It would also delay international projects and full Phase I operations for two to four years. In the last analysis, it would increase cost without much payoff and another time consuming exercise was put in the Space Station's back pocket.

## President Reagan's Space Vision – Impact on Space Station

At the beginning of President Reagan's last year in office his administration made a bold move. On January 5th he signed National Security Decision Directive-42 (NSDD-42). The result of a five month interagency review, that included NASA and DOD, it was a wide-ranging declaration designed to assure United States space leadership. The White House described its objective "to consolidate and update Presidential guidance on U.S. space activities to provide a broad policy framework to guide U.S. space activities well into the future."[15] Emphasizing national security, the first goal of the NSDD, "to strengthen the security of the United States," contained language that would reopen the controversy regarding the use of the Space Station for DOD projects. It stated that assured access to space is a key element of national space policy and stressed that the U.S. "will conduct those activities in space that are necessary to national defense." It went on to say that the U.S. "will pursue activities in space in support of its inherent right of self defense . . . including research and development programs." Although not called out by name, DOD had made it clear in earlier discussions and public comments that the Space Station was one of the space systems that should be available if needed by DOD. A fear that Fletcher thought he had calmed just a few months earlier with the international partners surfaced again. The new MOUs with the international partners were in the final stages of negotiations and language would need to be inserted to clarify what "national defense" meant.

Beside the national defense aspects, the NSDD included a comprehensive space policy and commercial space initiative that had three major components: "Establishing a long-range goal to expand human presence and activity beyond Earth orbit and into the Solar System; Creating opportunities for U.S. commerce in space; and Continuing our national commitment to a permanently manned Space Station." To achieve the goal of the first component, the President requested $100 million in his FY 1989 budget to fund "Project Pathfinder" that would develop a broad range of manned and unmanned technologies for exploration beyond Earth orbit. For the second component, he announced a fifteen point commercial initiative that included prohibiting NASA from developing ELVs as adjuncts to the shuttle leaving that role to the private sector. One of the more encouraging parts of this component was the promise that the President would establish a National Microgravity Research Board chaired by NASA to encourage and coordinate research in microgravity environments. For his third component, the Space Station, he would request $1 billion in FY 1989. (The actual request was $967 million.) In addition, the President would ask Congress for a three year appropriation commitment of $6.1 billion but directed "NASA to rely to the greatest extent feasible on private sector design, financing, construction, and operation of future Space Station requirements."[16]

Thus, the NSDD contained both good and bad news for Andy Stofan and his troops. In the first meeting of the Space Station Advisory Committee (SSAC) in early February 1988, he outlined how he desired the Committee to help his office and Moser's Space Station Program Office (SSPO). He

---

15. The White House, Office of the Press Secretary, For Immediate Release, February 11, 1988, Fact Sheet, Presidential Directive on National Space Policy.
16. The text and quotes in these two paragraphs are an amalgam of information and author's interpretation from Note 16 above and a similar White House release, same date, titled "The President's Space Policy And Commercial Space Initiative To Begin The Next Century."
17. Author's notes from the first SSAC meeting of February 12, 1988. The SSAC, as previously described, was established as a result of a request by Congressman Nelson. It started with 16 members selected from industry, academe, and other government agencies. Many had previously worked at NASA. The first Chairman was John McLucas, Chairman of the Board Questech, Inc., who also served on the NAC. However, two months after the SSAC began its work he was selected to chair the NAC and decided to continue serving on the SSAC as a member. The chairmanship was transferred to Laurence J. Adams,

reviewed the health of the program, generally good, but warned of troubles ahead. Even with the additional unspent $100 million from FY 1987, he could run out of funds by June. Congress had asked for a "rescoping" of the program, and a new schedule and his office was in the final stages of responding. There would be changes.[17]

Stofan's warning was based on several factors. At the end of December the joint Conference Report on Appropriations for HUD and Independent Agencies directed NASA to rescope and reschedule Space Station activities consistent with the revised budgets. The Appropriation Committees had been playing games with the Space Station budgets beginning with FY 1987. For FY 1988, NASA had planned its activities based on an obligation plan of $513 million that included $121 million appropriated in FY 1987 but not obligated; $100 million of that sum was unavailable because of conditions imposed on its release by the Appropriation Committees.

Eventually the conditions were relaxed and most of the $100 million was released and this is where the monies came from that Stofan added to the Phase C/D contracts. But the meddling did not stop there. New restrictions were placed on the FY 1988 appropriation. NASA was not permitted to obligate $225 million of the FY 1988 appropriation until June 1988; the reason that Stofan had told the SSAC that he could run out of funds by June. The FY 1988 Authorization Act contained language that the Space Station development budget requests could not exceed 25% of the total NASA budget and that operations costs, including utilization, could not exceed 10% of the total. (Until this point, the Space Station never exceeded 4.4% of NASA's budget but would increase to 14% by FY 1990.) Sound complicated? It was. The constantly shifting budget ground made it difficult for Space Station managers to run a coherent program. Many of the contracts awarded spanned several years and contained milestones or deliverables based on measured continuity of the contracted effort. Changing the level of effort in midstream because of a lack of funds or other restrictions guaranteed wasted effort or caused schedules to slip.

Moving ahead a few months, language in the NSDD did not translate into approval of its terms by Congress. Congress cut $67 million from the FY 1989 Space Station request of $967 million. In a speech to the New Orleans Chamber of Commerce in March, 1988, Fletcher reviewed recent public opinion polls that showed a strong majority of Americans supported the space program. He went on to say that despite this support there were continuing efforts in Congress to eliminate funding for important programs, and "though unsuccessful, such moves could resurface in future years, for, as we all know, while Congress does well at distributing the 'pork', it often ignores the beef."[18] As he forecast, the events spelled out above came to pass.

Toward the end of February 1988 a flurry of announcements were made related to the Space Station. The National Microgravity Research Board, announced in NSDD-42, was established. A Commercial Development of Space Industry Advisory Group was formed reporting to the NAC. And an Astronaut Science Support Group was created at JSC to provide direct interaction with prospective experimenters who would use the Shuttle and Space Station. Stofan took another action in February establishing a panel to review the TMIS contract.[19] If TMIS did not live up to expectations, the program could run

---

retired Corporate Vice President, Martin Marietta. The SSAC's Executive Secretary was Dr. William P. Raney, special assistant to Andy Stofan. At the first meeting, Stofan asked the Committee to focus on top level technology, management and policy issues and, in particular, how to provide stability to the program and assure its utilization. That is how the Committee began its work, but it soon started to study specific problems, influenced by the backgrounds of the individual members. Besides, it was more exciting to get involved in trying to solve problems that the Program Office acknowledged could use some help. As a result, the Chairman soon split us up into seven panels in order to address these problems that included: reviewing how SET&V was functioning, what was the threat and how to protect against damage from orbital debris, and how to insert advanced technology into the program in future years. As we were few, we usually served on more than one panel. It would require panel meetings and meetings of the whole every two to three months followed by homework and report writing. Until disbanded in 1994, the SSAC continued to operate in this fashion reporting the results of our efforts to Stofan and his successors. I am sure that all the SSAC members would feel that they made useful contributions to the program.

18. Excerpts from Remarks Prepared for Delivery: New Orleans Chamber of Commerce, New Orleans, LA; March 11, 1988, Dr. James C. Fletcher, NASA Administrator.
19. Report to the Administrator, Space Station Program Technical and Management Information System Review, NASA, March 8, 1988.

into serious problems maintaining oversight and control of the multiple, diverse pieces under design by four contractors and their subs and managed at four NASA centers. History had shown that the centers were not known for willingly sharing information. The Panel knew that because JSC did not put such a system in place for the shuttle, it had been difficult and costly to track life cycle costs for the shuttle reconfiguration. The Panel concluded that the scope of the TMIS contract had been properly defined and was needed by the program. The first phase had been successfully completed and the Panel recommended that funding be released for the next phase, developing the communication / computer system infrastructure that would include cost price modeling and configuration management. It was noted that the Phase C/D contractors had already costed and staffed their contracts assuming that TMIS would provide automated capabilities that they could plug into.

At the end of February 1988, Stofan released the first Space Station Program Requirements Document (PRD).[20] At NASA, all programs were managed by PRDs that established the highest level program requirements. For the Space Station program it was the official Level I controlled set of technical and management requirements, budgets, and milestones that Level II (SSPO) management would use for guidance. Based on the PRD, Level II would generate and implement the more detailed requirements Level III would follow. Release of the PRD was a major accomplishment as it provided Space Station management with a foundation on which to control the Phase C/D contracts and the deliverables.

### Rescoping – Rescheduling

On April 12th, NASA responded to Congress on how the Space Station program would be "rescoped and rescheduled" based on the revised FY 1988 and projected 1989 budgets.[21] In the cover letter forwarding the response, Fletcher stated that NASA had "reviewed various means of accommodating the funding reductions – further descoping of the Station, changes to the program management approach, and slipping major program milestones." (Note the reference to "further descoping." The program had already been descoped because of earlier budget cuts, in case the Committees had forgotten.) He reminded them that, "The current configuration results from 4 years and over $600 million worth of definition analysis by government and industry." Fletcher went on to say, "It is my conclusion that the only sensible way to accommodate . . . the funding profile is to slip the major program milestones." He concluded that it would entail a one year slip in first element launch (to first quarter 1995) and warned that "any further reductions . . . would result in further delays in the major program milestones." He closed his letters to House Chairman Boland and Senate Chairman Proxmire by expressing his appreciation for their "continuing support" of the Space Station; the obligatory, bite-your-tongue, bow to higher authority knowing that their support was ephemeral at best.

In the body of the report, NASA provided details of how the budget cuts would be accommodated. In addition to slipping first element launch by twelve months, the cuts impacted the Phase C/D contracts. NASA directed the four work package contractors to constrain hiring personnel and incurring costs from their subcontractors. As a result work planned to support the scheduled preliminary requirements review (PRR) was deferred and the PRR was slipped three months. Full scale detailed design activities would also be delayed. Similar hiring constraints were placed on Grumman's support contract, the TMIS contract and the software support environment contract. Thus the three SSPO support contracts, and especially the Grumman contract, would be behind schedule. Finally, NASA deferred some long-lead procurements and test hardware six to twelve months. What the full impact of these deferrals would be was not discussed. But NASA ended its analysis by advancing the argument usually heard when a program's budget was in jeopardy of losing funds; "If you cut my budget, it's going to cost you more in the long run to complete the program." Congress wasn't buying that argument, after all some members wanted the program to disappear.

---

20. Space Station Program, Program Requirements Document, February 24, 1988, NASA, Washington, D.C. 20546. The PRD was the successor to the Program Approval Document (PAD) that had been used by NASA from the earliest days as a first level management tool.
21. Space Station Program Response to the Fiscal Year 1988 and 1989 Revised Budgets, submitted to the Committee on Appropriations, U.S. House of Representatives, and the Committee on Appropriations, U.S. Senate, NASA, April 1988.

The Administration just did not have enough muscle, or perhaps desire, to override a position that was being pushed by a few powerful chairmen, with the rest of the members following meekly behind. Stofan recalls that in conversations with Congressional staff he was told that unless the Administration got behind the program, the budget cuts would stand. It was well known that within the Administration Casper Weinberger at DOD and William Casey at CIA did not support the Space Station, and OMB Director Miller was not a big fan.[22] The support never came and the cuts stuck.

The FY 1988 NASA Authorization bill introduced a new requirement on Space Station bookkeeping. Up until this point, Space Station cost estimates included only the research and development costs which was why the original $8 billion estimate was such a low figure. After FY 1988 NASA was directed to include other costs, such as marginal costs incurred to modify the shuttle for assembly flights, ground test facilities dedicated to Space Station, tracking and data services that would be required for the assembly flights and for the elements left in orbit, and the development costs for the Flight Telerobotic Servicer. By including these costs, the estimates took a big jump to over $19 billion dollars. The cost of each shuttle launch still was not included in the runout cost estimates.

### James Odum Replaces Stofan – Progress Continues

These were trying times for the program complicated by Stofan's decision to leave. When Fletcher first approached Stofan in June 1986 requesting that he transfer from his position as Director of the Lewis Research Center to lead the Space Station effort, Stofan turned him down. Having served in Washington as an Associate Administrator earlier in his career he was not enamored about returning. Center directors had more clout than AAs and operated almost independently. Fletcher called him at home several times after his initial refusal and he finally agreed on one condition, as soon as the Phase C/D contracts were underway he would resign and move on. In April he reminded Fletcher of their agreement. Fletcher tried to talk him out of leaving but to no avail. He offered Stofan a position at Ames Research Center but Andy turned it down and after thirty years at NASA he became eligible for an early-out and retired. He soon accepted a position at Martin Marietta and James B. Odum was appointed to be the new Headquarters AA for Space Station.[23] Odum transferred from MSFC where he had managed the Hubble Space Telescope and the Science and Engineering Directorate but had not been involved with MSFC's Space Station Work Package 1.

At the May 1988 meeting of the SSAC Odum made his first presentation to the committee. A new Space Station Management Plan[24] was distributed. Levels I and II would develop requirements and manage the program, Level III and the prime contractors would design, develop and fabricate the Space Station. He indicated he would delegate authority to the lowest level practical commensurate with demonstrated, real, accountability. Even though budget cuts would delay this development, the program would depend on TMIS as "the key management tool."

For the past month Moser and his staff had been reviewing the multilevel Space Station management plan with the objective of producing a new plan that would carry the program through Phase I. They had relied on successful NASA management concepts employed in past manned programs to form the foundation of their plan and then went at least one step further in recognition of the "increased challenge posed by the technical and organizational complexity of the Space Station." From Level 0, the Administrator, downward to Level IV, the contractors, management lines of responsibility were defined. How management would interact with the international partners and potential commercial users was also defined. The diagram illustrating the many interactions that were required to implement the program would give any manager pause, there were almost too many to count (Figure 18). If you took time to carefully study the management complexity, you would be justified in wondering if it would ever come together and SSAC members were quick to point this out.

---

22. Based on a interview with Stofan, July 10, 2004.
23. Ibid.
24. NASA, Space Station Management Plan, Summary, May 15, 1988. The Plan was submitted by Thomas L. Moser and approved by James B. Odum.

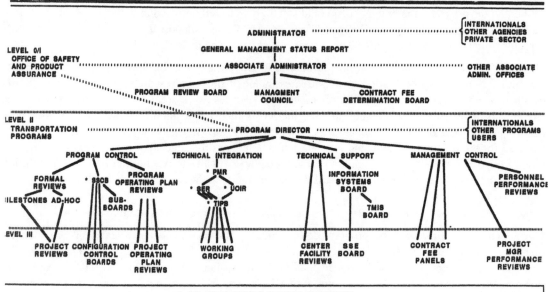

Figure 18. Space Station Management Approach. (Moser brief)

Looking at the plan from the bottom up, reporting status, problems, and progress would begin with the contractors and work their way up culminating with the monthly General Management Status Reviews (GMSR) presided over by the Administrator and senior Headquarters staff. The GMSR was brought about as a result of Bob Seaman's dissatisfaction with NASA's decision making process and was established soon after his departure in January 1968. It consolidated the functions of the Administrator's reviews and monthly status reviews. Although not designed originally to be a decision making forum, it became one.[25] The GMSR served NASA well by providing a carefully choreographed format to review all the programs. It allowed the Administrator to keep a close eye on the programs and made it difficult for program AAs to hide problems. What it did permit when problems arose, as they always did, was a collegial response and solution. A few years after the Space Station Program adopted its management plan, the GMSR system would be abandoned and program oversight deteriorated. Odum reviewed the budget situation for the SSAC. The picture had not improved since February. In addition to the cuts already described, the House had eliminated $15 million from the construction of facilities (CofF) line item that had been earmarked for the Space Station processing facility at KSC. This was considered a major set back, but as schedules stretched out it would be added back at a later date with no impact. Perhaps the most interesting chart showed a runout of expected budget requests through FY 1996, the budgets that would allow the program to achieve PMC. The line labeled Development included all the items costed in the original estimate of $8 billion. It now totaled $18.894 billion. In 1984 dollars it was said to equal $12.8 billion, a 60% increase. The total of all past appropriations and future requests through FY 1996, including the definition phase ($595.8 M), operations, and the Flight Telerobotics System, items not included in the original $8 billion estimate, was $24.297 billion in real year dollars (Figure 19). Shuttle launch costs were not included. In summing up his budget issues and concerns, the column labeled FY 1989 Appropriation was blank except for a large, bold, question mark. The Senate was still deliberating the HUD and Independent Agencies appropriation and the conference committee had not met. (Eventually

---

25. For further discussion of how the GMSRs evolved and other management functions, such as the NASA Management Council, instituted during the 1960s, see Managing NASA In The Apollo Era, NASA SP-4102, and Aiming At Targets, NASA SP-4106.

the House markup of $900 million was accepted.) Odum showed the latest version of the assembly schedule containing more detail for each flight payload and the slip of one year in first element launch to accommodate the budget reductions. He ended by reviewing the status of negotiations with the international partners and predicted that the MOUs with Canada and ESA would be signed by August. The signing date with Japan was unknown as it depended on action by the Japanese Diet.[26]

## SPACE STATION
## FY 1989 BUDGET REQUEST
## (RY $ M)

|  | 87 and Prior | 88 | 89 | 90 | 91 | 92 | 93 | 94 | 95 | 96 | Total |
|---|---|---|---|---|---|---|---|---|---|---|---|
| Development* | 147.0 | 386.8 | 935.4 | 2035.2 | 2755.8 | 3243.0 | 3074.5 | 3047.3 | 2257.1 | 1121.8 | 18983,9 |
| Flight Telerobotics | 30.0 | 21.5 | 20.0 | 45.0 | 45.0 | 50.0 | 45.0 | 45.0 | 55.0 | 50.0 | 406.5 |
| Transition Definition | 0 | 4.0 | 12.0 | 25.0 | 32.0 | 42.0 | 44.0 | 46.0 | 49.0 | 51.6 | 305.0 |
| CERV Definition | 3.0 |  |  |  |  |  |  |  |  |  | 3.0 |
| Operations | 0 | 0 | 0 | 25.0 | 79.7 | 199.1 | 545.9 | 794.0 | 1134.0 | 1225.0 | 4002.7 |
| Total | 180.0 | 392.3 | 967.4 | 2130.2 | 2912.5 | 3534.1 | 3709.4 | 3932.3 | 3495.1 | 2447.8 | 23701.1 |
| Definition | 595.8 |  |  |  |  |  |  |  |  |  | 595.8 |

\* *Current Development Estimate In 1984 Dollars Is $12.88*

Figure 19. Space Station FY 1989 Budget Request (real year dollars in millions. (Odum brief)

At this meeting of the SSAC Moser addressed the issue of hiring staff, a concern raised by the SSAC at its first meeting. In 1986 the Ross Committee, after reviewing the equivalent responsibilities of other large NASA programs (the Apollo Washington office, JSC's Shuttle Program Office, and JSC Space Station Level B staff whose responsibilities had been transferred to Level II at Reston) had recommended a staff of 300-400. The new SSPO would have similar responsibilities to the offices reviewed, and more. Moser's latest estimate was a need for 382 positions, a reduction of 96 from his December 1986 estimate. Nineteen months after taking office, his staff was 153 and he described overworked personnel spending 10-12 hours a day, six to seven days a week, in the office. Those who had agreed to come to Washington from the NASA centers on a temporary basis that he had hoped to convince to stay, were turning down the honor. To make matters worse, he was being asked by his management to develop a staffing plan that would have only 213 equivalent people at full complement. His final point made to the Committee was: "To reduce manpower below 382 and still maintain a medium risk program is not a challenge, it is impossible."[27] Committee members urged Odum to request relief from the hiring target. More on this subject later.

During the two day meeting, Moser and his staff made thirteen presentations on a wide variety of subjects.[28] All the presentations represented the results of studies not undertaken for the benefit of SSAC, but for programmatic reasons. Some were required by Congress. However, after hearing of his staffing woes, the Committee felt guilty about using so much valuable staff time to prepare for its presentations. This concern served to accelerate the Committee's desire to be more than a "top level" review and put its expertise to work on problems of Moser's choosing. At his request, seven panels were formed to examine problems including program requirements, system evaluation-test-and-verification (SET&V), communications, and how to balance short-term and long-term program goals.

---

26. Based on author's notes and handouts from SSAC meeting of May 19-20, 1988.
27. Program Implementation Responsibilities and Requirements, Thomas L. Moser, Director Space Station Program, April 27, 1988. Briefing handout to SSAC at May 19-20 meeting.
28. Rather than write about each of the presentations, which might result in the reader learning more than may be desired, I have listed just the subjects. In addition to Odum's opening overview, the thirteen topics were: Space Station Management Plan; Program Implementation Responsibilities and Requirements; Space Station Assembly Sequence; Master Verification Plan and Its Implementation; Space Station Evolution; SSOFT Update; Space Station Commercial Activities; Advanced Automation Study; NSTS Integration and Operations; Orbital Maneuvering Vehicle; Assured Crew Return Capability; Advanced Solid Rocket Motor; and Commercially Developed Space Flight Facility, the latter a presentation by OSSA.

One other study with repercussions for the Space Station closed out the month of May. The NAC Life Sciences Strategic Planning Committee reported the results of an eighteen-month study reviewing the Life Sciences programs within NASA. Chaired by Dr. Fredrick C. Robbins, Case Western Reserve University, the Committee consisted of members from clinical and medical institutions, private companies, and NASA. The Committee stated: "From the point of view of learning what is necessary so that man can exist safely for extended periods in space . . . the Space Station becomes crucial." They took note of the flight of Soviet cosmonaut Yuri Romanenko who had just completed 326 days in space, and suggested his experience highlighted the need for well designed experiments to study the physiological and psychological effects of prolonged exposure in space. Other recommendations included: flying an augmented series of Spacelab missions, dedicating suitable facilities on the Phase I Space Station for life sciences research, augmenting life sciences capabilities during the post-Phase I Space Station, and establishing a combined national and international life sciences research facility on the Space Station. One of the facilities recommended was a "variable-force centrifuge."[29]

By the time the SSAC met again in September Odum had made several important changes. Working with the SSPO support contractors, the four NASA work package centers, KSC, and prime contractors, he and Moser established a program integration process to simplify and reduce program risk. As he had promised in May, the process would establish the lowest level of accountability for problem resolution. This would be done by closely defining the interfaces between the work packages and achieving, where possible, a high degree of commonality of how the contractors would report and schedule flight and ground deliverables. They believed this would reduce contractor paperwork but preserve NASA quality control. It was expected that the contractors would agree to this new way of doing business, termed "Associate Contractor Relationships," and contract modifications reflecting this change would be completed by October.

The Grumman support contract was reviewed including the background and experience of the personnel (146) that Grumman and its subs would transfer to Reston. The Level III NASA centers were complaining that the Grumman statement of work (SOW) was too broad and could interfere with their technical authority and control of their work package contractors. They noted that Fred Haise, a former astronaut, was now working as Grumman's contract director. They suggested that his office located in Washington, close to top NASA managers with whom he had past close working relations, would try to siphon off a larger share of the work for Grumman.[30] SSAC Panel 7 met in Reston in August for two days to work on their assigned SET&V task. Reporting at the September meeting, it found that the Grumman staff was too small and inexperienced to do the job assigned. The Panel also received the impression that the Level III centers did not recognize the degree to which their former SE&I responsibilities now resided at Level II and were unnecessarily duplicating this work.[31] This issue would continue to complicate SSPO relations with the centers in the years ahead and was another example of lack of communication or mistrust between the centers and NASA offices in Washington.

Between the last meeting of the SSAC in May and September 1988, the Space Station had been given a new name by the President, "Space Station Freedom." Odum's overview of the program described his lack of freedom in managing the program. In addition to cutting the overall request by $67 million, Congress had transferred $26 million from the development line to the Flight Telerobotic System (FTS), in keeping with a letter sent to Fletcher in February signed by Representatives Boland, William Green (D-NY), and Senator Proxmire. The letter demanded that the FTS be developed on a schedule that would permit a demonstration flight in 1991 and a engineering test flight in 1993.[32] In response, Fletcher agreed to this interference in the program content. In further tinkering with the appropriation, Congress mandated that only $385 million, less than half the appropriation, could be spent before May 15, 1989 when the FY 1990 budget would be debated and the results of congressional and presidential elections known. A clear message to NASA as to who was in charge regardless of which party controlled the White House.

---

29. Exploring the Living Universe, A Strategy for Space Life Sciences, A Report of the NASA Life Sciences Strategic Planning Study Committee, June 1988, Washington, D.C., NASA.
30. Based on the Grumman Program Support Contract Review, June 17, 1988.
31. Based on the author's notes and handouts at the meeting.
32. The letter, addressed to Dr. Fletcher, is dated February 25, 1988. Fletcher's response is dated March 22, 1988.

Explaining the predicament in which the budget shenanigans had put him, Odum said that "an already oversubscribed budget is significantly reduced and constrained." It had greatly restricted his flexibility and forced him to drop his three month reserve, a small cushion he tried to maintain in case of unforeseen problems. In addition, staffing levels had to be frozen and there would be a delay in adding major subcontractors until the May deadline passed. The centers were forced to review the work package contracts to delete or delay items until later years. Adding to the budget problems, the FY 1989 Authorization language required him to submit seven new reports; two were due by October 1st.

Next, he discussed the first major milestone for the work package contractors, the Program Requirements Review (PRR) that had been completed in June. During the course of the past year, the NASA centers had identified 6600 Review Item Dispositions (RIDs). A RID was NASA terminology for various types of discrepancies that would arise during the course of a contract and might range from minor documentation errors to more serious schedule problems or specification errors. Every RID had to be worked off in one fashion or another and the PRR was the first opportunity to review their disposition. The results were then reported to the Space Station Control Board (SSCB) chaired by Tom Moser. As of September 4200 were still being processed and some potential program impacts were anticipated. Other subjects covered showed that if the schedule could be maintained, the next major milestone was the Preliminary Design Review (PDR) for PMC scheduled for the third quarter of CY 1990. The final problem area he covered was civil service manpower. A "bottoms up" analysis had been completed in May for all the participating NASA organizations. The result showed a deficit of 726 bodies and no relief in sight.[33]

Included on the SSAC agenda was a briefing on the results of the June meeting of the Space Station Science and Application Advisory Subcommittee (SSSAAS). Its report was wide-ranging covering general and discipline specific recommendations. The Subcommittee pointed out potential challenges to conducting life science and materials sciences experiments simultaneously in the lab modules. They recommended establishing a clear management process to achieve isolation and environment requirements for microgravity disturbance, bioisolation and toxic isolation. They endorsed the need for uniformity of requirements to reduce cost and to be able to rapidly add new experiments as attached payloads. Finally, the Subcommittee expressed the need to establish multiple classes of data security and to develop alternatives in the event of loss of the Tracking and Data Relay Satellite (TDRS) for data acquisition.

There was one piece of good news. Margaret Finarelli reported on the status of the MOUs with the international partners. A signing ceremony was scheduled for the following week. All the points of contention had been resolved, perhaps not to everyone's complete satisfaction in view of the configuration and schedule changes forced by budget considerations. In her presentation, she described the "Fundamental Deal" that included: (1) All partners would provide flight hardware and supporting ground elements; (2) User space could be exchanged for resources enabling use; (3) All partners would participate in management; (4) All partners would provide crews; and (5) All partners would share operating costs. This last item was the most difficult to define and would be raised again in the next round of MOU negotiations. NASA had responsibility for overall planning and would provide the Space Station Control Center, the Payload Operations Integration Center, and the control center for NASA's polar platform. Part of ESA's responsibility included providing control centers for their polar platform and the Man-Tended Free-Flyer (MTFF). A Multilateral Coordination Board would be established and all decisions would be made by consensus. In the event consensus could not be reached the Chairman, NASA's AA for Space Station, would make the final decision.

On September 29, 1988, a signing ceremony was held at the State Department. Four documents were signed: a multilateral agreement among the U.S., Japan, Canada, and the nine ESA member states participating in the Space Station Program (Belgium, Denmark, France, Italy, the Federal Republic of Germany, the Netherlands, Norway, Spain, and the United Kingdom); an Arrangement on Interim Cooperation between the U.S., ESA, and Canada; and two bilateral MOUs between NASA and ESA, and NASA and Canada. The bilateral MOU signing with Japan would take place next spring. ESA's contribution would be the *Columbus* attached laboratory, the *Columbus* MTFF laboratory, and the

---

33. Based on the author's notes and handouts at the meeting.

*Columbus* Polar Platform. Canada would provide the Mobile Servicing System and a simulation facility. Japan's contribution would be the multipurpose Japanese Experiment Module (JEM). With the agreements, each partner provided an organization chart with key individuals identified. Dale Myers spoke on behalf of the U.S. Government and NASA congratulating everyone involved in finally reaching this historic moment and presented each representative a framed, artist's rendition of how Space Station Freedom would appear when fully operational (Figure 20). Perhaps portending things to come, only ESA identified their contribution as being a part of an "International Space Station (ISS)."[34]

Figure 20. Artist Vincent di Fate's rendition of the Space Station.

## New Administration – Same Congress

The November 1988 elections did not result in any significant changes in the fortunes of the Space Station Program. A Republican Administration was reelected and both the House and Senate were still firmly under the control of the Democratic majority. At least another four years ahead of gamesmanship and gridlock. Vice President Bush became the forty-first President and it was expected that he would continue the programs initiated by President Reagan, including the Space Station.

Progress inched along for the Space Station in the last quarter of CY 1988. The four work package prime contractors, working up to this time on letter contracts, now had final contracts and accepted the Associate Contractor Agreements. OSSA released the Announcement of Flight Opportunities for Space Station attached payloads. SSPO was developing specific guidelines, criteria, and procedures for evaluating commercial proposals related to Space Station Freedom. On the manpower front, OMB approved staffing that would allow Levels I and II to staff to their full request of 444. That was the good news. The bad news was that Congress had yet to approve how NASA would pay for the additional staff so the new positions would be in limbo until this little matter was resolved. Hiring for

---

34. Transcript of Space Station Program International Signing Ceremony, held at the U.S. Department of State, Washington, D.C., September 29, 1988. For additional background on how the international partnerships evolved, see "Together in Orbit, The Origins of International Participation in the Space Station," John M. Logsdon, NASA History Division, Monographs in Aerospace History #11, November 1998.

positions that had been approved earlier and for which salaries were appropriated, was still proceeding slowly. Only 29 of the approved 54 positions had been filled. A new concern, two important positions, the Chief Scientist on Odum's staff and the SSPO Program Scientist had similar job descriptions as the contact points with the user community. Vacancies and turnover at these positions, along with a lack of aggressive leadership in the years ahead, would result in problems as facilities for the users were defined and strong user advocacy was missing at Levels I and II. User and experiment interfaces would become more and more complicated as the program tried to incorporate strict, Apollo-like, requirements. At the end of December it was announced that E. Ray Tanner would replace Moser as Director Space Station Freedom Program Office; Moser would continue as Odum's Deputy wearing just one hat.[35] Before his appointment, Tanner was the MSFC manager of Work Package #1. He was replaced at MSFC by George Hopson. All in all, a mixed bag of minor accomplishments and changes to start the new fiscal year.

At the next meeting of the SSAC on January 24, 1989, Odum brought the Committee up to date on personnel changes and progress made since September. He listed his modest goals for the forthcoming year. At the top of his list was: "Continue to ensure Space Station Program has solid support in new Administration and new Congress." There would be some new faces in the Bush Administration but the day-to-day players at OMB would remain the same. The State Department was usually supportive of NASA's efforts to include international partners in space programs but the Transportation and Commerce Departments, with their new mandates to increase private sector involvement in space, and the National Science Foundation's traditional competition for research funding could not be counted on as strong supporters. A ray of hope for strengthening ties to the new Administration was Bush's decision to reestablish the National Space Council. This could provide an advocacy voice close to the President. Vice President Dan Quayle would chair the Council but the names of the senior staffers he would bring in were unknown. Trying to improve relations with the new Congress was a noble goal but Odum would be dealing with the same committees whose positions vis-à-vis the program were well known. New chairmen had been selected but it was too soon to know if their committee's positions would change. Odum felt that the biggest challenge the program faced was: "To design a permanently manned (his emphasis) facility that is safe, productive, and affordable."

In his first briefing to the SSAC, Ray Tanner provided the status of closing the RIDs first reviewed at the June PRR. The total number reviewed had dropped from the 6600 estimated in September to 6442; some had been withdrawn. By January, 4161 were closed. Of the 2281 remaining open, 1533 required additional study before they could be closed to determine if they would effect cost or schedule or had the potential to lead to modifications to the work package SOWs. Such a large number of open items was causing concern at SSPO because the date to complete the sign-offs had not been fixed by the centers and progress had been slow.

Based on activity of the Space Station Science and Applications Advisory Subcommittee (SSSAAS), OSSA presented an ambitious mission requirements briefing covering long-range planning for Space Station experiments. Referring to the Ride Report as justification for the type of experiments that the Space Station should support, the briefing concentrated on payloads and facilities that would be required to carry out those experiments and the studies recommended in the report of the NAC Life Sciences Strategic Planning Committee. A wide variety of attached payloads and free-flying platforms were under study. Facilities being planned for on board the Space Station included 1.8 and 6-10 meter diameter centrifuges. Where such facilities could be located was a TBD. For the Earth Observation System (EOS), OSSA was studying how to coordinate experiments on five polar orbiting platforms (two U.S., two provided by ESA, and one Japanese), various attached payloads, and one free-flyer sponsored by NOAA. A complement of forty-three experiments had been identified to make use of the different platforms. SSPO welcomed this effort on the part of OSSA and the international partners and planning was underway to understand all the requirements. Attempting to accommodate one or more centrifuges became an important design consideration as the program progressed.

On the second day of the meeting the committee was briefed on the status of the polar orbiting platforms managed by GSFC. General Electric, the WP #3 contractor, had completed Phase C/D

---

35. NASA News Release: 88-175, Tanner Named Director Space Station Freedom Program, December 29, 1988.

definition and the Level II program requirements review for NASA's polar orbiting platform (NPOP-1). The PDR was now scheduled for May 1990. Design of NPOP-1, with a total weight of approximately 30,000 pounds, was being baselined for launch on a Titan IV. Experiments would be accommodated by grouping them in orbital replacement units (ORUs) and configuring the platform for specific missions such as astrophysics or ultra low G materials processing. ORUs, small modules designed to fulfill many different functions as the Space Station was assembled and operated, would have a standard design and interfaces. They could operate indefinitely with robotic or shuttle EVA servicing, or changed out as their missions required. Each ORU was estimated to weigh 7700 pounds including the housekeeping functions and the experiment(s) they would carry.[36]

As was the custom, after delivering his first State of the Union Address, President Bush sent his Administration's FY 1990 budget request to Congress. The request honored Reagan's NSDD-42 commitment and included $2.050 billion for Space Station. However, how the FTS would be funded had become an issue. Based on the NSDD objective of increasing involvement of the commercial sector in space technology, the Administration had developed a list of seven projects that would be privatized. FTS was on the list but the new administration had signaled that it would be removed and funded solely by the Space Station Program, a decision attacked by the House Space Science and Applications Subcommittee. This created a dilemma for Fletcher. Either way he had a problem but supported privatizing as the least impact on his budget, a potential savings of $30 million. Other projects on the privatization list included the Neutral Buoyancy Facility at JSC, the Processing Facility at KSC, the Extended Duration Orbiter (EDO), and the Space Station Docking Module. Except for the EDO, which was being negotiated with Rockwell, NASA doubted that the other projects would be attractive to the private sector.

And so the budget wars were joined for FY 1990. OMB had a new Director, Richard Darman. His position on the Space Station was still unknown. His boss' position seemed clear. In a conversation with the crew of space shuttle *Discovery* orbiting overhead shortly after his inauguration, President Bush said: "The space program, especially Space Station Freedom, is an investment in our future. We're living in tough budgetary times but I am determined to go forward with a strong, active space program."[37]

The NAC Aerospace Safety Advisory Panel (ASAP) completed a review of the Space Station program in 1989. It praised the program for progress made in the last year but then criticized the program organization as "very complex and at times . . . unmanageable." It warned that "the situation is fraught with opportunities for safety hazards to occur." It was especially critical of the SSPO stating that it had "not utilized its system engineering and integration support contractor effectively, is currently understaffed and appears to be encountering some difficulty in effectively directing and monitoring the work at the centers." The list of specific deficiencies found by the Panel was long and detailed but in reality did not uncover any unknown issues. Problems with the Grumman support contract were well known but were difficult to resolve because of budget restrictions and the inability to attract qualified personnel, with an emphasis on "qualified." The ASAP report just reinforced the need to solve the problems that Odum had been addressing since his appointment. He needed help from his management and the new administration. Without it the problems could not be solved.

---

36. Discussion of the January SSAC meeting based on the author's notes and handouts.
37. Some of the above information extracted from Space Station News and Advanced Technology, vol. 3, no. 7, March 27, 1989.

## — Chapter 4 —
## More Management Changes – Design Continues To Evolve
### (February 1989 - December 1990)

The elections of November 1988 brought some changes to the 101st Congress and the House and Senate Committees with NASA oversight. Five of the six had new chairmen. Bill Nelson (D-FL) kept his position as Chairman of the Subcommittee on Space Science and Applications of the House Committee on Science, Space and Technology, and had four new members out of twenty-six. Also on the House side, Robert Traxler (D-MI) was the new chairman of the Subcommittee on HUD and Independent Agencies of the Committee on Appropriations with two of nine new members. Leon Panetta (D-CA) chaired the Committee on Budget with twelve of thirty five new members. In the Senate, with all new chairmen, Albert Gore (D-TN) chaired the Subcommittee on Science, Technology, and Space of the Committee on Commerce, Science, and Transportation, with three out of ten new members. Barbara Mikulski (D-MD) was the chairperson of the Subcommittee on Veterans Affairs, HUD, and Independent Agencies of the Committee on Appropriations with three of eleven new members. Finally, James Sasser (D-TN) chaired the Committee on Budget with three of twenty three new members. Six committees, each with their own special interests, requiring kid glove treatment and awareness of their differing approaches to NASA programs and twenty-seven new members to educate. Odum's goal of improving relations with the Congress would be a daunting task.

The process of obtaining a new fiscal year appropriation at government agencies and departments rivals the most Byzantine procedures devised by man. For some high visibility agencies, such as NASA, the process is an unending struggle with the ground rules constantly changing. Before one budget is put to bed by an administration, the discussions are already underway for the next year. Usually an agency's guidelines for the next fiscal year would be received from OMB in the spring. They might include major cuts, status quo, small increases, or in rare cases, higher targets. Each agency begins by adding up the commitments already on the books and then, if there is room left over, propose expanded or new programs. If commitments exceed the OMB target number, the agency is forced to make cuts in ongoing programs or completely terminate programs. Either way a painful exercise allowed by a clause in most grants and contracts that permits the government to terminate or scale back a program at its discretion.

Assuming this latter exercise is not required, here is how the game is played. After receiving its initial mark, an agency sends back to OMB its request covering ongoing programs and usually containing new or expanded programs that may require more funds than the original OMB mark, but close to OMB's direction. Negotiations then take place and a final budget number is agreed to that would then appear in the budget submitted to the Congress shortly after the President's State of the Union address. If an agency has strong objections with the negotiated number, as we saw with Culbertson's appeal in 1985, there is an opportunity to go directly to the President. This is a risky course to take and could backfire if the grounds for the appeal are not strong enough to win as you are telling the President, in affect, OMB had given him bad advice. If a new fiscal year submission coincides with a change of administrations, as may occur every four years, a budget will be submitted that essentially follows the last administration's out year forecast. A modified budget may follow once the new administration gets its act together. Sometimes these modifications are significant if the new administration's goals differ greatly from the preceding.

Now the budget request is in the halls of Congress. Again, for the high visibility agencies, preliminary hearings might have been held before the budget was received. They might be friendly, they might be very contentious, but in either case they are important and consume enormous amounts of time for senior agency management. It is at this point that the opening salvoes are heard from the various committees and subcommittees either supporting the programs or laying down the gauntlet for the battles still to come. And, although there might be agreement between House and Senate committees on some or most aspects of an agency's programs, there are always differences. This is most apparent and makes for lively debates and strong words if either or both the House and Senate are controlled by a party different than the administration in the White House. That has often been the case since the end of WW II.

Hearings can continue throughout the remainder of the fiscal year, often requiring compromises on the part of the administration. When each committee finally agrees on a program's content and budget, it must then go to a House-Senate conference committee to iron out the differences in how the two view the agency budget. More compromises, more delays. If the issues are not resolved by the time the new fiscal year begins, another procedure is invoked, called a continuing resolution. It permits the agency to operate at the same level as the previous year, or in some cases at a reduced level, until a final appropriation is passed. The Space Station program was continually besieged and battered by the process just outlined from the day it was proposed by President Reagan. To make matters worse, NASA proponents might change their colors as the Space Station reached critical milestones and budgets were requested to move ahead.

The first session of the 101st congress convened on January 23, 1989. It was expected that the partisanship that closed the 100th Congress would grow over the next months with the White House and Congress still controlled by different political parties. Unlike previous years, the outgoing Reagan Administration had not held a prebudget summit with Congressional leaders that might have softened some of the usual rancor. Comity and any bipartisanship agreement was expected to disappear during the FY 1990 hearings. Budget bashing on some of the Bush Administration programs that would gain points for the "bashers" was inevitable.

Five preliminary hearings on the FY 1990 budget had already been held before President Bush's new NASA management team was installed. The first hearings had been orchestrated by Fletcher and Myers with Odum, Moser, and Tanner providing the detailed defense of the Space Station request. Odum appeared before the House Subcommittee on Space Science and Applications at the end of February to defend the Space Station request. Following the usual procedure, he reviewed progress to date and then described how the FY 1990 funds, representing a substantial increase over FY 1989 (more than $1 billion), would be allocated. To meet schedule, the design of the Space Station underway at the four work package centers required that the Space Station aerospace work force of 15,000 increase to 32,000. Of the $2.05 billion requested, $1.556 billion was earmarked to go to the work package centers to pay for the buildup of activities at the four prime contractors and their subcontractors. The remainder would be spent on a number of items such as the FTS, the design of facilities, and program support at Level II that contained funds in NASA's program management budget to hire 725 additional civil servants. Odum's testimony included a new request. He outlined the administration's proposal to authorize and appropriate $8.5 billion though FY 1992 in place of returning each year for new funding. He argued that it would increase program stability and improve cost effectiveness. He closed by saying: "Our budget for FY 1990 is just barely adequate. Under funding Freedom at this critical phase of the program would reduce its performance, and thus its usefulness. At the same time, it would increase program risk, and increase total long term costs."[1]

On the Senate side, at a hearing in mid-March chairman Gore of the Science, Technology and Space Subcommittee complained that NASA needed to present better budget, foreign policy, and technological arguments than he had thus far heard if it expected the Senate to approve a $2.1 billion Space Station budget. Some Senate staffers had attempted to bolster NASA's arguments by inviting the Work Package contractors to testify as to the importance of the program but the Subcommittee members were unimpressed with what they heard. None of the Work Package managers who testified made a convincing case much to the chagrin of NASA and its supporters.[2] In the next few months, the persuasiveness of Odum's arguments would be challenged.

### FLETCHER RESIGNS – NATIONAL SPACE COUNCIL ESTABLISHED

Adding to Odum's problems, on March 21, 1989 Fletcher submitted his resignation effective April 8. In the interim, while the Administration was interviewing candidates, Dale Myers was in charge.

---

1. Subcommittee on Space Science and Applications, Committee on Science, Space and Technology, House of Representatives, Statement by James B. Odum, Associate Administrator for Space Station, February 28, 1989.
2. Selected from *Space Station News and Advanced Technology*, March 27, 1989, Vol. 3, No. 7.

Fletcher was replaced in May by Richard H. Truly, rumored to be a friend of the new president and until that point the Associate Administrator Office of Space Flight. Five years, four different NASA Administrators. Every new administrator would rearrange the chairs and revamp the organization. The good news, Truly was an insider and knowledgeable, he would not have a steep learning curve. How he would adjust to this step up in responsibility remained to be seen. Dale Myers resigned in May. In July the Administration agreed with Truly's endorsement for his Deputy and MSFC Director James R. Thompson, Jr. was nominated and confirmed.

As described, budget season is a trying time under the best of circumstances. This year there was a new Administration player to contend with, the National Space Council (NSC). Suddenly, Administrator Truly with some experience appearing before Congressional committees in his former position as Associate Administrator, found himself thrust into the hearings in midstream describing and defending the whole NASA budget. Truly's honeymoon would be short.

Actually, the NSC had little impact on the FY 1990 budget debates as it was still hiring staff and defining its turf. However, it did get into the loop as another approval sign off needed for NASA testimony to make sure everyone was on message before agency management testified to Congress. As required by the FY 1990 NASA Authorization Act, the NSC would be chaired by the Vice President and would have a staff of not more than seven persons headed by an executive secretary appointed by the President. The first appointee to this position was Dr. Mark Albrecht who, until his appointment, was a legislative assistant to Senator Pete Wilson (R-CA). Although a late starter in the NASA budget debates, the NSC charter was very broad covering space policy in the civil, commercial, and national security sectors. Its membership of nine senior Presidential appointees, plus the Vice President, included the NASA Administrator, the Secretaries of State and Defense, and the Director of Central Intelligence.[3] Six of the nine members represented the same agencies included on the Senior Interagency Group for Space – SIG (Space) – during the Reagan years and carried over the same opposition or tepid support for the Space Station as the original SIG (Space) members. The major difference was that with the Vice President as chairman, the members were the top appointees of their departments rather than deputies as was the case for SIG (Space). Another difference was that the NSC included OMB's Director whereas that position was not included on SIG (Space). The National Space Council would soon begin to imprint its charter on the Space Station and NASA programs as a whole.

Panel 7 of the SSAC met at the end of March to review the status of the program's SET&V functions. Some progress had been made, a new organization structure was in place but the System Engineering and Integration (SE&I) Group Director and deputy positions had not been filled. Richard Carlisle, one of the managers who had been assigned to the program since the early days, was leading the Group on an "acting" basis. The Reston Level II office had been functioning for two years and SE&I responsibilities transferred from JSC were still undermanned. Of the 222 civil service positions identified for the Group, only 74 were filled. As the meeting progressed, the impact of both the civil servant and contractor staffing shortfall was discussed.

Some tasks that should have been under the direct purview of the Program Office were in the process of being reassigned to the program support contractor, Grumman. PDR planning was behind schedule. Agreements on over 50% of the Phase I contractor deliverables, including some specifications, were unresolved. There was a likelihood that risk management program goals would not be met. A configuration management control system was just being implemented and verification documentation had recently been restructured. For those who have not been closely associated with complex programs, this may sound like engineering double talk, but for a program at this stage of maturity, the above list showed a serious deficiency in management control and oversight. All Panel 7 could do after these briefings was express alarm to NASA management. But without funding and staff to overcome these deficiencies, all that the Level II managers could do was hang on and hope that help would soon arrive.

---

3. NSC membership was Vice President-Chairman, Secretary of State, Secretary of Defense, Secretary of Commerce, Secretary of Transportation, Director Office of Management and Budget, Chief of Staff to the President, Assistant to the President for National Security Affairs, Director of Central Intelligence, and Administrator of NASA. Other senior officials would participate as appropriate.

At the SSAC meeting in early April, Odum reviewed the status of the program in light of the recent upheavals and new bosses at all levels. Budget hearings were still ongoing however he expected that "the full budget request of $2.05 billion would not be approved." There were many unknowns associated with the several commercial projects being studied. Nelson's subcommittee had already indicated it would recommend cutting over $200 million from the Space Station budget earmarked for projects that might be turned into commercial ventures. Committee staffers warned that the hearings would become more contentious in the weeks ahead as the various factions, including Republican members, lined up their arguments to support pet projects.[4]

Since the last SSAC meeting, a new line item had been added to the FY 1990 - FY 1992 budget runouts to develop a radar to track orbital debris. Now acknowledged as a potential concern, the new radar had a total price tag of $20 million. How it would fare in the hearings was also unknown. One of the SSAC panels, led by MIT professor Edward F. Crawley, had been assessing the orbital debris problem and identified it as a serious and growing danger with implications for design and maneuverability. There was evidence that space debris had collided with the shuttle during several missions and good statistical data was expected as soon as the Long Duration Exposure Facility (LDEF) was recovered during an upcoming shuttle mission. Space Station design had been predicated upon using the U.S. Space Command optical and radar tracking network to receive warnings in enough time to alter the Station's orbit or orientation to avoid collision with space debris. However, the resolution of the system was not sufficient to pick up pieces of debris smaller than 10-20 centimeters in diameter that might be crossing Space Station orbits with impact velocities exceeding 3.5 kilometers per second. Debris below the network's resolution had enough mass and velocity to penetrate module hulls or damage solar panels and other structures.[5] Although the Space Station had a lot of open space, the total frontal area would be larger than a football field in size; it would be a large, clumsy target. As the debris problem was brought into sharper focus, design changes would be studied.

By May Vice President Quayle and the NSC had settled in. Preparing for a major space announcement by the President, NASA was directed to identify an exploration goal and the resources required to permit significant milestones to be achieved early in the 21st century. In his first major assignment since becoming administrator Truly quickly put together a team to respond. Using several past studies for background, three options were identified: robotic missions to the planets, a manned lunar outpost and then on to Mars or, bypass the Moon and send astronauts directly to Mars. Now he would have to wait and see which option the new administration would choose as NASA's future, or none of the above.

### TRULY MAKES MANAGEMENT CHANGES – PROBLEMS PERSIST

Two weeks after his confirmation, Truly made some major management changes in Space Station program management. Odum resigned and in his place Truly installed fellow astronaut William B. Lenoir as Associate Administrator for Space Station. Richard H. Kohrs was also transferred from Houston and as number two in the management chain given the title of Director. Headquarters managers with backgrounds at MSFC had been replaced by JSC managers. As had Stofan before him, when Odum agreed to come to Washington it was on the condition he would stay no more than two years and then move on so the timing was convenient for Truly to make the changes. Ray Tanner was made Deputy Director for Program and Operations. A few other chairs lower in the organization also were rearranged and new offices appeared. The critical Level II position of Director of the System Engineering and Integration Group remained vacant.

When the SSAC met at the end of June 1989 at Woods Hole, Massachusetts, for a joint meeting with the SSSAAS, they were briefed on the new management organization. Although there were some new faces in Kohrs office, they inherited the same old problems, staffing and funding shortfalls and other uncertainties. Bill Raney, who gave this briefing, discussed the preliminary planning underway to

---

4. The Odum quote and background come from the March 27, 1989 issue of *Space Station News and Advanced Technology*, Vol. 3, No. 7.
5. For those interested in understanding the space debris problem in more detail, see "The Dynamics of Orbiting Debris and the Impact on Expanded Operations in Space," V.A. Chobotov and M.G. Wolfe, *The Journal of the Astronautical Sciences*, Vol. 38, No. 1 January-March 1990, pp. 29-39.

merge the Office of Space Flight and the Office of Space Station similar to the recommendation made by the Phillip's Committee back in 1986. Two years in gestation; no date was given for its establishment but it appeared it would soon be born.

James M. Sisson, Deputy Program Director, reviewed the status of the technical audit initiated by Odum before his departure. The audit focused on three elements: prime contractor activities, supporting development efforts, and the capability to support operations. The goals of the audit were to simplify the development effort and assure that implementation plans were reflected in the design and expected performance. A large number of issues (223) were identified by the review team and prioritized for resolution ranging from qualification testing, to EVA, to emergency procedures. Actions were assigned to resolve the issues, some required immediate attention and some with lesser urgency would need additional study and review with the Space Station Management Council. The new management team would have its hands full resolving the issues identified during the audit plus the still unresolved RIDs found one year earlier. Adding to the workload and with first element launch holding for March 1995, three important PDRs were scheduled in CY 1990: the FTS in January, the polar platform in May, and the Space Station "manned base" expected to be completed by August.

Sisson also reviewed the latest assembly sequence schedules and plans for backup hardware. The SSAC and the 1987 NRC study had been critical of the program's philosophy for addressing the problem of spares and some changes had been adopted. Four approaches were now planned. The first was called "pipeline redundancy." The Space Station configuration, in general, consisted of pairs of elements launched at different times, two PV panels, two similar trusses, two habitation modules, and so on. If one of the elements should fail, for whatever reason, the second part of the pair, either already at KSC or soon to be delivered, would take the place of the failed element. This would entail modifications to some elements as they were not always completely identical. Assembly schedule might also change depending on which element had a problem. However, this approach would buy some time to fabricate a replacement. The second part of the plan was to accelerate the procurement of selected "operational spares" such as the PV panels that would then complement the "pipeline" sparing. Another source of spares could be the testbed articles already in the program. For testing purposes they would be manufactured as "flight-type" hardware and they could be fully converted to flight hardware. Finally, funding for long lead item spares not included in any of the above fallback positions would be added to the four prime contracts. It was estimated that this would add $311 million to the estimated cost; a large sum that the program had been reluctant to include in the overall Space Station runout costs but now there was agreement it was a necessity.

This was the first joint meeting of the SSAC and SSSAAS. It was an opportunity for SSAC members to listen to and join in "user" discussions. Since the last meeting of the SSSAAS in January a number of positive steps had been taken. As the result of the response of the scientific community to two OSSA Announcement of Flight Opportunities, NASA had selected a large number of principal (P.I.s) and co-investigators who would conduct experiments on Space Station Freedom. This was the first step toward making the investigators eligible to receive NASA funds to define and design their experiments. The experiments chosen were grouped in two categories: attached experiments (14 selected) that would be mounted on elements of the Space Station structure during assembly and outfitting, and "concept studies" (13 selected) for experiments that would be conducted after the assembly phase. Examples included: Laser Communications, X-ray Astronomy, Cosmic Dust Capture, Lightning Imaging Sensor, Tropical Rain Mapping Radar, and a Laser Atmospheric Wind Sounder.[6] Many of the newly selected P.I.s also were in attendance at Woods Hole.

At the five day SSSAAS meeting two main subjects were reviewed: experiments to be conducted on free-flying platforms and as attached payloads, and experiments that would be performed within the pressurized laboratories. Attendees broke up into a number of splinter groups working during the day and at evening sessions to complete the heavy agenda. A major concern was data load and information sharing. These concerns affected all of the experiments. To alleviate the first concern, data storage and

---

6. Attendees at the meeting were given two NASA Press Releases: NASA Selects Science Experiments for Space Station Freedom, and Space Station Attached Payload Principal and Co-Investigators, both dated June 29, 1989.

tape recorders were discussed and viewed as poor alternatives to direct transmission. There was a real problem on how to permit the rapid return of high volume data from both free-flyers and onboard experiments. TDRS' inability to handle the high data rates was potentially a bottleneck. Information sharing added another problem. Some experiments required critical information on such items as the ability to accurately measure Freedom's attitude (in x, y, and z) and the internal and external ambient environment and contamination levels. Depending on what was going on at the Space Station, electrical noise levels might exceed the sensitivity of some instrumentation and measurements were needed of the "weather" outside the Station when venting of various liquids or gases, would take place. The experiments conducted externally and internally would not be performed in a pristine environment.

To address these and other concerns, it was agreed to establish a User Accommodation Panel consisting of representatives from SSPO, OSSA, OAST, and OCP. Its task, to address the questions: (1) What are the rationale and objectives for the Space Station? (2) Will the Space Station be physically equipped to meet the objectives? and (3) Is the manifesting and integration process being devised so as to ensure easy access? All very basic questions and, surprisingly, still unresolved at this late date. Undoubtedly, these deficiencies reflected the continuing absence of a Program Scientist, a position that had been vacant for two years, and the minimal interactions with users by the Chief Scientist on the Associate Administrator's staff. The Panel promised to produce a report by midwinter.

One other briefing and discussion at the meeting addressed another important problem. What would be NASA's reimbursement policy for Space Station users? Responding to a requirement in NASA's FY 1988 Authorization Act to have a policy in place by PDR, SSPO and the Office of Commercial Programs made an attempt to clarify this issue. They had been studying the question for one year since NSDD-42 announced that there would be a commercial space initiative policy. Two types of users would be recognized, U.S. government users, and non-government users. For the first category NASA sponsored users would generally not make reimbursements. For other U.S. government agencies that would sponsor users, there would be reimbursement subject to negotiations between NASA and the other agencies. For the second category, non-government users, it was anticipated that the users would be primarily domestic private companies. Reimbursement would be required, again based on negotiations. For foreign companies, the international partners would negotiate reimbursement based on the allocation of their space and resources. It was anticipated that the cost reimbursement for domestic and foreign companies would be comparable. All the above was caveated by a number of rules and assumptions. However, it was noted that: "There is insufficient detailed information regarding operations costs and total demand for Space Station resources to warrant making a final decision at this time." The Report promised to release an initial policy for "early use in 1990."[7]

## PRESIDENT BUSH ANNOUNCES A LONG-RANGE SPACE PROGRAM

On July 20, 1989, the twentieth anniversary of the *Apollo 11* Moon landing, President Bush, addressing a large crowd gathered on the north plaza of the Smithsonian Air & Space Museum, revealed his vision for the nation's space program. "I am not proposing a 10 year plan like Apollo. I am proposing a long-range, continuing commitment." Filling in with a minimum of specifics he said, "for the 1990's Space Station Freedom [is] our critical next step in all our space endeavors. And next, for the new century, back to the Moon . . . and this time back to stay. And then . . . a manned mission to Mars."[8] Bush justified his Space Exploration Initiative (SEI) by stating it would carry out the National Space Policy goal proposed a year earlier by President Reagan to expand human presence and activity beyond Earth orbit and into the solar system.

Responding to the President's speech Administrator Truly created a task force, chaired by JSC Director Aaron Cohen, with the goal of understanding the technical parameters of conducting such an ambitious exploration initiative. He directed Cohen to convene a 90 day study to examine the technological requirements, science opportunities, international considerations, and to estimate needed resources.

---

7. Reimbursement Policy for the Space Station Freedom Program, Submitted to the Committee on Science, Space and Technology, U.S. House of Representatives and the Committee on Commerce, Science and Transportation, U.S. Senate, Office of Space Station, NASA, May 1989.
8. Quotes from President Bush's speech given at the Air & Space Museum, July 20, 1989.

Cohen's task force drew primarily upon the many studies conducted in past years and summarized the requirements. Among the task force's many conclusions: major investments in challenging technologies would be required, current launch capabilities were inadequate, Space Station freedom was an essential component, and a long-range commitment and significant resources would be required.[9] (On this latter point, the total cost was not estimated by the task force but some critics suggested that it would take at least $300 billion to carry out all the goals.) Frank Martin who had been Stofan's and Odum's Deputy was made Assistant Administrator for Exploration, a new office with the responsibility to identify and pull together all the pieces that would be needed to carry out the President's new mandate.

In a briefing to the NAC in August, Martin outlined how his office viewed the challenge.[10] He had two goals: (1) Provide recommendations and alternatives for an early 1990s national decision on a focused program for human exploration of the solar system, and (2) Steer agency investments on a practical, year-by-year basis toward providing defined choices in the early 1990s. He had a tight schedule. In order to influence the FY 1991 budget preparation the first studies had to be completed by September. There would be three phases to the plan. First, robotic exploration would take place in the period 1995 to 2010 and would be a precursor to humans returning to the Moon and then on to Mars. A lunar outpost would be established to provide experience for working on another planetary body and as a research platform for space-based astronomy and other disciplines. The dates for occupying the lunar outpost were 2001 to 2010. Human Mars expeditions would begin after 2010. Space Station Freedom would serve as an operations base in the 1990s. NAC was told that to accomplish such an ambitious program would require a major, national, long-term commitment, and that NASA receive authority to hire additional personnel, restructure the agency, develop heavy-lift launch vehicles, and streamline the procurement system. International partners would be invited to join with an objective of announcing new agreements by 1992, the International Space Year. Once again NASA was undertaking an exercise that eventually would lead nowhere and jeopardize ongoing programs.

Before the SSAC would meet again, the House voted in August to cut $400 million and the Senate $200 million from the FY 1990 Space Station request. Based on warnings of the pending cuts from committee staffers, the program convened a Configuration / Budget Review Team on July 5th to evaluate the impacts of either reduction. On September 25th SSFPO briefed Truly on options that would be available to save $1 billion ($400 million in FY 1990 and $600 million in FY 1991) in design and construction. Changes were proposed for almost all elements and systems including assembling the Space Station to accommodate a crew of four rather than eight and cutting the power in half. These changes would reduce the number of assembly flights to achieve man-tended operation and subsequent logistic flights but severely limit the amount of science that could be done. Truly decided to wait before selecting any of the options until the House-Senate conference committee met and agreed on a final number. In the meantime, it was reported that the international partners were "livid" upon hearing of possible downsizing. They had not been invited to take part in the first meeting to decide how the program might be adjusted to reflect new funding restrictions. Subsequently, they were brought into the process but one official said that their relationship with NASA "has been damaged to a certain degree."[11]

## NEW SPACE STATION MANAGEMENT TEAM

When the SSAC met in mid-October the consolidation of the Office of Space Flight and Space Station had been completed. Bill Lenoir was announced as the Acting Associate Administrator for Space Flight (the Acting label would soon be removed) and Richard Kohrs, reporting to Lenoir, was the senior Space Station manager. Former astronaut Robert L. Crippen was promoted from Deputy Director NSTS Operations to Director of the Space Shuttle Program. Arnold D. Aldrich, who had been NSTS Director and successfully returned the shuttle to flight after the *Challenger* accident, was named Associate Administrator for the Office of Aeronautics and Space Technology. Truly's new Headquarters team was now in place.

---

9. Excerpted from a briefing given to the NASA Alumni League, March 1990.
10. Briefing to NASA Advisory Council, Franklin D. Martin, Assistant Administrator for Exploration, August 16, 1989.
11. The quotes are from an article by Kathy Sawyer appearing in the *Washington Post* on September 26, 1989.

Dick Kohrs first briefing to the SSAC brought the Committee up to date on the results of the Configuration / Budget Review Team's analyses. The Team had held three meetings and the international partners and science users participated in the last two in August and September. Their conclusions: a minimum of $1.85 billion was needed in FY 1990 for a "stable and sustainable" program and even at this level important capabilities in crew size, power, and data services would have to be deferred. Kohrs presented five charts detailing the impact on the baseline program. Most of the major systems would have to be modified or reduced in capability. He stated that maintaining first element launch in 1995 was a "line in the sand" that the program would maintain unless directed by Congress to slip the date. At $1.655 billion, significant schedule slips of six months to a year would result for MTC and PMC and for the launch of the JEM and ESA modules. Assembly complete would be delayed 18 months. Grim forecasts. A meeting of the Level I and II Management Board was scheduled for the end of October when the final budget would be known and then a new baseline program would be established to fit the new numbers. As another cost cutting measure and to resolve the problem of getting enough experienced staff to move from the field centers to Washington, the program was examining transferring some of the Level II management tasks back to MSFC and JSC.

In spite of press reports of "livid" international partners, many joint activities were taking place. While the SSAC was meeting, the 2nd International Evolution Working Group was meeting in Paris hosted by ESA. At the meeting, the partners expressed concerns on delaying launch dates and the probability of reduced user capabilities, however, they were continuing to work on developing the program documentation and management plans. A Systems Operations Panel and Utilization Operations Panel were being established to formalize guidelines for operations and utilization activities after assembly was completed. Proposals had been received from European companies to utilize the Space Station and they were being evaluated. Truly and senior NASA management were receiving a constant stream of dignitaries from the international partners and discussions had begun to have two Canadian astronaut candidates trained at JSC.[12]

At the end of October, one month after the start of FY 1990, a final Space Station budget was still being debated in Congress. In preparation for more hearings, Chairman Roe had sent a series of questions (91, many with multiple parts) on October 18th to NASA requesting responses by October 30th.[13] Thus is business conducted between the Executive and Legislative branches of the government, often in adversarial exchanges, often in the form of excruciatingly detailed oversight requiring numerable appearances before cognizant committees.

Answers to the questions revealed a recognition on NASA's part that the future of the program would entail a reassessment "even if the program had received its full funding."[14] The general gist of the questions appeared designed to put NASA on notice that the program would be either stretched out or reduced in scope. In defense of the congressional concerns, one must remember that five years after President Reagan gave his approval approximately $2.1 billion had been spent and not one rivet had been hammered home on a piece of flight hardware. And, congressional staffers knew that there remained a number of unresolved issues. Cost estimates continued to be on the low side and did not include many costs directly associated with the program. Following the conference committee's compromise position, Congress appropriated $1.7496 billion, a reduction of over $300 million from the Administration's FY 1990 request, halfway between the House and Senate markups. Bill Raney informed the SSAC membership in December that "we had pretty well completed the new baselining of the phasing changes to the program." (The rephased baseline was approved at the end of November.) He reported that changes to the work package contracts required by the rephasing were underway and that NASA was working with OMB on the FY 1991 budget request that would reflect the rephasing.[15]

---

12. Information on the international partners is based on the briefing provided to the SSAC by Lynn Wigbels, SSFPO International Program Group.
13. Letter to the Honorable Richard H. Truly from Robert A. Roe, Chairman of the House Committee on Science, Space and Technology, October 18, 1989. To show that there were no hard feelings on the part of the Chairman, he had written "Richard" over the Dear Admiral Truly: salutation, and in his signature block had written "Best wishes" over Sincerely.
14. Quote in the answer to question I.2 from enclosure to NASA letter to the Honorable Robert A. Roe signed by Lynn W. Henninger for Mary D. Kerwin, Director, Congressional Liaison Division, October 30, 1989.

## Program Rephased

Rephasing the baseline resulted in a number of major changes. Among the most extensive was the switch back to launching the U.S. laboratory and habitation modules before the ESA and Japanese modules; once again the partners modules would be launched after PMC. The power system would be changed from AC to DC. Other changes included reverting to a hydrazine propulsion system, and deciding to use shuttle EVA suits through the first phase deferring, indefinitely, development of a new constant volume design.[16] This latter change would become a major obstacle toward accommodating the mushrooming EVA hours needed for assembly and maintenance. A flight demonstration of crew and equipment translation aids (how to move around safely during EVAs using attached holds) was now scheduled for a shuttle flight in November. In addition to the baseline changes, a number of deferrals were also required in order to live within the reduced budget. Perhaps the most significant was not having the full 75 kilowatts of power available until the end of 1997, a nine month delay.[17] Several new studies would also be required to satisfy questions raised by users including location and size of the centrifuge, knowledge and accuracy of payload pointing, and developing a mapping system to understand acceleration forces at different places on the structure and in the pressurized modules.

When the SSAC met in mid-February 1990, Kohrs reviewed all the actions taken by the program in the last four months. The rephased baseline changes and deferrals were described and a new milestone schedule distributed. FEL still held for March 1995 but MTC would slip six months. PMC would be delayed seven months and assembly complete would be eighteen months later than originally planned. Assembly of the JEM and ESA modules would be delayed eight and nine months, respectively (Figure 21). Compared to the last assembly sequence provided to the SSAC in 1988, the number of assembly flights to achieve MTC was now seven versus four, the same number of flights (13) to achieve PMC, and assembly complete would require eighteen flights versus nineteen. Thus, the impact of the budget reduction was closer to that projected in the summer for the original low-ball House vote. PDR was now scheduled over a ten month period in 1990 and would consist of incremental PDRs of distributed systems, elements, and ground systems. The first system (data management software) would be reviewed in February and the integrated system review, completing the PDR, would take place in November and December, a slip of four months from the date provided in June 1989. CDR was scheduled for mid-1992. Some members of the SSAC expressed concern about such an ambitious schedule in view of uncertainties with the FY 1991 budget request. Kohrs agreed that it would be a challenge to meet the schedule.

---

15. From a letter, with enclosures, sent to all SSAC members from William P. Raney, Executive Secretary, December 14, 1989.
16. The decision to defer the development of the Space Station Mark III 8.3 psi space suit — saving a not insignificant plus or minus $350 million — was one of the bad decisions SSFP was forced to make by budget cut backs. The MARK III was the outgrowth of twenty years of technology development reaching back to Apollo. If available, it would reduce the astronauts required prebreathing time by three to four hours when using the shuttle EMU to exit from the Space Station on each EVA, because of the difference in operating pressure between the Space Station and the shuttle EMU. Because the Mark III was a constant volume suit, all EVA tasks would be easier to perform, the astronauts would not be working against suit pressure to move the suit joints. Glove technology was better making the astronauts more finger-dexterous and less fatigued when holding equipment, and because it could be dismantled into many pieces, the Mark III was much easier to don and stow. After deferral, NASA contracted with Hamilton Standard Div. of United Technologies Corp. and ILC Dover, Inc. to study ways to improve the operation, durability and ease of refurbishment of the shuttle EMU. See for example: SAE Technical Paper Series, Shuttle Extravehicular Mobility Unit (EMU) Operational Enhancements, July 1990. Reprinted from NASA SP-830, Space Station and Advanced EVA Technologies, 1990.
17. The total list of changes in addition to those mentioned were: reducing airlocks from two to one, passive rather than active cooling of external payloads, reducing lab support equipment, deleting solar dynamic power generation, and reducing polar platform unique hardware and transferring the program to OSSA for management. Beside the power deferral, other deferrals included some of the crew habitability amenities, such as washers, dryers, and a freezer, until 75 kW of power was available, and the closed-loop oxygen and carbon dioxide system (resulting in more change-outs of expendables), and supplying ultra-pure water for users.

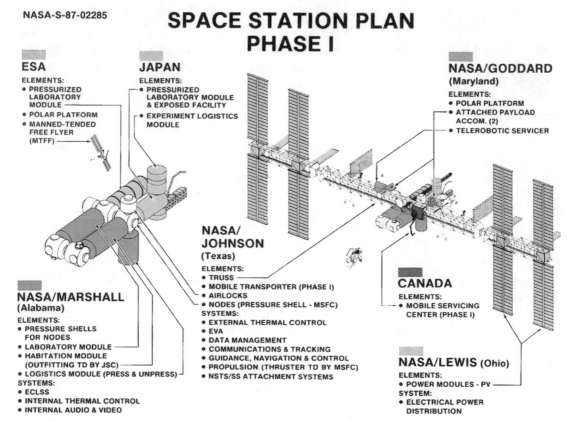

Figure 21. Space Station Plan – Phase I.

The Committee was briefed on the status of the in-house ACRV studies initiated in FY 1986 and completed in FY 1988. Two approaches were addressed: having the ACRV space based, where it could respond in minutes to an emergency, or ground based where response time might be weeks or months. A space based ACRV was selected for the next round of studies. Two contractor teams led by Rockwell and Lockheed had been selected early in February for negotiations to conduct six month Phase A studies. The studies would start in April with an option to go directly to Phase B. If the Phase B option was exercised, it would begin in October with the objective of having enough information to request a new start on an ACRV in FY 1992.

Kohrs reported that agreement had been reached with international partners on a complex issue, how Space Station resources would be allocated based on each partner's contribution (Figure 22). For the Manned Base, the U.S. could utilize 71.4% of the total resources, Canada 3%, ESA 12.8%, and Japan 12.8%. Utilization of discreet elements might deviate from the above percentages. For example, it was agreed that the U.S. could utilize 46% of ESA's *Columbus* module, Canada 3%, and ESA 51%. Resources for the two polar platforms would be shared on a balanced, reciprocal basis between the U.S. and ESA with Canada allowed to use its standard 3%.

### Amount of EVA Raises Concerns

One month after the SSAC meeting the results of an EVA analysis conducted at JSC were made public. Freedom's design would require an enormous amount of EVA time for preventative maintenance after only 60 to 70 percent of the first elements were assembled on orbit. The analysis showed that the current design life of over 5,500 electronic devices and other parts mounted outside the pressurized modules would require 2,200 hours of EVA each year to replace components as they began to fail. The PV panels were especially troublesome because the cells would slowly degrade and the system would lose about 1 kilowatt of output per year. Putting the 2,200 number in perspective, it would mean on average more than three two-man EVAs per week with each EVA of about 6.5 hours in duration. Complicating the high number of EVA hours needed for maintenance was the ground rule that EVAs

## SPACE STATION UTILIZATION ALLOCATIONS

| MANNED BASE | U.S. | CANADA | ESA | JAPAN |
|---|---|---|---|---|
| 1. Utilization Resources | 71.4% | 3.0% | 12.8% | 12.8% |
| 2. User Accommodations | | | | |
| a. NASA Lab Module | 97% | 3% | | |
| b. NASA Attached Payload Accommodations Equipment | 97% | 3% | | |
| c. ESA Attached Pressurized Module (APM) | 46% | 3% | 51% | |
| d. ESA Man-Tended Free Flyer (MTFF) | (25%) (Option to use) | 100% | | |
| e. Japanese Experiment Module (JEM), consisting of a pressurised module, an exposed facility & experiment logistics modules | 46% | 3% | | 51% |
| **PLATFORMS** | | | | |
| a. NASA Polar Platform | * | 3% | * | |
| b. ESA Polar Platform | * | 3% | * | |

\* Shared on a balanced reciprocal basis. Users may propose specific splits on actual payloads.

Figure 22. Space Station Utilization Allocations.

would only be conducted while a shuttle was docked to the Space Station. When EVA hours were added to the time spent prepping, suit donning, pre-breathing of several hours, and EVA close-out activities, each EVA period would consume over twelve hours of an astronaut's time, a full, hard day's work. How the shuttle space suit would stand up to such high usage was under study and there were concerns that the suit design was never intended to allow many repetitive EVAs without major refurbishment back at JSC. At this point, NASA had a total of 400 hours of EVA experience, including the Apollo missions, on which to base projections of potential complications that might arise with such a high work load.

The need for maintenance EVAs was not unexpected; the magnitude of the numbers that became evident by the end of the Phase C studies was the show stopper. During the definition phase in 1986 it was recommended that as much as possible of the electronic equipment and other components be located inside the pressurized modules rather than mounted outside on the trusses. In this way maintenance could be conducted without resorting to EVAs. But as the design matured the need grew to mount more and more equipment outside. Volume inside the hab and lab modules was needed for experiments, support equipment and some spares not to mention all the furnishings that were required to make the Space Station a home away from home. If EVAs were permitted from the manned base when shuttles were not present, EVA suits alone would take up a lot of stowage space, another penalty created by deferring the constant volume suit that could be disassembled into small components for easier storage. Data management, guidance and navigation systems, communication antennas, and a host of other components could only be accommodated by hanging them on the trusses. David M. Walker the astronaut manager for Space Station assembly was quoted as saying that, as presently designed, the Space Station probably could not be built.

Kohrs met with the contractors in January to discuss how to reduce the amount of EVA. His goal was to lower the number of maintenance EVAs to one per month. Whether or not it was attainable remained to be seen. A number of options were possible and recommendations from the External Maintenance Task Team, studying the problem, were expected by July.

Word of the increased requirement for EVA had not escaped the notice of Congress. Lenoir was summoned to appear before Chairman Roe and his Subcommittee at the end of March to explain this new problem. He was asked to address the question: "Did the Space Station need to be redesigned in light of the heavy demand for EVA during assembly and maintenance?" Lenoir flatly stated: "The

Space Station does not need to be redesigned." He told the Subcommittee that the analysis was still in an early stage and that "it has been well known that EVA hours for maintenance could grow to an unacceptable number if not controlled." And that "NASA plans to keep EVA time as low as possible." Lenoir also promised that robotics and artificial intelligence systems would decrease reliance on EVA.[18] Although politely received, he failed to convince the Subcommittee that all was well with the program, as you will read.

At the next meeting in May the SSAC was brought up to date on the status of a number of the problems described above. Orbital debris studies had been assigned to the WP #2 contractor. An analysis had been done based on the current design of how impacts of various sizes of debris could effect critical elements (Figure 23). Several different shielding concepts were being studied. New design criteria were expected to be ready in the next two months. To arrive at possible solutions a test schedule was established to conduct hypervelocity impacts on various prototype Space Station elements such as the truss structures, airlocks, and module windows. Testing would begin in June and continue over the next year.

**ORBITAL DEBRIS ENVIRONMENT:
ISSUES FOR SPACE STATION FREEDOM**

| Debris Size / Range | Issue | Potential Solutions | Concerns |
|---|---|---|---|
| less than 1 cm | • Loss of critical elements due to direct impact<br>• Damage to non-critical elements due to direct impact and secondary ejecta | • Shielding<br>• Maintenance<br>• Redundant systems | • Weight limitations<br>  – New materials<br>  – Add-on shielding<br>• Additional EVA |
| 1 cm to 10 cm | • Loss of critical elements<br>• Damage to non-critical elements due to secondary ejects from direct impact of non-critical areas<br>• Potential loss of Station from catastrophic collision | • Increased Ground tracking capabilities<br>• On-board collision warning sensor<br>  – Active shielding<br>  – Heavily shielded shelter<br>  – Directed energy diversion<br>• Materials which minimize secondary ejecta | • Technology development |
| Larger than 10 cm | • Loss of Station from catastrophic collision | • Collision avoidance using ground tracking | • US Space Command limitations<br>  – Accuracy of tracking<br>  – Completeness of data<br>• Frequency of maneuvers |

Figure 23. Orbital Debris Environment: Issues for Space Station Freedom. 4/26/90 SSAC briefing

Laboratory support equipment was being studied by a joint team consisting of SSFPO and OSSA. At present the equipment list was extensive (44 items) and included autoclaves, cutting and polishing systems, freezers, glove boxes and a wide variety of measuring devices (Figure 24). Agreement had been reached with the international partners on standardizing the interfaces for payload racks such as the locations within each laboratory, and common instrumentation, controls, valves, sensors and other hardware elements. Studies on where the centrifuge would be located and its size, were nearing completion and a decision was imminent. The favored position was to place it in one of the nodes if its size could be accommodated. Initial recommendations to have more than one size centrifuge had, for the moment, been cast aside.

---

18. Committee on Science, Space and Transportation, House of Representatives, Statement by Dr. William B. Lenoir, Associate Administrator for Space Flight, March 29, 1990.

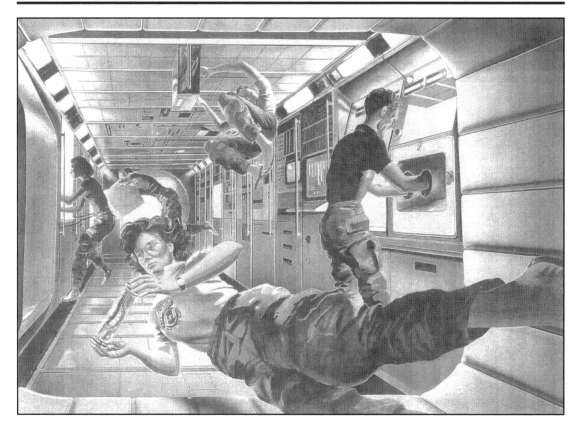

Figure 24. Artist Harold Smeicer's drawing of interior of Space Station Freedom laboratory element. NASA 90-14C-S49.

Progress had been made on addressing some of the EVA maintenance problem even though the latest estimate had climbed to over 3500 hours. Many of the 5,500 plus components that would require maintenance were located in what were termed "greedy" orbital replacement units (ORUs). Twenty percent of the ORU inventory was discovered to account for almost 80% of the EVA time and the study team was now concentrating on the "greedy" ORUs to determine how their maintenance requirements could be reduced. The team also was examining how the estimates for mean-time-between-failure (MTBF) for each component had been calculated and it appeared that the contractors had not used consistent criteria. Also, JSC was conducting tests in the water immersion facility to better estimate the time needed to accomplish the change-out of ORUs. The tests included redesign of tools, standardizing equipment and procedures, and determining if some of the tasks could be carried out by robots. Perhaps the Space Station could be built after all.

Members of the SSAC SET&V Panel sat in on the program verification review given in June. Marc Bensimon briefed on the results of six months of study to develop recommendations on critical Space Station verification functions in order to bring them more in line with previous NASA manned programs. The participants in the review included Level II and III managers as well as the work package contractors. Recommendations coming out of the review were numerous. They were based on providing a program-wide capability for integrated physical and functional interfaces for the flight hardware. In simple terms, make sure everything fit together and functioned when mated. Two facilities would be needed, one at JSC and the other at KSC. The JSC facility would be responsible for verifying the integrated truss assembly flight hardware including unpressurized logistics carriers and pallets, the PV modules, and the Canadian MSC. KSC would verify the pressurized modules, all the nodes, and the JEM. The integrated truss assembly would be moved to KSC for final sustaining engineering and end-to-end testing at what would be called the Stage Integration Facility.[19] This was

---

19. Space Station Freedom Program Verification Review, Presented to Deputy Director, SSFPO, June 8, 1990, Marc Bensimon, Deputy Manager, SSFPO.

the facility that Congress had deleted from NASA's FY 1988 construction budget. As discussed earlier, integration and verification of Space Station elements before launch would be at a scale never before required by a NASA program. The program's success depended on getting these procedures right.

Wrapping up its analysis of the program's verification plans, SSAC Panel 7 submitted its report. A number of deficiencies were noted. High on the list of concerns was the continuing lack of resources at Level II to provide the necessary systems engineering and integration oversight that only NASA could provide. The Panel reported: "The fact that there is no up-to-date or adequately detailed Level II Verification Plan is a manifestation of the lack of necessary resources . . . Level II philosophy and requirements need to be established urgently." The Panel noted that the Space Station will have many in-orbit configurations as it evolved to final assembly complete. Each configuration would be an individual spacecraft that would require verification. A detailed analysis was needed of each in-orbit configuration to define what can be verified prior to launch and what can only be verified in space. The Panel also endorsed the need for the software and hardware integration facilities that Bensimon had described.[20]

On July 19th the External Maintenance Task Team presented the results of their study to Administrator Truly. In some respects the problem had worsened. Based on the current design, the Team had inventoried all the ORUs including those that would be attached to the elements supplied by the international partners. The grand total – 8,158. The good news was that after assembly complete the EVA hours were manageable. The problem occurred prior to assembly complete. They calculated that during the twenty-four months from FEL to assembly complete there would be 811 maintenance actions, requiring a total of 5250 EVA hours and 437 two-man EVAs. On average, 4.2 two-man EVAs per week over the entire 24 months.[21] This latter number did not tell the whole story because the shuttle would only be docked to the Space Station for short periods each month. To conduct both assembly and maintenance during those periods appeared to be more than the shuttle time on orbit and available man-hours would allow.

The Team made fourteen recommendations. Summarizing just a few: develop the ability to maintain the ORUs either by astronauts or ground controlled robots, move a large number inside, and redefine the role of the Space Station to be a "facility" with regularly scheduled downtime for maintenance and refurbishing. Many of the recommendations dealt with the use of robotics and the Teleoperator Maneuvering System (TMS) to do the servicing. How the recommendations could be carried out was left open. For instance, moving many ORUs inside had already been studied and largely rejected because of a lack of space. One solution, examined early in the study to increase MTBF with more reliable components, was rejected as being too costly (The higher the required reliability of a component, the more it would cost). In spite of the obvious problems, it was recommended that the solutions the Team proposed be incorporated in the PDR. In reality, the EVA problem still had not been solved.

With all the problems surfacing in the Space Station program in the summer of 1990, as well as embarrassing deficiencies in the manufacture of the Hubble Telescope mirror, continuing shuttle maintenance problems, and Congress not showing any signs of embracing the President's space exploration goals, the Administration established a blue ribbon committee in August. Its charter was to advise NASA on approaches it could use to implement the U.S. space program in the years ahead, a procedure used in years past to garner support when NASA programs were in trouble. Norman R. Augustine, Chairman and CEO of the Martin Marietta Corporation, a well known and admired space pioneer, was selected to chair the study. His committee was made up of eleven other notables including Tom Paine who had chaired a similar study four years earlier. The Committee was requested to complete its deliberations within 120 days; in time to influence the debates on the FY 1992 budget.

---

20. Extracted from the Panel report submitted to the NAC September 28, 1990.
21. Numbers in this and next paragraphs extracted from: External Maintenance Task Team Management Review, Presentation to NASA Administrator, William F. Fisher & Charles R. Price, July 19, 1990. The Space Station Freedom External Maintenance Task Team Final Report, July 1990, was a two part volume, hundreds of pages in length. The Space Station Freedom External Maintenance Solutions Team Final Report, July 19, 1990, 54 pages in length including appendices.

As the dog days of the summer of 1990 wound down, on August 14th the SSAC panel, chaired by MIT professor Jack L. Kerrebrock, that had been studying EVA and robotics made its report. After reviewing a number of studies conducted by NASA and contractors, and visiting GSFC, the WP #3 manager, the panel's conclusion was that the current Space Station plan to minimize assembly and maintenance EVA in favor of using robotics "may not be viable." Current plans to use six different teleoperated systems for these functions appeared overly ambitious and risky in view of the limited experience with such systems. Also, many of the tasks that the teleoperated systems would perform still needed astronaut EVA to oversee operations. The Panel recommended that the program reexamine the decision to stop development of the SSF space suit and external maneuvering unit (EMU). In the long run, if available, they would reduce the cost of assembling and maintaining the Space Station far more than the cost to develop the units. Although there were no firm plans to use semiautonomous robotic systems during assembly and maintenance, the Panel recommended that this be carefully studied and that the functions and capabilities desired for these systems be better defined and coordinated with a single point of oversight at Level II.[22]

## NEW PROBLEMS ACCENTUATE CONGRESSIONAL OVERSIGHT

Phase C studies revealed a new concern; the total weight of the Space Station after assembly complete was growing and was estimated to be over 655,000 pounds. This number included the weight of both the U.S. elements and those to be supplied by the international partners but did not include approximately 15,000 pounds of government furnished equipment (GFE). The original allocation established for the prime contractors at the start of the Phase C/D contracts was 512,184. Of the four work package elements, WP #4, the power system, was the most overweight, WP #3 was under allocation and the international partners had done a good job and were essentially on target. Since June, when the problem was first addressed, a weight loss program had been instituted and succeeded in shaving over 69,000 pounds. If the new weight of 586,488 pounds could not be further reduced, it would mean that two additional shuttle flights would be required assuming that the overweight elements could be packaged in a manner to fit shuttle payload constraints.

The waning days of summer also saw the rapidly approaching new fiscal year without Congress agreeing on the FY 1991 Space Station funding. The Administration's request of $2.451 billion was in trouble in the House committees where there was a proposal to cut $195 million from the request. In the Senate, a cut of $70 million was being debated but by September essentially no actions had been taken by the Senate subcommittees. Once again, the program was left slowly turning over the fires of Congressional indecision. Almost 80% of the budget was earmarked for the four work packages and the FTS contract. If the House passed the proposed cut, more deferrals and schedule slips would be necessary. Concerns about the amount of EVA time, weight growth, and escalating costs, were fueling the Congressional critic's arguments to cut the program back or cancel it. Their concern – did NASA really have control of the program?

---

22. Since last reported in Chapter 2, NASA continued to submit required semiannual automation and robotic Progress Reports to the Congress under the auspices of ATAC. After the September 1988 Progress Report 7, ATAC modified its reporting procedures. Among the research studies coordinated by ATAC, two are especially interesting: Space Station Freedom Program Capabilities for the Development and Application of Advanced Automation, Steven E. Bayer, December 1989, Sponsor: NASA/OSS Contract No.: NAS9-18057, conducted by the MITRE Corp. McLean, VA 22102. This study provided a reference document describing all the advanced automation engineering and research activities (numerous) under way for SSFP at all the centers. The second report is: Space Station Freedom Automation and Robotics – An Assessment of the Potential for Increased Productivity, March 1990, Sponsoring organization: Advanced Development Program, Space Station Engineering, Office of Space Flight, NASA Headquarters. This study is important because it consists of a compilation and analysis of lessons learned from Skylab, Spacelab, space shuttle missions and other programs, such as Soviet Space Station, that would influence how advanced automation and robotics could be applied to SSF. The consensus of the 32 astronauts interviewed for the study was that automation and robotic systems would improve their productivity during EVAs. You may remember that Gordon Fullerton in 1986 had questioned the ability to use the Canadian MSS during Space Station assembly EVAs, believing that it would be too difficult. ATAC would complain in their reports to Congress that NASA budgets did not have enough funds to pursue high-value automation and robotics development that would alleviate some of the Space Station EVA needs. But little ever came from the complaints and Congress, which had mandated ATAC's formation, never provided new funds except for FTS.

On September 27, Chairman Nelson's subcommittee held a hearing to obtain answers on the questions his subcommittee and others had been debating during the last five months. SSAC members were invited and Jack Kerrebrock would be a witness to report on his SSAC Panel's findings. Nelson opened by asking a series of four questions: Should SSF be redesigned? Can we afford a permanent manned presence in space? Can NASA do the proper kinds of ground testing to assure that all the pieces would function correctly on orbit? Does SSF still have public support? The subcommittee's positions on the questions appeared to lean to: yes, no, we don't want to fund a facility at KSC, perhaps not. Lenoir and Kohrs addressed all of Nelson's questions attempting to turn around the subcommittee's leanings. How persuasive Lenoir's and Kohrs' arguments were in countering these positions would soon be known.[23]

All of Congressman Nelson's questions reflected the overt and covert dissatisfaction with the program. New estimates of the cost of the Space Station indicated it could be as much as $38 billion. The Planetary Society with membership consisting of present and former industry, government, and academic individuals was on record as questioning if the Space Station was the right program for the future of manned space exploration versus moving on directly to the Moon and Mars. Hearing the rancor and complaints from vocal constituents, professional societies, and lobbyists, it was not surprising that the Congressional committees were raising pointed questions.[24]

Finally, in October, Congress ended all pretense of nibbling around the margins and the appropriation conference committee, going far beyond the earlier proposed reductions, cut the FY 1991 request by $551 million. Congress then directed NASA to develop a plan within 90 days to shave $6 billion from the program in the next five years. This meant approximately a forty percent reduction below the budget profile on which the program had been based. In the 90 days NASA would have to cut in half a program that had ramped up over six years to spend $2 billion per year. In the month of October the program's spending rate was $170 million per month and growing in anticipation of the future, larger funding requests needed to maintain content and schedule. At the next SSAC meeting, the second week in November, Dick Kohrs described the Congressional order and the initial look at how the program might respond. Committee members raised the question as to whether the Space Station could be salvaged with the continuous downsizing and changes, and if NASA might recommend canceling the program. At this point, NASA had the opportunity to lay down the gauntlet and stick to its earlier declared position by telling the Congress that program goals could not be met and redesign would not solve the budget impasse. NASA backed away from the confrontation.

## Management Adjusts to Congressional Demands

Bill Lenoir and his Space Station team decided to soldier on while admitting it would be a challenge to reduce the program as Congress requested and still maintain a useful program. Lenoir stated, "If we cannot come up with what we consider to be a reasonable program . . . we would have to readdress whether we can go forward with the Space Station."[25] With so many unknowns but drastic reductions in capability a certainty, the international partners were reported to be "uniformly disappointed." Probably the understatement of the year. Congress, wearing its instant-engineering hard hats, ordered NASA to break the program into a series of self-sufficient phases starting with a platform that would remain in orbit and be visited several times a year by shuttle astronauts in the extended duration orbiter (EDO). Research on the platform would concentrate on the effects of weightlessness and material processing. Each shuttle mission might last up to four weeks depending on how long the EDO could remain on orbit and provide the crew with their life support and other needs.

In a memo to all the Space Station offices, Kohrs laid out the ground rules for conducting the restructuring.[26] The assessment would take place over three months and, based on the existing

---

23. Based on author's notes made at the hearing.
24. *Aviation Week & Space Technology* carried an excellent summary of the difficulties the Space Station was facing in its October 8, 1990 edition, written by correspondent James R. Asker: Space Station Redesign Likely; Contractor Hiring Freeze Ordered.
25. Quote from an article written by Kathy Sawyer in the *Washington Post*, November 8, 1990.
26. NASA Memo: To: Distribution, From: M-8 / Director, Space Station Freedom, Subject: Space Station Restructure Assessment, November 2, 1990.

program structure, determine what deletions, adjustments, or schedule deferrals, would be necessary. Level II would be in charge with the involvement of Level III work package project managers, contractors, the international partners, and users. All resulting changes would then be implemented by new budget allocations, and formal directives from his office would document any schedule changes. The new budget targets were $1.9 billion in FY 1991 (the Congressional appropriation), $2.1, $2.3, and $2.5 billion in fiscal years 1992-1994, and $2.6 billion in each of the next three years. However, he also instructed that as the assessment went forward, the integrated system PDR would continue on schedule to be completed at the end of CY 1991.

Attached to Kohrs' memo were six pages of additional, detailed ground rules. They included instructions on how to develop the phased approach ordered by Congress starting with a man-tended capability and ending with an eight-person full-time crew, and a list of candidate areas that could be considered for deletion, deferral, or simplification that ranged from the lab and hab modules to communications and tracking. In essence, no element might be spared scrutiny. His schedule to complete the restructuring had a preliminary review date of December 10th, a final review for the Administrator on January 14, 1991, submittal to OMB on the 17th, and submittal to Congress on the 22nd. A very tight schedule considering the amount of analyses that would be needed. Office lights would burn late at night during the next three months.

Behind the scenes, and soon to be made public, the work package contractors, facing a $6 billion reduction in funding for the program over the next few years, began to lobby to preserve their pieces of the Space Station. Each of the prime contractors had been asked by NASA to participate in the restructuring and each was beginning to come up with its own, unique response that would assure maintaining their share of the new design. These alternative designs were presented to the SSFPO in Reston in November. Kohrs and his team would have to select which, if any, would be accepted.[27]

## AUGUSTINE COMMITTEE REPORT

On schedule, and having full knowledge of Congress' order to restructure the Space Station, the Augustine committee submitted its report on December 17th.[28] As expected, the Report contained very broad recommendations on all aspects of NASA management and programs. The Committee recommended that space science programs deserved the highest priority "above space stations, aerospace planes, manned missions to the planets, and many other major pursuits which often receive greater visibility." It endorsed the "Mission to Planet Earth" and a "Mission from Planet Earth" to be the major programs to receive the greatest support with the latter focusing on robotic exploration of space, but also including human exploration. A bit of a mixed signal with these last recommendations as they strongly supported human exploration of Mars and returning to the Moon. They included the advice that rather than proceed on a rigid schedule, the nation should decide to "go as you pay" and seek and obtain the funding needed to proceed taking into account national economic factors.

There were specific recommendations for the Space Station. Recommendation 6 stated that NASA and its international partners should reconfigure, reschedule, and reduce the Space Station's size and complexity and then wondered if it could be done in the 90 days Congress mandated. Aside from their question on whether or not the restructuring could be completed in 90 days, this recommendation, echoing the Congressional order, seemed to be a cave-in to the Congress' direction to restructure the program. Until this report, all the studies or reviews conducted by commissions outside NASA had gone down the line supporting the established Space Station program and would recommend how the baseline program could be improved. The Augustine committee, composed mostly of senior industry managers, some of whom were directly involved in the program, surely saw the many problems looming for the Space Station and decided it would be easier to go along with the Congressional order. Certainly the lack of enthusiasm by the committee to continue the baseline program seemed to confirm the uninspiring testimony given earlier to Congress by the work package contractors that had left Congressional staffers perplexed. To put some possible context on the committee's recommendations,

---

27. For this controversy, see for example *Space News*, Contractors Jockey for Position in Space Station Redesign, November 26-December 2, 1990.
28. Report of the Advisory Committee on the Future of the U.S. Space Program, December 1990.

one must consider that committee members at one time or another had appeared or expected to appear before Congressional committees on this or other business. Its recommendations were good politics that might be remembered when trying to appeal in the future to a sympathetic ear.

Recommendation 11 was to initiate a design effort so that manned activity in the Space Station could be supported without the presence of the shuttle, and that a crew recovery capability be immediately available based on launching from an expendable launch vehicle. This last aspect was also covered in Recommendation 6, where it was stated that assured crew return in an emergency should be in place prior to human occupancy. As this recommendation did not fall under the "go as you pay" advice, how a crew return capability could be paid for, in light of the budget restrictions, was not addressed.

Among the many other recommendations directed to the Space Station were those to concentrate on two missions, life sciences and microgravity research; improve ground testing and verification; reduce launch requirements; reduce EVA; reduce cost; honor the international commitments; procure a fifth orbiter; and work with OMB and appropriate Congressional committees to establish a reasonably stable share of NASA's total budget to develop advanced technology. A laundry list of recommendations, some motherhood and obvious, and some beyond the ability of NASA to carry out without the support of future administrations and Congress. One recommendation, to have NASA and DOD share the cost of a new $10 billion launch vehicle to put payloads of 150,000 pounds into low Earth orbit, was accepted by the Administration and details were being worked on. It would soon fizzle out when it reached Congress. In one respect the Augustine Committee accepted the Congressional direction to reduce Space Station budgets but then turned around and made recommendations that, if accepted, would have required much larger, and unrealistic, NASA budgets.

If all the recommendations were implemented, they would benefit several of the companies with representatives on the committee. This observation is made to indicate how difficult it is to form a committee of knowledgeable individuals to provide needed advice on government programs, and how easy it is for critics inside or outside government to claim conflict of interest when such committees deliberate. The Report, in Appendix III, attempted to head off such criticism and contained a legal compliance disclaimer that included the statement: "It was the determination of the appointing authority (the President) that the private interests of the individuals appointed to the Committee were not so paramount as to impede their objectivity or integrity as members of the Committee." Adhering to the disclaimer, for example when the Committee debated procuring a fifth orbiter where Martin Marietta had a strong business interest, Augustine recused himself from the discussions. Although quoted frequently in the months following its submission, and at times today, the Augustine report had little influence on NASA's future beyond supporting the Space Station down sizing. Another commission, another report to be filed.

In spite of all the confusion that pursued the Space Station in the latter half of the year, the program completed the element and system PDRs on schedule in December. Several open items (exceptions) remained including verifying the final assembly plan, analyses related to safety, reliability, maintainability and quality assurance (SRM&QA), and certification of the ESA and Japanese contributions. Open questions remained as to how useful the PDRs would really be if new designs were needed and how much of the program would survive the restructuring? Some of the elements reviewed in the PDR were sure to be deleted and the hope was that those that remained would not be greatly altered and the time spent on the PDRs wasted. All of these questions would have to be answered as quickly as possible if the Space Station program would survive in 1991.

## — CHAPTER 5 —
## RESTRUCTURING THE SPACE STATION
### (JANUARY 1991 - DECEMBER 1992)

Once again the new year started with the Space Station program in disarray. The 90-day response time to tell Congress how NASA would rephase and restructure the program proved to be much too short and an extension was requested and granted. Actually, two extensions were requested, the first for 30 days, the second for an additional 34 days. Congress agreed to both extensions. As a result, the original submission date of January 22, 1991 stretched out into March. The new schedule that Kohrs and his team put together to meet the revised deadline reviewed the response with the Administrator at the end of January, followed two days later with briefings to OMB and NSC. One week was added after meeting with OMB and NSC to accommodate any new direction that might be received. After that followed another round of reviews with the Administrator and NSC, and submission to Congress in March, with a final report on April 8th.

In the meantime, more details became known of the Congressional directions received by NASA for the restructuring. Congress required that NASA reinstate development of the solar dynamic power system, to the tune of $10.8 million in the FY 1991 budget (at the request of the Ohio delegation, home of WP #4 at LeRC.) Congress also required (at the request of Senator Mikulski and the Maryland delegation) continued development of the FTS in WP #3 at GSFC, earmarking $106 million in FY 1991. Congressman Nelson's Subcommittee insisted that a fully equipped microgravity laboratory, provided by the U.S., be in orbit before launching the hab module. His Subcommittee members believed microgravity research should be the primary focus of the Space Station. They also had ideas on life sciences research and required that the centrifuge be located outside the U.S. lab module. The reasoning for this requirement is a little obscure, but undoubtedly linked to the concerns raised by the SSSAAS during the Woods Hole meeting; the life sciences community was also a powerful Congressional lobbying group. A few other Congressional requirements included specific power levels that would be available to users, and shuttle launch rates. Kohrs and his team found themselves between the proverbial rock and a hard place. On one side were all the Congressional requirements not in the baseline, and opposing them was NASA's desire to keep the Space Station tracking as closely as possible to the original baseline program. Both could not be accomplished within the new budget profile that Congress had dictated.

When the SSAC next met in March, Kohrs was able to describe how the restructuring was proceeding. Beyond any question, the restructured program incorporated major changes to the baseline program NASA had been following since CETF and the Phased Program modifications adopted in 1987. In the new phases, man-tended capability would be followed by an extended period of man-tended operations, then a permanent manned capability and finally permanent manned operations. Follow-on phases could come later. This might sound very similar to the previous baseline program but the details were quite different.

One detail that illustrated the difference was the nuance between capability and operations. In the two man-tended phases, capability meant that the Space Station would be operating the majority of the time as a free-flyer and astronauts would not be present. The operations phase meant that astronauts would be spending more time at the Space Station with visits every 45 to 90 days but still dependent on the shuttle for life support. When or if the extended duration orbiter would be available to allow attached missions to expand from thirteen to twenty eight days was unknown. The EDO was only on the drawing boards. Similarly, when the permanently manned phase was achieved, the ability of the Space Station to support science and engineering experiments would gradually grow from a limited capability to a full range of operations.

Based on the first look at how the restructuring would effect the shuttle manifest, to arrive at the permanently manned configuration would require seventeen assembly flights and nine logistics flights through 1999 with FEL scheduled to start no later than the second quarter of CY 1996. This compares to the old baseline manifest that required eighteen assembly flights and sixteen logistics flights through the same time period, a sizable reduction. Crew size would be four at PMC, two of whom would be available to conduct research. Eight person crews remained a commitment but would not be possible until after 2000 when an additional hab module would be attached. New ground rules required a crew rescue vehicle be available at PMC. This requirement introduced a big unknown because the ACRV Phase A/B studies had not begun and its cost was only an educated guess. Complicating this unknown even further, Congress' restructuring did not allocate any funding for the ACRV. NASA ignored this direction and asked for $6 million in FY 1991 and projected an additional $25 million in FY 1992. The program still held out hope that first element launch could take place before the end of CY 1995. The FEL redesigned payload now would consist of the first truss section with a PV array and other ORUs attached (Figure 25).

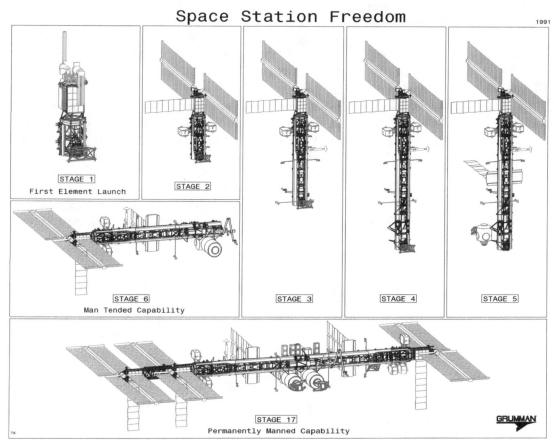

Figure 25. Space Station Freedom. Grumman artist concept of progressive stage build up during assembly flights to PMC.

Other significant changes included reducing the length of the hab and lab modules from 44 feet to 27 feet (diameter would remain the same), and the length of the truss sections. The new design, called a pre-integrated truss, would have a segment completely built and checked out on the ground and then launched as a fully functional "spacecraft." Previously, the trusses would be built "stick-by-stick" and checked out on orbit. With the new design, each truss would be joined to other truss segments that had gone through the same process and then all the plumbing and electrical connections would be made between the assembled segments. All elements would be redesigned in this way to permit better end-to-end ground testing prior to launch. On orbit assembly would be simplified, consisting of mating each of the spacecraft. As a result, it was estimated that the amount of EVA required for assembly would be reduced 50% and maintenance 40%. With the reduction in the length of the trusses and the size of the pressurized modules, the overall length of the Space Station shrank from 493 feet to 353 feet. A corresponding reduction in weight would occur, but the new weight had not yet been calculated.

Restructuring had several negative consequences for potential experimenters. Power available for housekeeping and users was reduced from 37.5 kW to 18.7 kW. This meant that only 11 kW to 13 kW would be available for users at man-tended capability. At PMC power would be reduced from 75 kW to 56.25 kW with a minimum of 30 kW for users. With all the other downsizing it was hoped that this amount of power would keep the users relatively happy. The number of experiment racks in the lab modules, as well as the amount of downlink telemetry dedicated to experimenters, also were reduced. (For a complete comparison of the restructured Space Station to the Space Station design prior to the Congressional mandates, see Figure 26.)

## PROGRAM PLAN
## CAPABILITY/CONFIGURATION COMPARISON

|  | Man-tended Operations | | FY 2000 Permanently Manned Operations | |
| --- | --- | --- | --- | --- |
|  | Before | After | Before | After |
| Power | 37.5 kW | 18.75 kW | 75 kW | 56.25 kW |
| Module Length | 44 feet | 27 feet | 44 feet | 27 feet |
| U.S. lab user racks | 28 | 12 | 28 | 12 |
| User research power | 10 kW | 10 kW | 30 kW | 30 kW |
| Logistic module capacity | 20 racks | 8 racks | 20 racks | 8 racks |
| Command uplink | 25 Mbps | 72 Kbps | 25 Mbps | 72 Kbps |
| Data downlink | 300 Mbps | 50 Mbps | 300 Mbps | 50 Mbps |
| Gravity level | 3 µg | 3 µg | 1 µg | 1 µg |
| Attached payload accommodations | 1 | Deferred | 2 | Deferred |
| U.S. assembly and logistics flights | 7 | 6 | 26 | 14 |
| Permanent crew size | 0 | 0 | 8 | 4 |
| Dedicated crew for research | 0 | 0 | 6 | 2 |
| Utilization flights | 2-3/year | 2-3/year | N/A | N/A |
| Truss | Nineteen 5m bays built and checked out in orbit | Segments built and checked out on ground | Built and checked out in orbit | Segments built and checked out on ground |
| Length (Total) | 312 ft. | 158 ft. | 493 ft. | 353 |
| Pressurized equipment nodes | 1 | 1 | 4 | 2 |
| International hardware: Canada | Mobile Servicing Center | Mobile Servicing Center | All | All |
| Japan | None | None | All (5 U.S. User Racks) | All (5 U.S. User Racks) |
| Europe | None | None | All (11 U.S. User Racks) | All(11 U.S. User Racks) |
| Life Support | Recycled air, recycled water | Shuttle Provides | Recycled air, recycled water | Recycled water |
| Flight Telerobotic Servicer | Yes | Transferred to Office of Aeronautics, Exploration and Technology | Yes | Transferred to Office of Aeronautics, Exploration and Technology |

**Quality maintained**
**Quantity can be increased in future years**

OSS-15739
3/15/91

Figure 26. Program Plan, Capability / Configuration Comparison.

The restructured program presented to the SSAC made several assumptions that remained to be accepted by NASA management, and Congress. To stay within the Congressional runout costs required SSFPO to transfer one of Senator Mikulski's favorite programs, the FTS, to the Office of Aeronautics, Exploration and Technology (OAET, formerly known as OAST) thus removing it from the Space Station budget line. GSFC's contract with General Electric to design, develop, and test the Attached Payload Accommodation Equipment (APAE), with a value of $555 million, had been canceled in February. The APAE element, that was being designed to ease the assembly of large external payloads, was deferred indefinitely. As a result, although still developing the FTS, all of GSFC WP #3 would be eliminated from the Space Station budget line. Whether this sleight of hand with the budget would fly with the Senator was unknown. Also, for budgetary and schedule reasons it was proposed, once again, to delete the solar dynamics power system. The Ohio delegation would strongly influence the outcome of this maneuver. If all these changes were accepted by the Congress, the Administration was said to be on board. The program was projecting a buildup to a constant runout amount in FY 1995 to FY 1997 of $2.6 billion per year, budgets that were in accordance with Congressional direction. Total cost to PMC was estimated at $30 billion. These changes were advertised to save $8.3 billion.[1]

Turning the clock back seven months and then forward to February 1991, NASA finally responded to a question asked by Senator Frank R. Lautenberg (D-NJ) at a hearing in July 1990. Apparently laboring under the impression that the international partners would be contributing $8 billion in cash to the program, he had asked: "Has any of that money been provided?" Now, after a long delay, NASA responded to the Senator: "Our agreements with the international partners provide for each partner to design and develop specific hardware elements, regardless of the ultimate cost. The partners do not financially support the development of hardware for which we are responsible and we do not financially support the development of hardware for which they are responsible. The current estimated value of the three partner's hardware contribution is approximately $8 billion in real year dollars." The $8 billion was broken down for the Senator: $1.0 billion for Canada, $4.5 billion for ESA, and $2.5 billion for Japan.[2] These same numbers were included in the report to Congress on how the program addressed the restructuring.

Some in the private sector, taking note of how NASA was struggling to preserve the Space Station program and compete with the Soviet *Mir* station, proposed alternative approaches for an early, cheap space station. External Tanks Corporation (ETCO), located in Boulder, Colorado, was one of two companies to propose to NASA to build a space station from the NSTS external tank that normally reenters after a shuttle launch and burns up. The ETCO concept was to strip a shuttle to the lowest weight possible, not jettison the external tank after the fuel was depleted and carry it still attached to the shuttle, into low-Earth orbit. Once in orbit, the external tank would be demated and then refurbished and changed into living quarters and a laboratory by later flights. The proposal was similar to the Skylab program that used an Apollo S-IVB stage and modified it to be our first space station in 1973. The estimated cost of the ETCO version of a space station was $10 billion. ETCO, with board members including former NASA managers Jim Fletcher and Phil Culbertson, had approached several aerospace companies to join their proposal. Their high powered team would include Teledyne Brown Engineering, managed by former astronaut Owen K. Garriott; Boeing; and Martin Marietta. ETCO's chairman of the board, John McLucas, a former NAC chairman, when interviewed about his company's plan, said it was probably too late to make such a proposal, and as things progressed, he was right. The ETCO space station never got off the ground. For many, throwing away the external tank on each shuttle launch, a multimillion dollar piece of hardware, seemed a terrible waste of money. Trying to use the external tank would generate other ideas, but none ever came to fruition. Perhaps one day someone will make an offer too good to refuse.[3]

## EFFECTS OF THE RESTRUCTURING

At the end of March, 1991, the National Research Council's Space Studies Board weighed in with their assessment of the effect of the restructuring on science.[4] It was blunt and to the point. Louis J. Lanzerotti, Board chairman, wrote to Truly: "In the judgment of the Board . . . the present stage of redesign does not meet the basic research requirements of the two principal scientific disciplines . . . (1) life science . . . and (2) microgravity research and applications." NASA had no means to rectify these concerns; the budgets simply would not permit the original promised capabilities to be developed. Over the years it would attempt to improve the ability of the Space Station to accommodate science, but the program never fully recaptured the early vision.

In April, three weeks after NASA submitted its plan for restructuring the Space Station, Lenoir appeared before the House Subcommittee on Science, Space and Technology. He described the progress made in 1990, including completion of the PDR, and the results of the many analyses to reduce EVA and weight.

---

1. The details of the restructuring are from the Kohrs' briefing to the SSAC on March 7, 1991, and the Report to Congress on the Restructured Space Station Program, March 20, 1991.
2. NASA Space Station Freedom Facsimile Transmission, from Gary DuBro, Chief International Policy Branch, Policy and Plans Division, Office of Space Flight, to Dan Omahen, GAO, February 9, 1991.
3. From material provided by Phil Culbertson and a *Space News* story: "Company Offers Cheaper Station Using External Tanks," April 22-28, 1991.
4. Letter from the National Research Council Commission on Physical Sciences, Mathematics and Applications, Space Studies Board, addressed to Vice Admiral Richard H. Truly (Ret.), signed by Louis J. Lanzerotti, Chairman, dated March 29, 1991.

Then he reviewed the new Program Plan described above and provided the reasoning behind the two controversial decisions regarding the FTS and solar dynamic power. As a result of the restructuring, the FTS would no longer be needed until after PMC. However, he acknowledged that some of the FTS technology continued to be of interest, particularly the end-effector systems that had potential application for the Space Station and other NASA programs. A flight test was still planned to validate the technology as Congress had directed. Solar dynamic power did not fare as well. He stated that the level of effort to advance this technology was not in line with program resources and no further development would be done. OAET would continue to study thermal energy storage materials that a solar dynamic system would require. On another contentious subject, changes in schedule would allow NASA to delay or scale back construction of some Space Station processing facilities at JSC and KSC, and in some cases rely on existing shuttle facilities as substitutes. Another saving that was realized as a result of the restructuring was a reduction in the number of flight spares. With fewer elements, and by adopting end-to-end ground verification and testing, fewer spares were needed.

Because of the funding cutbacks and hardware changes, layoffs and transfers had begun at the prime contractors and their subs. WP #1 contractor, Boeing, expected that its subs would layoff about 500. WP #2 contractor, McDonnell Douglas, estimated it would reduce its work force by 160 and its subs would lose another 470 jobs. Until OAET determined how it would proceed with the FTS, the impact on WP #3 contractors was unknown. No layoffs were expected at WP #4, as eliminating the solar dynamic system affected only LeRC civil servants, who would be reassigned. At Reston, about 30% of Grumman's support contractor work force would be let go.

Lenoir testified that the international partners had been fully involved in the restructuring and "have written to Administrator Truly strongly endorsing the restructured program plan and urging him to proceed with its implementation." (The international partners had been briefed on the changes at the Multilateral Program Coordination meeting at the end of January.) He stated that the resulting changes would be in place by May, and a new stage design review would take place early in the next fiscal year (1992) to assess the impact of all the changes. He failed to mention the concerns of the Space Studies Board.

Turning to the FY 1992 request, Lenoir was particularly blunt, echoing warnings from his predecessors. He said that, based on Congressional guidance, a new, well-balanced program was in place to achieve a manned orbiting laboratory. However, he warned: "Any further reductions in this program's budget can come only through schedule slips." He went one step further when summarizing his testimony by stating: "However successful this restructuring assessment was, it was not completed without cost. This and past program reconfigurations resulting from budget reductions have adversely affected the morale and momentum of the teams working on the program. It is now more important than ever that this program receives the support of the Congress to avoid any further delay in its implementation."[5] Strong words for his Congressional bosses, but words that had no real bite. The program would continue to be at the mercy of further Congressional second guessing, and changes brought about by his NASA bosses, and a soon-to-be-installed new administration.

## SPACE EXPLORATION INITIATIVE AND SCIENCE COMMUNITY ISSUES

May witnessed the submission of one more study report. Established by a Vice Presidential Directive in December 1989, the Synthesis Group was asked to evaluate and recommend alternative ways to implement President Bush's Space Exploration Initiative (SEI) that he had announced in July 1989. Chaired by former astronaut Lt. General Thomas P. Stafford, the Synthesis Group pulled together the work of many organizations, including studies done by NASA, the Rand Corporation, AIAA, DOD, DOE, aerospace industries, and others. It was a monumental task that was conducted over a period of eighteen months. The result was a thick, glossy report of 189 pages containing beautiful color photos and paintings that illustrated what America's space future might hold. It included the obligatory recommendations on all aspects of space exploration: solar system missions, returning to the Moon and then on to Mars, advancing beneficial technologies in many fields, and the commercialization of space. Altogether, it was a hugely ambitious forecast with scheduled activities extending out to 2020.

---

5. Extracted from statement by Dr. William B. Lenoir, Associate Administrator for Space Flight, presented to the House Subcommittee on Space, Committee on Science, Space and Technology, April 11, 1991.

Interestingly, the Synthesis Group essentially omitted any contributions Space Station Freedom could make, as opposed to other earlier reports that made it the key component and hub for any future space activities. Its role was confined to developing precursor information in the fields of microgravity research and life sciences.[6] The Synthesis Group report recommendations broke some new ground suggesting that there were different ways to approach space exploration. But like all the commission reports that came before, it did not generate wide support, especially in Congress. It did make a colorful addition to place on someone's coffee table.

In anticipation of the Synthesis Group Report, the Office of Technology Assessment (OTA) circulated in April a review "Exploring the Moon and Mars" that was seen as an attack on SEI. It contained many pejorative statements rather than a impartial assessment, which is not surprising since OTA was an arm of Congress and usually supplied analyses to support the majority's positions. By now that position for SEI was well known; it was going nowhere in the hearing rooms and offices of Congress. In May GAO comptroller Charles A. Bowsher testified before the House Subcommittee on Governmental Activities and Transportation on the redesigned Space Station.[7] Bowsher reviewed the program's history and renewed GAO's earlier disagreements with NASA on runout costs of $30 billion to achieve permanent occupancy. GAO now had a much higher estimate. By adding some costs not included in the NASA number, GAO estimated it would cost $40 billion to achieve PMC. GAO then estimated the cost to keep the Station operational after PMC would be at least $78 billion versus NASA's estimate of an additional $54 billion. GAO's position on the program's cost expressed in 1991 would be repeated in the next few years, with the differences continuing to grow between its estimates and NASA's.

Support for the Space Station from the scientific community was a love-hate relationship from the moment that President Reagan announced his decision. When NASA decided how it would respond to Congressional direction to restructure the program, the response, mostly negative, flowed in from all directions. From the American Geophysical Society: "it is likely that support for the ongoing development of the Station may drastically impact the unmanned program at NASA." The American Physical Society: "scientific justification is lacking for a permanently manned space station . . . the potential contributions of a manned space station to the physical sciences have been greatly overstated." And perhaps most damming, as already noted, was the National Academy of Sciences – Space Studies Board's report that: "Space Station Freedom, at the present stage of redesign, does not meet the basic research requirements of the two principal scientific disciplines for which it is intended." Representing a minority position, support came from the American Chemical Society: "a balance between space station funding and other science-related funding is particularly important . . . science is not the only objective of a space station, and very good science might be accomplished even if it is a 'tertiary' priority."

To drive home the discontent felt in the scientific community, a news conference was held in Washington, D.C. on July 9th at the National Press Club. More than a dozen scientific societies attended and followed up by sending letters to the congressional appropriations committees with some of the above specific statements attached.[8] NASA had not been doing a good job keeping its needed allies, the user community, happy for either the Space Station or its many other programs. And Congress, intentionally or unintentionally, was making matters worse. Of course, one could ask: Can you ever give the science community enough money from the public coffers to keep them all happy?

Throughout the summer of 1991, Kohrs and his team tried to pull all the loose ends together in the face of increasing criticism of the escalating program cost estimates. The budget runout presented to Congress with the FY 1992 testimony contained a table that showed the FY 1994 request would be just

---

6. America at the Threshold, Report of the Synthesis Group on America's Space Exploration Initiative, May 3, 1991.
7. Statement of Charles A. Bowsher, Comptroller General of the United States before the Subcommittee on Government Activities and Transportation, House Committee on Government Operations, delivered on May 1, 1991.
8. Media Alert: American Physical Society, Office of Public Affairs, July 3, 1991. The *National Journal*, July 6, 1991 and the *Washington Post*, August 18, 1991, both ran stories under the heading "Lost in Space." Both articles briefly reviewed the history of the Space Station and other NASA programs and the problems NASA and the Congress were trying to resolve. See also: *Time Magazine* cover story, August 26, 1991, Crisis in The Labs.

under $2.5 billion and for FY 1995 more than $2.7 billion. Projected costs through 1999, when PMC would be achieved was $29.6 billion. After PMC, NASA was projecting mature operations would cost $2.0 billion per year in FY 2000 dollars. If the Space Station remained operable for the full twenty-five years expected by the science community, this would add another $40 billion or more to the program's price tag.

At the July SSAC meeting, again a joint meeting with SSSAAS, Bill Raney, standing in for Kohrs, described the growing opposition in the scientific community and asked for suggestions to counteract its biases. SSAC chairman Larry Adams promised to look into how the Committee could help. A new round of PDRs (Delta PDRs), based on the restructured designs, had been scheduled with the three remaining work package teams. They would be completed by mid-August. Following these PDRs, a Level II PDR for the MTC phase was scheduled for October. On the surface, the program appeared to be moving ahead in an orderly manner but everyone knew that there were many unresolved problems over which the program had little control. Perhaps the most troublesome was the status of the ACRV. Without a way to have the ACRV operational by 1999, PMC could not be achieved. The Space Station would only be able to operate as a unmanned free-flyer until the shuttle was docked and man-tended operations would become possible.

### More Management Changes – Progress and Problems

During the summer, more changes in the Space Station management structure were being contemplated by Truly and Lenoir. Arnie Aldrich, recently appointed as AA for OAET, was being considered as Associate Administrator for a new organization, the Office of Space Systems Development. In this new organization, he would have two Deputy AAs, one for Space Station Freedom and the other for Flight Systems. Programs under the latter deputy would include ASRMs and the New Launch System (NLS). When the Office of Space Systems Development was finally approved at the end of the year, there were few changes in Freedom's organization. Kohrs was still in charge, and Bob Moorehead remained at Reston as Deputy Director SSF Program and Operations. Dr. William W.L. Taylor, a JPL employee who had joined Kohrs staff in July, was appointed Chief Scientist and John-David Bartoe became the Director SSF Operations and Utilization, the contact with the user community. Bartoe had a counterpart office at Reston with the same title. Meanwhile, OSSA was also restructuring its organization to improve Space Station utilization and better reflect its seven budget line items that provided funding directly supporting Space Station Freedom.

To further improve and understand customer needs, a customer support team was established headed by Kohrs and former astronaut Robert Parker, Director of Policy and Plans in the Office of Space Flight. This team, composed of members from Headquarters' offices, the centers and some principal investigators (PIs), was charged with interviewing former and current NASA managers from Apollo, Viking, Skylab, and Spacelab for lessons learned from those programs in terms of how users and PIs were accommodated. This included how the users were involved in the design process, payload integration and ground processing at KSC, flight operations, and post-flight support and analysis. Although some attention had been given to these issues as the program progressed, it was finally trying to tap the expertise and knowledge from NASA's successful past that could be applied to the Space Station's largest challenge, how to assure that the Space Station would be a productive facility.

The ACRV new start had been denied by Congress in the FY 1992 budget, and negotiations between NASA and OMB on the FY 1993 budget indicated that it could not be added until FY 1994, at the earliest. If the ACRV was given the go-ahead in the FY 1994 budget, and if the schedule was maintained and all elements for PMC were on orbit at the end of CY 1999, there would be less than six years from the start of Phase C/D studies until the ACRV would be needed docked to the Space Station. To develop a new, man-rated vehicle in six years would be cutting it close. Prior to its November meeting, Kohrs asked the SSAC to establish a subcommittee to review ACRV requirements, examine possible system concepts, provide cost and schedule estimates, and recommend a course of action. Jack Kerrebrock volunteered to chair the subcommittee and the first briefings were held at JSC at the beginning of November. JSC had not been sitting on its hands since it was decided that an ACRV would be required at PMC. A number of in-house studies had been conducted of a capsule based on the Apollo command module, a scram jet vehicle and, going far afield, the Soviet Soyuz. The Phase A/B study contracts with Lockheed and Rockwell, mentioned in Chapter 4, had not progressed very far, due to budget constraints. The FY 1992 budget included $3 million to continue low-level studies of the extended duration orbiter and Soyuz, and an additional $3 million to continue other studies. Kerrebrock and his small subcommittee had taken on a difficult assignment. It might result in recommendations requiring a compromise between opposing factions at JSC.

When the SSAC met again on November 6th, the Delta PDRs and the MTC PDR had been completed. A number of RIDS remained to be worked off, but the good news was that the total number of RIDs, compared with the PDRs held in 1990, was significantly reduced. Kohrs reported that several GAO studies had been requested by members of Congress, covering a wide variety of SSF subjects, including reviewing the restructured design, software standards, risk management, and surveys of the work package contractors. These types of surveys were becoming more onerous, as any member could initiate a request for GAO to investigate real or imagined problems. Eventually, some requests were seen as harassment; keep NASA putting out small fires rather than managing the program and resolving real problems. The GAO studies were added workload requiring staff to spend enormous amounts of time educating GAO staff and then responding to Congressional questions when GAO submitted its report. However, there was some good news. Once the program had adjusted to the $551 million FY 1991 cutback and submitted the restructured design, the Congressional budget conference committee agreed to appropriate $2,028.9 million for Space Station in FY 1992, the amount requested by the Administration.

A new requirement was being staffed at this time, SSF's environmental impact statement (EIS). The first round of EIS submissions had been completed by March 1991. Writing a EIS for a large program is a task seemingly without end. For some statements, the process takes on a life of its own when finally released for public and official comment. The Space Station EIS would fall into this category. In its April 23, 1990 response to the draft Tier 1 EIS, EPA found that it did not include sufficient information "based on potential impacts associated with . . . the launch activities"[9] and recommended that the Space Station EIS include a summary of the environmental impacts of the "referenced" space shuttle EIS. This shortcoming was addressed in the final Tier 1 EIS, and jumping far ahead, the Tier 2 EIS was finally approved by EPA in 1996, after seven years of bureaucratic wrangling. Of special note, the Tier 2 EIS addressed the problem of end-of-life for the Space Station. It was finally decided that rather than break it into small pieces for reentry, or boost it into a higher, long-lasting orbit, it would be deorbited intact and burn up such that the pieces that survived reentry would splash down in the ocean.[10]

NASA started 1992 by releasing "The NASA Strategic Plan" a response to the agency's many critics that NASA lacked a vision and plan for the future.[11] In the Plan's opening message Administrator Truly described his vision for the agency: "NASA exists to inspire and better the lives of all Americans, young and old, through our achievements as the world leader in space and aeronautics research." The Plan (41 pages long), claiming to be a "living road map to the future," briefly placed in context all the major ongoing "missions" such as Mission to Planet Earth and Aeronautics. Space Station Freedom was categorized as an "Enabling Capability," along with space transportation, space technology, communications, and NASA's human and physical resources. These capabilities would support the commercialization of key technologies. The Space Station was called the "critical next step for advancing the human exploration of space." Considering the behind the scenes conflicts that would soon become known, the Strategic Plan outlined a rather unusual view of how all NASA's pieces fit together. Unveiling the Plan at this time was probably Truly's last attempt, in light of increasing criticism of his leadership, to convince the Administration that he could lead NASA toward its unknown and uncertain future.

---

9. Letter from EPA addressed to Mr. Richard H. Kohrs, signed by Richard E. Sanderson, Director, Office of Federal Activities, dated April 23, 1990.
10. Tier 2 Environmental Impact Statement for the International Space Station, Space Station Program Office, Office of Space Flight, NASA, Washington, D.C. 20546, May 1996.
11. Vision 21, The NASA Strategic Plan, January 1992.

## SSAC Panel Reports on ACRV and Program Verification

Kerrebrock made his initial report on the findings of his ACRV subcommittee at the next SSAC meeting in February 1992. His major recommendations were to structure the program to include life cycle and embedded operations costs, not just vehicle cost alone, and that the contract contain an incentive to achieve the ACRV performance specifications at the lowest cost to the government. This type of contract, called performance-based, is not typical for a project that would be breaking new ground, even if the final configuration selected mimicked the Apollo command module. ACRV specifications applied to a command module-like design would result in an entirely new spacecraft in terms of size, and the ability to remain docked at the Space Station for extended periods (months) of time.

At this meeting Kohrs provided a status on the FY 1993 budget negotiations with OMB. The NASA budget request sent to Congress included $2.250 billion for the Space Station, an 11% increase over FY 1992. It did not include a new start for the ACRV, but did have $15 million for the Phase B studies. Kohrs stated that he hoped OMB would put $10 million in the FY 1994 budget request to complete Phase B and a new start in mid-1994. Beside the budget and ACRV problems, he discussed a laundry list of continuing concerns: orbital debris, program verification, location and size of the centrifuge, launch weight margins, construction of facilities, and on and on. Always part of his opening briefing to the Committee, the list continued to grow, not decrease, as one would have expected in a maturing program.

Concerns for how, when, and where program verification would be carried out appeared to be no closer to full resolution. It was planned that verification would begin with prototype hardware as a learning process and then those elements would be retained as restorable spares. Responsibility had been established for certifying the end-to-end tests, but final launch package coordination was still being studied. Test philosophy and test protocols had not been agreed upon because, among other problems, most of the facilities were still under construction and it was uncertain when they would be fully operational. Completing construction and outfitting of the facilities at JSC, MSFC, and KSC was wholly dependent upon when Congress would allow NASA to increase its facilities budget. SSAC's Verification Subcommittee, chaired by MIT Professor Stanley I. Weiss, had met at JSC two weeks earlier to review the status of this critical process and reported its findings. In general, the Subcommittee agreed that progress was being made, but many open items remained. Of particular concern was verification of the software and data management interfaces between elements, and between the shuttle and the Space Station, when the shuttle would be docked. Grumman seemed to think that they would be responsible for certifying each element prior to launch, but this was not confirmed in discussions with the other work package managers, an obvious communication failure or discrepancy.

The micrometeoroid-orbital debris problem also was still being studied. The program now considered that debris above 10 cm (softball size) would be detected by ground assets and could be avoided by maneuvering the Space Station. Debris smaller than 10 cm would continue to be a risk and shielding was being designed to reduce or avoid damage to critical components. Shielding, of course, would add weight. What the penalty would be was a TBD. Kohrs declared: "When the Space Station flies, it will be safe."

As mentioned earlier, debate was ongoing in the life sciences community on the size (diameter) of the centrifuge. Congress had dictated that it would not be in the U.S. lab module. The proposed solution: put it in a node. In this way the node would be launched after PMC as the eighteenth spacecraft, fully equipped to conduct the many experiments contemplated for the centrifuge, and then attached to a lab or hab module. The centrifuge, the critical device for life sciences experiments, was considered by the microgravity community to create major problems for their research. A massive spinning wheel whose spin rate would vary from time to time, attached to a relatively flimsy structure holding their experiments, would induce a microgravity environment that would compromise their work. For this reason and others, in the next few years the centrifuge would experience a troubled evolution.

With the restructuring, the international partners had modified the design of their elements. The *Columbus* module, like the U.S. hab and lab modules, had been reduced in size. The ESA free-flyer would no longer be serviced from the Space Station but by an ESA launched mission. Design changes had been made to the Canadian MSS and Japanese experiment module (JEM). Negotiations were underway with the Italian Space Agency to provide a new element, a mini-pressurized lab that, among other capabilities, could also house the centrifuge if desired. Kohrs showed a new major milestone chart with PDRs scheduled for JEM in mid-1992, for the MSS in March and September 1992, and for ESA's *Columbus* module in March 1993. CDRs for various U.S. elements were scheduled from mid-1992 to April 1993. It appeared that the program was getting back on track, but trouble was brewing.

## TRULY RESIGNS – SEARCH FOR NEW ADMINISTRATOR

Three days after the SSAC meeting ended, Dick Truly resigned. In the trade press it was called a forced resignation.[12] The main rumor circulating at the time was that the Vice President and his NSC staff were behind Truly's departure. Another rumor had Quayle and his staff, desiring to put their stamp on the agency, trying to ease him out back in December by requiring him to change some of his priorities, but Truly refused to knuckle under. Although he had been appointed based in part on his friendship with the President, his strong advocacy to buy more shuttles and support the Space Station program had placed him in conflict with the NSC, and eventually with many members of Congress.[13] Truly's differences with the Administration added to the growing gulf between the Administration and Congress. The Administration, now steered by the NSC, wanted to develop a new launch system and press on to the Moon and Mars. Congress, with its long infatuation with the development of robotic and artificial intelligence systems, wanted to use those systems to explore the planets and solar system, and focus on the "Mission To Planet Earth." Support for the Space Station had been declining in the White House for two years, despite the outward appearance that it was a favorite program. The Administration had not fought hard to maintain the program's FY 1991 budget during the debates in Congress.

Truly's resignation (he was quoted as saying he was fired) at this critical point – defense of budget time before House and Senate committees – created a top management vacuum that would be hard to fill. J.R. Thompson, Truly's Deputy, had resigned in November 1991, leaving NASA without a number two manager for the last four months. Aaron Cohen, JSC Director, was quickly transferred to Washington on February 19 to hold the fort as Acting Deputy Administrator. President Bush, expressing the obligatory praise for a departing administration appointee, immediately started a search for Truly's successor using the NSC staff to lead the search.

Two candidates were considered front runners for the job: Edward C. "Pete" Aldridge, President and CEO of the Aerospace Corporation, and James A. Abrahamson, Vice President of Hughes Aircraft Company. Abrahamson, a former Air Force MOL astronaut and NASA AA, was the clear favorite among senior NASA managers. He had an outstanding reputation as a manager and was well liked outside NASA and by the Congress. His major problem was that, as a former military person (Lt. General), he would need a special dispensation, as had Admiral Truly, to become NASA Administrator, a position that by law was not open to former high-ranking military officers. Senator Gore was known to be very concerned that appointing a former military officer would threaten to militarize the civil space program.

As often happens inside the beltway, those rumored to be front runners for high-level government jobs soon drop by the wayside. The Bush Administration's choice to become the next NASA Administrator, Daniel S. Goldin, caught everyone by surprise. Goldin was not an unknown in the aerospace world. When selected, he was Vice President and General Manager of TRW Space and Technology Group in Redondo, California, known primarily for its "black" (classified) projects although it also worked on civil space programs, including TDRS. Again, rumor had it that he was the choice of Mark Albrecht, NSC Executive Secretary, who knew Goldin when Albrecht worked on the "Hill" as a Senate staffer. Goldin, despite rumors of questionable management conduct at TRW, was soon confirmed by the Senate. Thus began a long, controversial, and contentious career. Goldin would become the longest serving NASA Administrator, working for three different administrations representing both political parties; an eventual resume almost unheard of in Washington.

At his confirmation hearings Goldin espoused the new directions that NSC and some in Congress wanted NASA to take. He said that NASA should focus on science and that he would support a larger number of faster and cheaper missions. NASA's "Great Observatories" projects had drawn criticism

---

12. "Truly Dismissed / Firing Shakes NASA Foundations," by Andrew Lawler, *Space News*, 17 February 1992.
13. See also: "Truly Ouster was Two Months in the Making," by Andrew Lawler, *Space News*, 17 February 1992.

because of cost overruns, and because they exposed the agency to unnecessary risk if the projects failed to achieve their objectives. For these mega-projects, requiring ten or more years of development before launch or data return, it meant that the scientists involved would devote a substantial fraction of their professional careers to a program, hoping that they would be successful. The Hubble Space Telescope was the latest example of placing so many resources in one project and then discovering it did not perform as advertised. Congressional concern was justified (Hubble had not yet been fixed), two other billion dollar plus programs, Cassini and EOS, were due to be launched within the decade. In regard to the Space Station, he said he would "look at this issue in depth." In his testimony Goldin acknowledged the difficult financial constraints that he faced and promised: "I will work with the Congress to realign our future programs to the budget realities." After his confirmation, Goldin quickly promoted within NASA his new agency mantra: "Faster, Better, Cheaper," a working philosophy supported by some and criticized by many, especially in the immediate NASA community.

Backing up a few weeks, at the end of February 1992 with Truly now on the sidelines, Arnie Aldrich in his new management role defended the FY 1993 Space Station budget request of $2.25 billion.[14] He went down the list of the past year's accomplishments and explained some of the changes that been made as a result of Congressional direction and reduced budgets. In addition to the restructured design, Space Station utilization had been consolidated in one organization in the Headquarters Office of Space Flight. This office would be responsible for coordinating both Space Station and Spacelab payloads. Many of the Spacelab experiments would be precursors for those that would eventually be conducted on the Space Station. Similarly, both Spacelab and Space Station payload operations would be controlled from MSFC, and all the individual Spacelab and Space Station payloads would be processed by the same team at KSC. In another move, the Expendable Launch Vehicle (ELV) program had been transferred to OSSA. One new line item appeared in the FY 1993 request, $35 million for Space Station operations, and the ACRV line included an increase from $6 million to $15 million. For the first time in three years Aldrich's testimony did not include any dire warnings of the consequences of not providing the full request as had become a standard closing statement by his predecessors.

Signals had already been received from the House and Senate Committees that there might be large reductions in NASA's budget request. House committees were discussing cuts of $500 million or more. The Senate, as usual, had smaller numbers in mind. Kohrs had declared that if the reductions under consideration in the House were passed, schedule slips of six to nine months in FEL would result and add 15-18 months to achieve PMC. It remained to be seen if the large cuts discussed would be agreed to and what the impact might be.

Leaving the Congressional battles behind for a brief reprieve, there was some good news. On the evening of May 7, 1992, *Challenger*'s replacement, *Endeavor*, lifted off on its maiden flight. The main task for *Endeavor*'s crew was to capture the INTELSAT satellite, stranded in a useless orbit due to a launch malfunction, mate it to a new perigee kick motor and release it. After this was accomplished (it was successful), the crew was scheduled to conduct two EVAs to demonstrate the ability to assemble and maintain Space Station Freedom. Although the EVAs were cut to one because more time was needed than planned for the INTELSAT work, the Space Station demonstration also was successful.

From the time the decision was made in 1986 to procure a new orbiter until it was ready to launch, five years elapsed. At *Endeavor*'s rollout, a NASA press release[15] extolled the effort to build the new shuttle. Its cost, $1.8 billion, was said to be below the original estimate of $2 billion.

### DANIEL GOLDIN TAKES CHARGE – ANOTHER NEW DIRECTION

As one of his first acts, and as he had promised during his confirmation hearings, on May 16th Goldin ordered the entire NASA management organization to conduct a one-month review of all programs. It was called the Red Team–Blue Team review. Essentially all activities came to a halt. The two Teams

---

14. Statement by Arnold Aldrich, Associate Administrator for Space Systems Development, to the Subcommittee on Space, Committee on Science, Space and Technology, U.S. House of Representatives, February 26, 1992.
15. NASA Press Release 91-62, Endeavor Roll-out, April 24, 1991. The press release included Endeavor's cost: $1.8 billion, $1.6 billion for the orbiter including some support items and GFE, and $200 million for the SSMEs.

were "locked up" in meeting rooms in Crystal City, in Alexandria, Virginia, until they completed their assignments. Each Team worked independently and developed their own criteria for the review, with the primary objective of finding ways to save up to 30% in major program funding. At the end of the month they were scheduled to come together, discuss their conclusions, and then Goldin would meld them into the new direction NASA would take. The next SSAC meeting scheduled for the end of May was postponed to allow the teams, which included senior Space Station management, to devote full time to the review.

In June Goldin asked Michael Griffin, the new AA for Exploration, to draft a one-sentence "vision" for NASA. Griffin's 31-member team proposed: "As explorers, pioneers, and innovators, we will boldly expand the frontiers of air and space for the benefit of all." A little Star Trekkie sounding (boldly go where no man has gone before), but it was almost impossible to encapsulate NASA's mission in one sentence. Goldin's plan was to require that all NASA employees sign on to this vision. One year later Griffin's team would still be arguing over the language and finally gave up saying only top management could set the vision.

Goldin also began training seminars for all employees called "Continuous Process Improvement" or as called by participants, "CPI Boot Camp." The seminar emphasized Total Quality Improvement, modeled after the management philosophy developed by W. Edwards Deming, that was credited with turning around Japanese industry. No one would deny that NASA needed some shaking from the top to change how the agency did business. But Goldin's heavy handed approach was creating more than a little grumbling, there was a strong undercurrent of discontent and revolt in the trenches.

An interview in *Science* magazine in October 1992 gave Goldin an opportunity to respond to his critics inside and outside the agency. He said that, although some might be afraid of change, "are we going to allow this nonsense and insecurity about change stop us from doing what we've been chartered with?" And he acknowledged that "there are some barricades in the street." The *Science* article quoted an anonymous NASA Division Chief as saying: "Goldin has the staff so tied up in boot camps and wrapped around the axle with this TQM nonsense that no one's doing any work."[16] Goldin's tough talk brought him supporters as well as critics.

Adding to all the uncertainties, 1992 was a presidential election year. Democratic candidate Bill Clinton released a position paper in the summer describing the space program his administration would support if elected. It contained some comforting words for NASA. He would "restore the historical funding equilibrium between NASA and the Defense Department's space program." He would also "maintain the Space Shuttle and continue work on the Space Station." For Space Station he would "base its development on the twin principles of greater cooperation and burden sharing with our allies."[17] Not a lot of detail, but his position paper seemed to indicate that he was an advocate of a strong civil space program.

Reviewing all of NASA's programs as Goldin requested was more than a one-month job. The Red Team–Blue Team reviews stretched out through the summer. When the SSAC next met at the end of July 1992, the Red–Blue Teams had settled on a 10% across the board spending reduction, but additional review of this recommendation was still proceeding as Goldin prepared his FY 1994 request to OMB. At this meeting the SSAC was briefed on the Long Duration Orbiter (LDO) studies (additional modifications to what was formerly called the extended duration orbiter) that would allow the shuttle to remain in orbit for thirty days, and perhaps as long as ninety days. To achieve ninety days would require that the Space Station supply power to the shuttle from its solar arrays, allowing the shuttle to power down its fuel cells.

The scenario being examined would have the LDO launched only during Space Station utilization flights, those flights when experiments were being conducted or serviced, as opposed to flights that were primarily used to carry elements for assembly. The LDO would be used to house the crews

---

16. News and Comment: Making Less Do More at NASA, Eliot Marshall, *Science*, Vol. 258. 2 October 1992.
17. Bill Clinton on America's Space Program. A two-page position paper in the author's files for the September 1992 SSAC meeting, no date, probably issued in midsummer, 1992. Under the heading "The Clinton Plan" the position paper lists seven major topics that include the statements described in the text.

before the hab module was on orbit, or as a complement to the hab module. The Red–Blue Teams had endorsed early implementation of the LDO and the Space Station program was proceeding on the assumption that it would be developed. A major drawback to configuring the shuttle as an LDO was the reduction in payload weight that the LDO could carry, perhaps as much as 15,000 pounds if ASRMs would not be available. The SSAC Utilization and Operations Subcommittee (SSAC Panel designations had been changed to subcommittee[18]) cautioned that this penalty needed to be carefully evaluated in terms of the beneficial or adverse effects on users. From information available, it appeared that life science research would benefit, but at the expense of microgravity experiments. And the Committee in general thought that the cost to convert a shuttle to an LDO was not only very high, but probably unrealistic in light of tight budgets.

### Russian Cooperation Explored

Aldrich provided the SSAC a status on ACRV activities. ESA had expressed an interest in providing the ACRV and discussions were at an early stage. The extra $3 million requested for ACRV studies in the FY 1992 budget was being held back by Congress awaiting the outcome of negotiations with the Russians to explore using a version of the Soyuz. A contract had been signed in June with NPO Energia to conduct a feasibility study. He reviewed the history of U.S. Soviet and Russian cooperation in space, starting with the Apollo-Soyuz Test Program in the 70s, through the 1987 Civil Space Agreement, and the most recent Human Space Flight Cooperation agreement, signed in June 1992 by Presidents Bush and Yeltsin. This latter agreement expanded cooperation in many scientific areas and explicitly mentioned human space flight.

Goldin led a multi-agency delegation to Russia in July to follow up on the June agreement. The delegation visited facilities and met with many Russian and Ukrainian space organizations. Of particular interest to the Space Station program were discussions on the possible use of the Soyuz-TM spacecraft as an ACRV, and cooperation on automatic rendezvous and docking systems – technology where the Russians were clearly ahead. The key difference in these discussions of cooperation with Russia, versus cooperative efforts with our other international partners where there was no exchange of funds, was that the U.S. proposed purchasing Russian goods and services, including maintenance of Russian facilities. A new Memorandum of Discussion included cooperation in human space flight involving flying Russian cosmonauts on the space shuttle, U.S. astronauts visiting the *Mir* space station, and joint rendezvous and docking missions between the shuttle and *Mir*.[19] The Memorandum of Discussion would lead eventually to a formal agreement in 1993.

### Space Station Budget Survives in Congress – More Progress

Keeping NASA in suspense until almost the last minute, the Senate-House conference committee reached agreement in the last week of September on NASA's FY 1993 budget. Calmer heads had prevailed and the cuts were not as bad as expected. NASA's appropriation included $1.939 billion for Space Station, only $311 million below the request.[20] Senator Dale Bumpers (D-AR) offered an

---

18. At this time the SSAC had six subcommittees: Intersite Deliverables, chaired by Richard M. Davis; System Integration, Test and Verification, chaired by Stanley I. Weiss; Utilization and Operations, chaired by Daniel E. Hastings; Natural and Induced Environments, chaired by Edward F. Crawley; Data Management Systems, chaired by John E. Miller; and Public Information, chaired by Donald A. Beattie. Most members served on two or more subcommittees. SSAC membership had changed since the committee was first formed, some members mentioned earlier in the text as panel chairman had resigned. The new Chairman of the SSAC, John Miller, had requested that NASA add additional members to lighten the workload, but this could not be done due to budget restrictions. This became a moot point as the SSAC was abolished 18 months later.
19. For details of the U.S. / Russian Civil Space Agreement and Goldin's trip to Russia and the Ukraine see: Presentation to NASA Advisory Council, U.S. / Russian Cooperation, Arnold D. Aldrich, Associate Administrator for Space Systems Development, August 24, 1992.
20. You may have noticed that I seldom refer to Authorization bills, the other half of the Congressional funding process. The reason is simple, although an agency's authorization bill might contain important language, such as rescissions or spending delays, the authorization process was becoming more and more controversial and in many years authorization bills were never passed. As of CY 2005 the last NASA authorization bill passed was in 2000. The appropriation bill is the key to proceeding in any program.

amendment for what had become his annual attempt to kill the program; it was defeated, but each year his amendment gathered more support. Conference committee language directed NASA to make no further architecture changes, give first priority to Manned Tended Capability, and stay on schedule. It also directed NASA to reduce "overhead costs." The new budget numbers were announced while the SSAC was having its next meeting.

Aldrich told the SSAC that Goldin considered the Space Station the top priority after the shuttle, and that the Red–Blue Team recommendations for Space Station had been accepted. Kohrs explained the impact of the funding reduction. Because funding for the ASRMs remained as a major issue in Congress, and the new appropriation was below the level requested that would have them flight ready at an early date, they would not be available to improve shuttle payload performance for early assembly flights. One additional assembly flight had been added to take up the slack in reduced shuttle payload capacity. The program also considered that a four-month slip in FEL was probable.[21]

Covering other actions, Aldrich said an Operations Phase Assessment Team was now working to develop a cost effective and sustainable operating capability, beginning at FEL. The Team, chaired by JSC's Eugene F. "Gene" Kranz, was composed of members from both the Space Station and shuttle programs. Its report would soon be available. Moving quickly after the ACRV contract had been signed with NPO Energia, the first technical meeting had taken place in Kaliningrad in August, and the contract tasks agreed upon.[22] The element PDR and CDR schedules were holding and the MTC Phase CDR, to show that all the elements would work together, was now scheduled for June 1993. Negotiations with OMB on the FY 1994 budget had not been completed, but the hope was that it would be close to the FY 1993 request of $2.250 billion. Aldrich said he was working with the various levels of Space Station management and after the CDRs he expected the Level I and II management structures would consolidate and the activities at the centers related to verification and operations would grow. JSC would be the center of focus for the man-tended phase and MSFC for the permanently-manned phase.

In spite of mounting criticism of the Space Station in the scientific community, the response to OSSA's recently released Announcement of Flight Opportunities to conduct experiments on the Space Station was encouraging. In October, from the 500 proposals received, 124 grants were awarded for microgravity research. The value of the grants was approximately $15 million. The investigators selected represented 60 universities, eight private laboratories, and nine government laboratories. Research was proposed in fluid dynamics, biotechnology, materials science, and fundamental science. NASA trumpeted: "The hardware and experiments developed from these grants could make Space Station Freedom a microgravity laboratory unrivaled by any other."[23]

One week after the release of the above news – which added 119 Space Station researchers (several received more than one grant) – the SSSAAS met in Huntsville, Alabama. During the four day meeting the attendees were brought up to date on the program changes that had occurred since their last meeting in February 1992, and there were many. Progress had been made to resolve or accommodate the issues and concerns expressed at that meeting, but a large number remained as open items. Data management, available data rates, providing active thermal control for experiments, the need for a nadir viewing port, mapping the microgravity environment, vibration isolation and preserving extended duration quiescent periods when the shuttle was not present, a better understanding of what elements were included in "commonality" for the attached payloads (data interfaces, utilities, etc.), and many more concerns were discussed. (A total of twenty recommendations from the "splinter groups" that reviewed specific aspects of the program were included in the Summary Minutes of the meeting).[24]

---

21. Presentation to the SSAC by Richard H. Kohrs, September 23, 1992.
22. The NASA Contract (NASW-4727) with NPO Energia contained a number of tasks. They included determining if the Soyuz could be carried safely in the shuttle cargo bay, and the minimum modifications needed to integrate it into NASA ground and launch systems, and perform the ACRV mission. The Soyuz was a three person spacecraft, so if it were to be used as the ACRV, the Space Station would be limited to a three person crew.
23. NASA News Release: 92-167. NASA Awards Science Grants for Space Station Freedom, October 6, 1992.
24. Summary Minutes, Space Station Science and Applications Advisory Subcommittee, Fall Meeting, Huntsville, Alabama, October 13-16, 1992. December 21, 1992.

Data management and software development had been a concern from the start of the program. The SSAC subcommittee on data management systems (DMS) had raised many of the same questions as the SSSAAS. At the September meeting John Miller, chairman of the SSAC subcommittee, summarized his Subcommittee's findings. Although many changes had been made in the DMS and NASA's management approach, the system did not have sufficient memory. It was too slow and too complex, and the program had given too much technical responsibility to the contractors. In a rapidly developing technology, problems were bound to occur, and computers and data management were at the forefront of such rapid change. The Subcommittee felt that the program was moving slowly and reluctantly in the right direction, but questioned if there was enough time to fix all the problems and recommended that the user community be given a larger role in designing the DMS and the interfaces with the payloads.

Another continuing concern was the external contamination that would be present from venting and thruster firings that might coat optics or interfere with other measurements. The SSSAAS asked that Ed Crawley, who chaired the SSAC subcommittee on external environments, provide a briefing on the findings of his subcommittee at the next meeting. Miller pointed out that the STS-52 mission, scheduled in October, would carry the Shuttle Plume Impingement Experiment to measure the amount and kinds of contaminants produced by shuttle firings. Results from this experiment would be a useful indicator of how bad the problem might be.

One new program, Expedited Processing of Experiments to Space Station (EXPRESS), was described and well received. EXPRESS was designed to reduce the time and cost of integrating small experiments and was a response to an earlier SSSAAS recommendation. Small was defined as an experiment with the following characteristics: weighed less than 50 kg, used less than 500 watts of power, and required uplink data rates less than 500 bps and downlink data rates less than 500 kbps. For such experiments, and it was anticipated that many would fit within these specifications, standard racks would be provided for pressurized payloads and attached payloads. What utilities would be included was still under study. The goal was to reduce payload integration requirements and to make the racks available to be integrated at either KSC or the experimenter's facility.

Representatives of the international partners also attended and reviewed the status of their programs. A Support Equipment Sharing Agreement with the partners had been negotiated but not finalized. The program was investigating how to involve Principal Investigators in developing the process for equipment sharing. Two pieces of equipment, a furnace and the centrifuge, were important candidates for sharing. The SSSAAS noted that the life sciences budget had been cut significantly in FY 1993 and that the centrifuge facility budget had been reduced 55%. In view of their understanding of the out-year budgets they feared the centrifuge, the centerpiece for space life sciences research, faced elimination.

## NEW ADMINISTRATION – NEW PROGRAM CHANGES

Three weeks after all the SSSAAS attendees went home, the November elections saw the changing of the guard. After eight years of on-again, off-again support for the Space Station from Republican administrations, there was a Democrat in the White House. Candidate Clinton's summer position paper on space would soon become administration policy. Over 100 new House members and fourteen new Senators would need to be educated on NASA programs. Stories circulating at NASA Headquarters described Goldin as working desperately behind the scenes in an attempt to get on the good side of the Clinton-NASA transition team, headed by former astronaut Sally Ride, in order to keep his job. While the transition was taking place, at the beginning of November, he appointed John R. Dailey as the Deputy Administrator (Acting), a position that had been vacant for a year. Designating Dailey as the Acting Deputy was the only alternative open to him, as NASA's Deputy would be selected by the new administration and required Senate confirmation.

Aldrich's forecast of potential management changes at the September SSAC meeting were confirmed at the end of 1992. Space Station Level I and Level II offices would be combined and operate at Reston. A joint vehicle integration team (JVIT) would be established at JSC consisting of the three prime contractors (Boeing, McDonnell Douglas and Rocketdyne) and managed by NASA. Grumman, the Space Station engineering and integration contractor, would also participate in the JVIT, but continue to be based at Reston. Kohrs would remain as Director but the Deputy Director for program and Operations would be transitioned to JSC. Aldrich, in the NASA Station Break newsletter, stated that these changes would "reduce overhead costs (as requested by Congress) and strengthen program execution and accountability."[25]

Summing up some of the progress made in 1992, a major milestone had been met. The WP #2 contractor was now "cutting metal" for some of the flight hardware, the trusses, that would be launched on the first assembly flights. Tests of assembly procedures had been successfully conducted at JSC in the neutral buoyancy facility. These tests were supplemented by EVAs during the *Endeavor* mission in April, again confirming that the design could be assembled by the astronauts during EVAs. The Alpha rotary joint bulkhead that would connect the solar arrays to the trusses was also in production. The rotary joint, not discussed previously, had been one of the most difficult design challenges of the Space Station. It had to be light weight (most electric power generation equipment is massive), firmly attach the large solar arrays to the trusses, and at the same time allow the arrays to rotate to a position that would point them in the direction to collect the maximum amount of sunlight. As the Space Station would always be moving relative to the Sun's position, this required that the arrays would be in almost constant movement to maintain the proper orientation. And then to add to the complexity of the design, the electric power generated by the arrays would have to be transmitted across the rotating joint, using slip rings, to the power distribution grid and the batteries. The final design was an ingenious mechanical / electrical device developed by LeRC and its WP #4 contractors.

According to the old Scottish saying: "The best laid schemes o' mice an' men Gang aft a-gley."[26] And so it was for the Space Station program at the beginning of 1993. Nine years of planning would be overtaken in an instant by the results of a national election. The NASA FY 1994 budget had been submitted by the Bush Administration, but it did not contain any detail for the Space Station. That type of detail would have to await for guidance from the Clinton Administration, and it was assumed it would be several months before the program would receive the new marching orders. Once again, at a critical juncture in the program, thirty six months to FEL, it was left to slowly twist in the winds of uncertainty.

---

25. *Station Break*, Vol. 5, No. 1, January 1993. *Station Break* became a NASA bimonthly publication of Code M-9, containing a compendium of NASA Space Station stories.
26. Robert Burns, "To a Mouse."

# — CHAPTER 6 —

## NEW ADMINISTRATION – BACK TO SQUARE ONE

### (JANUARY 1993 - MARCH 1994)

When the SSAC met in mid-January 1993 all the uncertainties created by the change in administrations were discussed. Aldrich described the transition as a "stalemate" between the two administrations. Not a good sign if you were a Washington manager, considering that the Clinton transition team was led by a former NASA astronaut. As a result, management changes announced by Aldrich in December were now on hold. The Joint Vehicle Integration Team, also announced at the end of 1992, had not been put in place and was still being studied.

Space Station managers had no alternative but continue to adhere to the last published schedules and move ahead until told otherwise. Eight shuttle launches were planned during 1993. In addition to the many experiments that would be conducted in Spacelab, three of the flights would feature EVAs to prepare astronauts for Space Station assembly and maintenance. The EVAs would require the astronauts to use large tools; move heavy objects in the payload bay, with and without foot restraints; and conduct other activities that would occur during assembly.

Two of the experiments that would be carried out during these missions were of great importance to Space Station design. One, to assist in developing SSF's orbital debris protective shield, would calibrate the ground assets that track orbital debris. Pairs of metal spheres of different diameters would be released to determine if the ground radars could accurately measure their size, and then the spheres would be allowed to have their orbits decay and reenter to calculate the life expectancy of orbital debris. The other experiment would test the design of Freedom's environmental control and life support system.

The SSAC meeting was held at JSC to allow Committee members to meet JSC managers and examine facilities and mockups. John Aaron, the WP #2 program manager, discussed the status of the WP #2 contract, and told the Committee everything was in "good shape" for the April CDR. Gene Kranz provided an update on his Operations Phase Assessment Team. The plan that was evolving would operate Freedom on the basis of an eight-hour work day, thus reducing ground operations costs. Also, significant savings could be made by moving some of the Level I and II engineering activities to Level III, consolidating some shuttle operational responsibilities. Cost savings also were possible with the support contractors that provided sustaining engineering and other services. Kranz's report was a harbinger of changes to come.

Two weeks after the SSAC meeting, more bad news hit the front pages. Marty Kress, the new Deputy Director for Utilization at NASA Headquarters, who had provided the Committee at JSC an upbeat assessment on how "user friendly" the Space Station would be, reported on his meetings with the contractors. He revealed that the program faced a $500 million dollar overrun above the costs last reported in October. McDonnell Douglas, the WP #2 prime contractor, was the guilty party. JSC, the work package manager, bore the brunt of the criticism that descended from all quarters. A different reading than Aaron's cheerful statement that WP #2 was in "good shape." Some increase had been expected but not anywhere near $500 million. This projected overrun was reported to Congress and OMB but the Clinton Administration, just finished celebrating its inauguration, had not made any public comment. It would not be long in coming.

### NEW ADMINISTRATION'S GUIDELINES – REDESIGN ORDERED

The first week in February, Executive Branch departments and agencies received guidance from the new administration on how to rework the FY 1993 budget submitted by the Bush Administration. NASA was the exception. Insiders, with knowledge of the debate on NASA's budget going on in the Clinton White House, reported that Leon Panetta, Clinton's OMB Director, had recommended canceling the Space Station. Panetta wanted to cut the total NASA request to $12 billion, and to do that meant dropping the Space Station. For the last four years the Space Station represented approximately 14% of NASA's budget. If it was eliminated, plus other items associated with the Space Station but carried in other parts of the budget, Panetta's goal would be achieved. Vice President Gore and Treasury Secretary Lloyd Bentsen, a former senator from Texas whose constituents had a lot to lose if the program was canceled, argued against cancellation.[1] A compromise was reached. NASA would be directed to redesign the Space Station (sound familiar), and depending on the outcome of the redesign, a new Space Station budget request would be sent to Congress.

Aldrich appeared before the House Subcommittee on Science, Space and Technology on March 2nd to explain the cost growth in the program and the steps that were being taken to reduce cost and prevent future growth. He listed six reasons for the unexpected cost growth and blamed most of the growth on the contractors' inability to achieve the productivity gains they forecast, management changes, and the 1991 restructured design. Another factor cited was the lack of overall design maturity that gave rise to the many change orders required as a result of the PDRs. He explained that in 1992 the program believed it had sufficient reserves to handle the projected growth, but the growth exceeded the planned reserves. He outlined the measures being taken at the Reston program office to better understand the reasons behind the cost growth and reduce the impact on the FY 1993 through FY 1995 budgets. He ended by stating, "we are working hard to fix the problem and to implement measures to prevent it from occurring again, and this new knowledge should benefit the redesign activities recently announced by the President."[2] Aldrich was shooting in the dark with his last statement as no one at NASA knew what the redesign would entail.

The hearing was described by the press as "contentious." Several members complained about NASA's management structure, in which accountability is diffuse, and that infighting among NASA facilities plagued the program. Bob Moorehead also testified, saying that he had warned top management about deficiencies at Houston (manager of the WP #2 contractor and largest culprit for the overrun) and had written letters about his concerns. Moorehead's testimony was very unusual, as he was telling the Subcommittee that his bosses had been derelict. Representative Dana Rohrabacher (R-CA) was quoted as saying: "What you have is a mishmash . . . It's not working."[3] All in all, a bad day on the Hill for Space Station managers facing an uncertain future.

On the same day that Aldrich testified, Goldin wrote to Subcommittee chairman Ralph M. Hall. He thanked him for taking the time to review the cost problem and wrote that he was following the Subcommittee's recommendation to establish an independent cost estimating activity. He promised to share the results that he expected would be available by mid-April.[4]

Goldin, still functioning as a holdover from the previous administration, had not been consulted by the new administration on the redesign. When he finally received his marching orders from the White House, he sent a memo to Headquarters offices and field centers on March 9th, detailing how the redesign process would be carried out.[5] He explained that the President wanted the "Space Station redesigned to be more efficient and effective, and capable of producing greater returns on our investment. The revised station program should strive to significantly reduce development, operations, and utilization costs, while achieving many of the current goals for long-duration scientific research." The word "many" in the last sentence was explained later in Goldin's memo. It covered "high priority research in materials and life sciences." High priority was not defined. It also included long-duration research "but not necessarily permanently manned" opportunities for international partners participation, and to achieve initial on-orbit research capability by 1997. Many other "constraints" were listed.

---

1. Clinton Urged to Cancel Space Station, Andrew Lawler, *Space News*, 8 February 1993.
2. Statement of Arnold D. Aldrich, Associate Administrator, Office of Space Systems Development, National Aeronautics and Space Administration, before the Subcommittee on Science, Space and Technology, House of Representatives, March 2, 1993.
3. Space Station Officials Queried on Overruns, Kathy Sawyer, *The Washington Post*, March 3, 1993.
4. NASA letter to The Honorable Ralph M. Hall, Chairman, Subcommittee on Space, Committee on Science, Space and Technology, House of Representatives, Washington, DC 20515, dated March 2, 1993.
5. NASA Memo: to Officials-in-Charge of Headquarters Offices; Directors, NASA Field Installations; Director, Jet Propulsion Laboratory. from A/Administrator. Subject: Redesign Process. March 9, 1993.

Goldin named Joseph F. Shea, Assistant Deputy Administrator, rehabilitated as a NASA manager twenty-five years after the *Apollo 1* fire, to lead the Redesign Team. He directed that the Team consist of 30 to 35 members, supplemented by representatives from the international partners. The Redesign Team eventually consisted of 60 members, including eight from the international partners, and support staff from the three NASA work package centers and LaRC. Senior Headquarters Space Station managers were excluded from the Team. In addition, the White House directed NASA to create a senior-level panel to assess the goals and redesign options that the Redesign Team developed. Goldin's memo did not indicate who would serve on this panel.

Still without a NASA FY 1994 budget request to Congress, the Administration needed to quickly arrive at an agreement on what the new Space Station program would be. The Redesign Team was given a tight schedule: provide an interim report by May 15th and complete their work with appropriate recommendations by June 1st. No big deal, just redesign a multibillion dollar program, nine years in development, in ten weeks! Shea and his Team began work on March 10th. Shea resigned on April 22nd for health reasons and was replaced by Bryan D. O'Connor. Congress, sitting on the sidelines, had yet to weigh in on the redesign directive.

An independent senior-level panel, called the Advisory Committee on the Redesign of the Space Station, was established the first week in April. Vice President Gore named MIT President Charles M. Vest to head the Committee. Committee members (16) were drawn from government, industry and universities. International partners also served on the Committee. Their job: review the Redesign Team report and make recommendations to the President in June.[6]

Members of Congress did not delay very long in expressing their displeasure with the ordered redesign. Some on the House Appropriations Committee clearly preferred the current baseline program. Representative Louis Stokes (D-OH), chairman of the House Appropriations VA, HUD and Independent Agencies subcommittee said: "I believe you could understand how it could strain NASA's credibility in the next three or four weeks if you come up with an option that suggests NASA in fact can build 95% of the existing Space Station for only $1.6 billion in 1994 and $9 billion in total when in January you couldn't do the job for $2.3 billion." Tom Delay (R-TX) whose district included JSC and Space Station contractor offices said, "After you have a Space Station that's scrubbed and scrubbed and scrubbed . . . to come back in and do this is irresponsible . . . I can assure you that I certainly don't like spending money for a half-baked idea."[7]

Grumbling was also heard from the international partners. By late April, based on their participation in the redesign, they had concluded that their roles were being given a low priority.[8] By the end of May the grumbling grew louder. George E. Brown Jr. (D-CA), Chairman of the House Science, Space and Technology Subcommittee, was quoted as saying he gave the Space Station a 50-50 chance "at best" of being funded in FY 1994, and gave the Administration scant hope of obtaining funding for any "harebrained" redesign options. When asked what were the chances of Congress passing Space Station funding on a scale of 1 to 10, he responded: "One." He also said that "The problem we are here to address today is not an engineering problem. It is a political problem. I do not believe that any design other than the Freedom-derived option (as you will read closest to proposed Option B) will carry the support of the House, and we risk losing the project altogether."[9] In a *Washington Post* article he was quoted as saying; "If the Space Station is killed, it could trigger a total unwinding of NASA." In the same article a Congressional staffer said any non-Freedom-based design would likely cost the project its winning coalition.[10]

GAO joined the chorus in May, providing a report on the Space Station's cost growth and program instability.[11] The Report briefly reviewed the program's history, the many redesigns, and the rescoping that had taken place. It faulted the program for not carrying sufficient reserves, and for how the

---

6. NASA News Release: 93-59, April 1, 1993.
7. *Aerospace Daily*, Stokes subcommittee skeptical over Station redesign plans, April 30, 1993.
8. *Aerospace Daily*, International partners dismayed at redesign proposals, April 28, 1993.
9. *Aerospace Daily*, Brown gives Station 50-50 chance on Hill, says he won't support redesign, May 21, 1993.
10. Space Panel Chair Predicts Freedom Alternates Won't Fly. Kathy Sawyer, *The Washington Post*, May 21, 1993.
11. GAO Report to the Honorable Tim Roemer, House of Representatives. Space Station Program Instability and Cost Growth Continue Pending Redesign, May 1993, GAO/NSIAD-93-187.

program kept its books. GAO reminded Congressman Tim Roemer (D-IN), who had asked for the report, that in 1991 it had estimated that the Space Station would cost $40 billion if items such as the ACRV, experiment equipment and shuttle costs were included in the total. By 1993, using the same program content, the total had risen to $43 billion. GAO was now adding additional costs after the year 2000: "$54 billion to maintain, supply, and operate the station for 27 years at $2 billion annually, and $24 billion . . . to bring it to the fully planned capability." GAO added other costs associated with funding experimenters and civil servant salaries. "When these costs are considered, the total space station life cycle cost estimate through 2027 is at least $121 billion." If you agreed with how GAO kept the books and wanted to scare someone, the $121 billion was probably not a bad estimate. NASA disagreed with GAO's total. Operational costs and civil servant salaries had never been included in the runout cost of previous programs.

## Redesign Team Reports

In the Introduction to the Redesign Team's Report,[12] Bryan O'Connor, the team's Director, explains that "Numerous concepts were presented . . . The team assessed all concepts within the framework of the guidance contained in the Administrator's letter (memo) and the existing international agreements." Technical briefings were also held with the Russians, but they were not formal members of the team. As a result, the team narrowed the field to three basic design options and "focused its efforts on defining those options *as thoroughly as time allowed*" [Author's emphasis] – a revealing admission in view of the many criticisms that soon would be leveled at the redesign options. It was a hurry-up job from the beginning.

The three options that the Redesign Team settled on were: Option A, Modular Buildup; Option B, Space Station Freedom Derived; and Option C, A Single Launch Core Station. To provide additional guidance and assist the Redesign Team in selecting the three options, the President's Office of Science and Technology Policy (OSTP) (the Clinton Administration did not continue the NSC), on April 30th, provided additional Space Station Program Objectives. OSTP's objectives, arriving two months after the Redesign Team started work, did not break any new ground but they did dictate that the Redesign Team had to incorporate four basic strategies to accomplish the objectives: (1) Build and operate a Space Station, (2) Provide a platform for technology development, (3) Provide an international laboratory facility, and (4) Integrate user outreach. Not too dissimilar to the direction the program had been going for the last nine years except for one major difference: there was no mention in OSTP's objectives of a permanently manned Space Station, the goal that drove Space Station Freedom and all the early designs and vision. The Redesign Team apparently chose to ignore this omission.

Option A was really two design options: A-1 and A-2. The major differences were that A-1 included Bus-1 (a new propulsion system for attitude control and reboost) that was not part of A-2, and a new orientation of the trusses and solar arrays. Both A-1 and A-2 were stated to be responsive to the overall redesign requirements and emphasized "programmatic and design solutions that result in a reduced size and cost station." The modular buildup for Option A consisted of four phases: a power station achieved after three launches; a human-tended capability after three more assembly flights and two logistics flights; international human-tended capability after six more assembly flights and two logistics flights; and assembly complete would be achieved with four more assembly flights and two more logistics flights, leading to permanent human capability (PHC) (manned capability had become human capability) with a crew of five as the primary goal. Compared to the earlier restructured design, it would be attained with two fewer flights. Many systems were deleted, such as nodes, several trusses, and the alpha rotary joints. It required modifications to the hab module, a long duration orbiter, the Canadian RMS, simplified radiator and data management systems, and many other changes, including the use of the mini-pressurized Italian logistics module. At PHC two Soyuz crew return vehicles would be attached that were not included in the restructured design manifest, presumably to be launched by the Russians.

---

12. Space Station Redesign Team, Final Report to the Advisory Committee on the Redesign of the Space Station, June 1993. The Report is 296 pages in length.

Option A could be placed in inclinations of either 28.8 or 51.6 degrees; the latter orbit was in the event that Russia would join the program and launch the Soyuz capsules. It was pointed out that the 51.6 degree inclination incurred a penalty of having only a five minute launch window for each shuttle flight. After PHC, Option A would require five to six shuttle maintenance flights per year for the 28.8 degree inclination and seven to eight flights per year at 51.6 degrees (Remember, at higher inclinations the shuttle payload would be reduced). Although a number of cost savings were listed for Option A, and the cost of some of the changes would be borne by the international partners, there were so many untested changes and assumptions that, intuitively, one would have to say it was difficult to understand how this option would save money. Fewer shuttle flights saved money but those costs had never been included in Space Station costs.

Option B used mature Space Station Freedom element design and operations to provide increasing capability during the assembly flights. The first stage, after two flights, would be a power station that would supply power to an orbiter carrying a Spacelab, thus allowing for extended missions. The next stage, human-tended capability, was achieved after six more flights and provided the U.S. lab for man-tended and long-duration untended experiments. This stage required periodic resupply and maintenance that would be done using the Canadian RMS. After nine additional flights, international human-tended capability would begin with both the JEM and ESA laboratories attached. PHC was achieved after three more flights. This last stage would support a crew of four. With additional flights, an eight-member crew would be possible and 75 kW of power would be available for users. Although based on SSF elements, there would have to be some modifications to the structure, systems and other elements, such as the DMS. Soyuz crew return modules would be included in the twenty flight manifest. Shuttle modifications that were needed for SSF, would also be required for the several Option B phases. Six logistics flights per year would be needed.

Option B, like Option A, could place the Space Station in either a 28.8 or 51.6 degree inclination orbit. The higher inclination would require the use of the STS lightweight, aluminum-lithium external tank being studied to increase payload capacity, otherwise the Freedom elements would be too heavy. This option took maximum advantage of the work already done, and cost savings would accrue from a reduction in the number of elements needed for assembly, and the number of shuttle flights would be reduced. Options A and B would have ORUs attached to the trusses, but fewer in number than Freedom. Other changes that simplified some of the Freedom systems were projected to reduce operating costs.

Option C was a completely different approach. Based on work that had been done earlier at MSFC and JSC for Shuttle-C, it required a single launch to achieve an initial operating capability. It would use known and proven shuttle systems, and development work done for SSF. The first element would be a fully integrated, 26,000 cubic foot space station requiring no assembly or EVAs. It also could be placed in either of the orbits described for Options A and B. Once on orbit, it would be completely functional for the first visit from a shuttle carrying the crew. Later flights would carry the Soyuz crew return vehicle and the JEM and ESA modules that would be attached using the Canadian RMS. A total of twelve flights would achieve PHC at the 28.8 inclination and eighteen flights for the 51.6 inclination. Option C would have almost all ORUs inside the pressurized modules. Maximum power available to the users at PHC would be 30 kW.

Each of the options supported users to varying degrees; however, it was difficult to assess, quantitatively, which would be the best. Each attempted to provide, as close as possible, the capabilities advertised by the baseline program. However, compromises were necessary because of changes in the lab and hab modules, assembly sequences, power availability, the number of ORUs and their location, crew size, and many more differences. In the Report Summary, a brief explanation was given: "The critical research requirements remained the same as those defined for Space Station Freedom. Others were reduced in an effort to contain costs, but with the potential for future growth." The "critical research requirements" that remained, and the "others" that would be reduced, were not listed.

The international partners made their own assessments of the redesign and these were included in the Report. ESA considered that Option B represented the lowest technical and programmatic risk. It had not been given detailed cost data, so it was not in a position to evaluate the cost impact. Japan's view was that Options A and B maintained its role as a contributor, but had several questions to be resolved.

Option C changed Japan's role to that of an "unwelcome user," because of the relatively small resource capability, and would impose an adverse cost impact on JEM. The Canadian Space Agency stated that its user requirements could be met by all three options at PHC, when the ESA and JEM modules would be assembled but with varying degrees of compliance. Its choice as the option with the best chance of success and least amount of risk was Option B. Finally, the Italian Space Agency, the newcomer to the table, indicated it could support all three options. However, it appeared to favor Option A. It was clear that the international partners were not very happy. Their unhappiness would soon grow.

The redesign report was submitted on June 7th to the Advisory Committee on the Redesign. In addition to descriptions of each of the options, it included a short discussion of possible Russian participation, and the required recommendations. Some of the recommendations involved streamlining management structures and acquisition policies, and cutting in half operations costs. But the bottom line was the estimated cost of each option. Annual funding targets to achieve PHC of $1.0 billion, $1.2 billion, and $1.8 billion had been included in the instructions to the Redesign Team. Ignoring all the caveats in the report, no option met the annual funding targets.

As one might have expected, the lowest cost option based on all the assumptions was Option C. The total estimate to achieve PHC for fiscal years 1994 to 1998 was $11.9 billion, an average of $2.38 billion per year, still higher than the highest target. Option A was $13.3 billion and Option B was $12.8 billion for fiscal years 1994 to 1998. The SSF baseline program was $17.6 billion or $3.52 billion per year. Post FY 1998 funding for each option was also estimated and, again, Option C was the lowest. Now it was up to the Vest Advisory Committee to make its recommendations to the President.

## PRESIDENT CLINTON SELECTS DESIGN OPTION – CONGRESS OBJECTS

On June 10th the Vest Advisory Committee made its report to the President. If nothing else, it sowed confusion and controversy. Seventy-eight pages in length, the Committee recommended Option A, but with some unidentified modifications. The President accepted the Committee's recommendation saying: "I am calling for the U.S. to work with the Congress, NASA and our international partners to develop a reduced cost, scaled-down version of the original Space Station." He then went on to say: "We are going to redesign NASA as we redesign the Space Station."[13] As part of its review, the Committee recommended a leaner management structure and, to cut costs, elimination of the Reston office. It also said that the civil space program "is in great need of stability of goals and budgets . . . and timely, clear and long-lived decision making" – probably the most cogent observation contained in the Advisory Committee recommendations.

Although Option A, the modular buildup, was chosen by the President as the design he and his advisors wanted, it was not exactly the Redesign Team's Option A. Insiders said Clinton called it the "A plus" option. On June 24th OSTP sent NASA a letter asking Goldin to "optimize" the Option A design.[14] Optimization was not defined but was understood to mean that some elements from Option B could be incorporated. Meanwhile, members of the Clinton transition team were at MSFC working with the Option A and B teams to complete the "A plus" design.

Controversy followed immediately after the President's decision was made known. Option A was not the favored option in Congress, as was mentioned earlier. However, both Senator Mikulski and Representative Brown, chairmen of their respective subcommittees, contrary to threats made earlier, welcomed the decision and said they would work to get bipartisan support to pass the funding needed. Chairman Brown had some "technical" reservations about the design choice, which was not surprising considering the many unknowns. In a "Commentary" published in the *Washington Times* in early July, Brown discussed the future of the Space Station and the recent close vote to terminate the program. Among other observations, he said: "it struck me as particularly disingenuous . . . to hear critics [in Congress] of the space station decry program stretch-outs, escalating budgets, and management disarray at NASA – delay, confusion, and costs for which they are at least partly responsible. No

---

13. The White House, Office of the Press Secretary, June 17, 1993. Space Station Redesign Decision Reduces Cost, Preserves Research, Ensures Int'l Cooperation.
14. Station Break: Option A – Described and Compared. Vol. 5 No. 5, July/August 1993.

federal program can long endure the kind of annual sniping, redrafting and retrenchment to which they have subjected the space station."[15]

The cost of Option A had become $10.5 billion over the next five years, as opposed to the Redesign Report of $13.3 billion. How this lower cost was achieved was not clear, in view of the fact that the design would incorporate some elements from Option B. But it was within NASA and the prime contractors that the largest dissension and acrimony surfaced. *Space News* reported that the program "appeared in turmoil."[16] *Space News* also reported that two NASA staffers who had been appointed by Goldin to redesign the program "were relieved of their duties after criticizing the effort as too secretive."[17]

At the next SSAC meeting on June 30 (and what would turn out to be the last), Kohrs gamely tried to put a positive spin on what was happening. He said the June CDR had gone well and, based on the maturity of the current design, the redesigned Space Station would start from a solid foundation. Committee members were not so sanguine. The CDR of the integrated design included all the work package centers, contractors and the international partners. The Committee was told there were a number of RIDs to work off, but no "show stoppers" had been identified. How many of the elements and systems that had passed the CDR (as part of an integrated design) would no longer be implemented was unknown. Kohrs' concern was that one of the goals of the redesign, to proceed under the announced target of essentially flat out-year budgets of between $1.8 and $2.1 billion, would not be sufficient to carry the program through assembly and initial operation. If NASA were forced to keep the books as GAO advocated, those amounts would not be close to enough to carry out the program.

Soon after the SSAC meeting, the trade press reported that there was growing dissension between Goldin and the Space Station management team that had been in place for more than four years. He announced that there would be a new team with Bryan O'Connor in charge of both the transition and the redesigned program. Working under O'Connor, heading up the transition team, was astronaut William Shepherd. Goldin projected that O'Connor's team would consist of some 300 NASA personnel, and that the civil service workforce at the work package centers would be reduced to stay within a target of 1,000, including the 300 on O'Connor's team.

### REDESIGN CHALLENGED – GOLDIN APPOINTS NEW MANAGERS – JSC AS HOST CENTER

Some staff from the old team were in open revolt. Bob Moorehead had written a letter to his management that was leaked to the press after the Redesign Team submitted its report. In it he said that the redesign was "inconsistent, self-contradictory, superficial and inaccurate."[18] As part of Goldin's plan to appoint "new thinkers" to the program, he had reassigned Moorehead to a Headquarters position with no Space Station responsibilities, but Moorehead refused the new assignment. NASA's local of the International Federation of Professional and Technical Engineers joined the fray and filed an unfair labor practices complaint. The union warned the Space Station civil service work force they were in danger of losing their jobs, especially those managers who were designated as Senior Executive Service (SES). SES managers, such as Moorehead, had some protection, but were susceptible to dismissal when programs were terminated. Not one of the NASA managers working under Kohrs and Moorehead had been assigned to O'Connor's new team.[19] John David Bartoe eventually was transferred to JSC.

Closing the books on Space Station Freedom, and in an act of symbolic defiance, SSF senior managers released a final report, almost 400 pages in length, summarizing the program's accomplishments and status at termination. The Preface begins: "As our bold endeavor on the Freedom program is brought to a close, we should all take pride in our accomplishments . . . No matter what future course the Space

---

15. Commentary, George Brown Jr., *The Washington Times*, July 11, 1993. Brown's article appeared shortly after he led the fight to defeat amendments introduced by Roemer and Richard Zimmer (R-NJ) to kill the Space Station in the FY 1994 NASA Authorization 216 to 215 and the Appropriation, 220 to 196.
16. Confusion Surrounding Station Redesign Intensifies, Andrew Lawler, *Space News*, July 12-18, 1993.
17. Ibid.
18. As reported in the Washington Edition / *Los Angeles Times*, Dispute Brews on Space Station Team, Contracts, Robert W. Stewart, July 1, 1993.
19. Ibid.

Station efforts take, there are facets of our work on Freedom which will endure the test of time . . . It is unfortunate that this team was not given the task to address the issues that overcame the Freedom program."[20] In the next years, rightly or wrongly, many of the Space Station problems would be laid at the feet of Freedom's managers; they had presided over the development of an overly complicated design, poor contractor management, and escalating cost growth. Even if true, they were small sins compared to what would follow.

In the meantime, to reduce civil service overhead, NASA was trying to win approval from Congress to structure an early buyout program of civil servants assigned to the Space Station Freedom program. Members of the SES, which included senior SSF managers, would be excluded from the buyout. Goldin was attempting to reduce the number of staff working on Space Station but at the same time maintain the total number of NASA employees allowed by Congress, the personnel level controlled by the annual appropriation. All of these questionable personnel actions caused Congressman James F. Sensenbrenner Jr. (R-WI) to remark: "Goldin has to be a little more concerned with people and their feelings, since morale at NASA is so low. What concerns me is that the program's institutional memory appears to be on the chopping block."[21]

As part of the management reorganization, Goldin announced on July 20 that Space Station Freedom was officially terminated, and that Kohrs would meet on a weekly basis with O'Connor during the transition. He also announced that the Reston office would be closed and most of its responsibilities transferred to either MSFC or JSC, and one would be designated the "host" center. As one of the conditions of the redesign, OMB ordered that the contractor work force, then running at about 14,000, be reduced by 30%. To do this the plan was to select a single prime contractor from the three work package prime contractors or the Reston support contractor, Grumman. With the possibility that they might not be chosen and their contracts reduced in scope by 30% or more, the prime contractors did what all contractors do under such conditions, they lobbied their Congressional delegations. Theoretically they were all in the running, but the grapevine had it differently. The Boeing Company was Goldin's favorite. The contract would not be competed; it would be a sole source selection with Goldin chairing the selection committee.

Members of Congress do not take kindly to having work and jobs in their districts spirited away based on the decision of some government bureaucrat. The Space Station program was a major job producer and, in addition, brought prestige and power to those members whose districts were fortunate enough to have a big piece of the pie. One member of Congress, Representative Leslie Byrne (D-VA), was especially displeased. Her district was in danger of two hits, the closing of the Reston office and the loss of 850 jobs by her constituents working for Grumman and several other support contractors. She was especially concerned with the plan to sole source the prime contract and said that it would cost more money than would be saved to move the Reston office to MSFC or JSC. She threatened to go to the Vice President and President with her complaints.[22]

By the middle of August the Space Station picture began to clarify. Goldin announced that he had selected JSC as the host center. The news release announcing the decision cited JSC's "experience, personnel and facilities to respond flexibly to the space station program needs. And secondly, JSC has a strong operations capability in terms of both the civil service workforce and its extensive facilities."[23] A little redundant but when making such an important choice one is obligated to expand the rationale. If one looked at the selection from MSFC's perspective, all the reasons given to choose JSC also existed at Huntsville. But by this time Goldin had replaced almost all of NASA Headquarters' top managers or moved them out of line management and into positions with few responsibilities. His senior staff was populated by ex-astronauts and his Special Assistant, George Abbey, was another JSC transplant still with strong ties to Houston.

---

20. Space Station Freedom Final Report, Prepared by Space Station Freedom Program Office, December 17, 1993. The Preface page was signed by Robert W. Moorehead, John T. Cox, and Marc Bensimon.
21. Union, Moorehead Fight NASA's Station Actions, Andrew Lawler, *Space News*, 26 July-1 August, 1993.
22. Slower Station Reforms Sought, Andrew Lawler, *Space News*, July 19-25, 1993.
23. NASA News Release 93-148: Space Station Host Center and Prime Contractor Announced, August 17, 1993.

JSC's primary responsibilities as host center and program manager were to manage the design, development and the physical and analytical integration of the Space Station as it evolved into operations. The new Space Station organization would employ about 1,000 civil servants. The program office at JSC would have 300 and the remaining 700 would be distributed to the other centers involved in the program, including the JSC work package office. MSFC and LeRC would continue to have major hardware roles, but would now report to JSC as the host center.

The third leg of the new organization was the single prime contractor. This plum, as some predicted, was awarded to the Boeing Defense and Space Group. Boeing's role was described as providing the Space Station elements necessary to sustain human life – the pressurized lab and hab modules, and the ECLSS. It would also be responsible for delivering the full-up Space Station assembly and for coordinating and integrating the U.S. elements with those provided by the international partners. McDonnell Douglas and Rockwell International, two of the three former work package primes, agreed to be subcontractors to Boeing and supply the specific hardware called for in their contracts. Grumman Aerospace also agreed to this arrangement, although what their new role would be was not clear. All three contracts would be modified to reflect their new relationships reporting to Boeing.[24] At the same time as these announcements were made, Kohrs informed his staff that he would retire, effective the end of September.

Often change is good or needed in an organization, especially in government bureaucracies. Sometimes change is bad, depending on how it is done. There was no question that NASA was getting old and stodgy in many of its activities, and bogged down with rules and regulations. It was not the vibrant agency of the 60s and early 70s. But the decision to move Space Station top program management from Washington to Houston was directly contrary to the recommendation made by General Phillips in 1986. Times change, but his recommendation to locate Space Station management near, but not in Washington, was based on a profound understanding of large government programs and NASA culture to which he had been exposed during Apollo. NASA centers are universes unto themselves, competing against each other and ignoring direction from Washington, if they could get away with it.

## Administration Courts the Russians

Immediately following the announcement of the management restructuring, the future of the Space Station program became murky again. The White House sent a team to Russia in mid-August for a one-day trip to discuss the ongoing space station studies, due to be completed at the end of the month, and the forthcoming visit of Russian Prime Minister Viktor Chernomyrdin to Washington. In addition to studying Soyuz as an ACRV, Russia decided to offer a range of hardware designed for use with its *Mir* program, including a large-core habitation module, a Russian designed ECLSS, a "space tug" propulsion module, and a solar dynamic power system. As a result of the trip and the studies, the Administration was now considering Russian participation in the Space Station as a major contributor, and in the words of a Russian Space Agency official "as an equal partner."[25] That official may have been jumping the gun a little bit, but the rumor spreading within NASA was that the Administration was preparing to announce during Chernomyrdin's September visit a wide-ranging pact for U.S. / Russian space cooperation.

At the end of August, the Working Group on Russian-American Cooperation issued its report.[26] The Introduction to the Report started on a high note: "Russia and the USA, being the leading Space powers of the world community . . . made it possible to go to new levels of Space exploration in the interests of fundamental science . . . and commercial utilization of Space assets." One hundred and thirty seven pages in length plus attachments, the Report included three main phases of cooperation: *Mir*-Shuttle, *Mir*-NASA, and an International Space Station combining Freedom and *Mir-2* programs. Even though

---

24. Ibid.
25. White House Considers *Mir* Option, Andrew Lawler, *Space News*, August 23-29, 1993.
26. Russian-American cooperation in manned space stations, Working group report, August 1993. The NASA Historians office holds the report titled as shown.

NASA had terminated Freedom four months earlier, it was still being referred to in an official document. As the old saying goes, there are always ten percent that never get the word.

The *Mir*-Shuttle phase was described as the necessary first step and would run from 1993 to 1995. This phase projected Russian astronauts on board shuttle orbiters in 1993 and 1995, and American astronauts on board the Soyuz-TM vehicle and *Mir* orbital Space Station. The Shuttle would dock to *Mir*, and NASA experiments would be carried out on board the *Mir* Station. The next phase, *Mir*-NASA, envisioned long-term missions of American astronauts on board *Mir* in 1996 and 1997. Research equipment would be supplied by NASA, and carried to *Mir* by the shuttle and Progress-M vehicles, and installed on the *Spectr* and *Priroda* modules. These two phases would lead to the main program of joint work on a space station combining the elements of Freedom and *Mir-2*, and would include the elements under development by the international partners. The 137 pages contained great detail on what could be done. It included an analysis that recommended that NASA purchase $2.5 billion in services from Russia over the next five years. If the program were to be approved as outlined in the report, the U.S. would launch the first Space Station element in 1997, one year earlier than the current plan.

On September 2nd, a space cooperation agreement was signed between the two countries incorporating some of the activities developed by the previously mentioned Working Group. One of the provisions of the agreement was that the U.S. would pay Russia $400 million for unspecified services. It did not contain any details of what cooperative efforts would be undertaken with the Space Station other than "the space station must be orbited at an inclination of 51.6 degrees to allow launch and assembly of Russian components."[27] Marcia Smith at the Library of Congress Congressional Research Service said in a speech that she learned that the joint space station being discussed with the Russians would begin with six to eight launches of Russian hardware followed by one or two U.S. launches. She said, logically, the station would be controlled, at least in the early phases, from Russia because most of the elements would be Russian.[28] Vice President Gore, an enthusiastic supporter of the partnership, said, "the agreement on space we signed represents the leading edge of what we are striving to accomplish, Russia and the United States together: from broad market access for Russian high technology goods to long-term projects to work together in complex, productive ways."[29] With such a glowing endorsement, what else might be in store for a joint U.S. / Russian program? The details were soon to be revealed.

## RUSSIAN PARTNERSHIP CHANGES PROGRAM – CONGRESS CONFUSED

The cooperative agreement seemed to require many changes to the Space Station design. Clinton's Option "A plus" became Option A plus, plus. The "optimized" design requested by OSTP was revealed as Space Station Alpha, but with many TBDs. Alpha's cost was under study, with the first estimates indicating that it could not be built and meet the Congressional $2.1 billion annual funding ceiling. Some of the proposed changes to Option A included replacing the Lockheed Bus-1 with the Russian functional cargo block (FGB) tug. The alpha rotary joints for the solar arrays were back in to reduce thruster firing to maintain proper sun orientation, and thus improve the microgravity environment and reduce contamination on experiment surfaces outside the pressurized lab modules. To maintain the scheduled number of shuttle flights carrying assembly and maintenance payloads required developing ASRMs and the lightweight aluminum-lithium external tank (ET). Neither development was guaranteed with the Congress tightly holding the purse strings. At this point, the planned orbital inclination was still 28.8 degrees and the "optimized" Space Station now included several Russian elements.

Apparently the terms of the recently signed agreement with the Russians that stated the inclination would be 51.6 degrees had not filtered down to NASA's working level. If the orbit was at 51.6 degrees, the shuttle payload problem had been carefully calculated by JSC and Boeing engineers. The

---

27. Ibid.
28. Before We Say "I DO," Marcia Smith, Specialist in Science and Technology Policy, Congressional Research Service, Library of Congress, based on a speech given to the Federal Bar Association, September 14, 1993, published in the AIAA National Capital Section Newsletter November 1993, Vol. 21 – No. 10.
29. *Station Break*, NASA Shows New Design, Explores Russian Options. Vol. 5 No. 6, September/October 1993.

shuttle would carry about 12,000 pounds less cargo to a 51.6 degree orbit. Assuming Congress would go along and approve building the lightweight ET, this weight saving would be approximately 6,000 pounds that could be put back into payload. That still left a deficit of 6,000 pounds that would have to be made up by reducing the weight of the elements, a redesign that had not been done or costed, or a guarantee that the ASRMs would be available.

With so much uncertainty and so many last minute changes, the Congress was having difficulty deciding how to structure NASA's FY 1994 budget, and the new fiscal year started in four weeks. The Administration had requested $2.1 billion for Space Station Alpha and an additional $313.4 million for ASRM development. If the ASRMs were not available by the time of FEL, the shuttle manifest and the elements scheduled as payloads would have to be modified or redesigned. Russia's future role added another dimension of uncertainty that could not be resolved by the time Congress had to pass the appropriation.

The Senate appropriations subcommittee took the only course available. It fudged. It recommended that the Space Station receive $1.946 billion and distribute the remaining $154 million to other budget line items. At the same time it reduced the overall NASA request by almost $650 million. The ASRM request was trimmed by $150.4 million, leaving $163 million for contract termination costs. If NASA didn't like this action, it would have to find a way to offset the reduction by cutting or terminating other programs. Message: NASA do you really need ASRMs? This was a typical hardball negotiation between Congress and an Administration and, surprisingly, both the antagonists represented the same political party. The Subcommittee then went one step further. Because of the uncertainties of Russian participation, $946 million, roughly half the appropriation, was "locked up" until January 1st of 1994. That gave the Administration three months to come back to Congress and tell it what was the "REAL" plan for Space Station. Many others also wanted to know.

Senator Mikulski, who chaired the appropriations subcommittee, explained the reasons for holding back the $946 million. "We support a cooperative effort with Russia, but we believe the space station effort, if approved by Congress, must be an American space station. They have to tell us (what we are buying), because this is like a floating crap game. I mean every time we turn around there's a new element to the Space Station." On the subject of ASRMs, Senator Phil Gramm (R-TX) said: "We're willing to fly ASRM if NASA wants it, but we're asking them to look at the new configuration which is based on fewer flights, look at the new Russian partnership – and the one thing the Russians do very efficiently is launch things – and come back and tell us do they want the ASRM program."

Mikulski said she and Gramm had given the Administration three criteria that must be met to get continued support from the Congress: "It must do significant science; it must be fiscally achievable on a sustained basis; and it must meet the criteria for the continued involvement of our international partners." "Significant" was defined as science equal to the science that would have been possible on the baseline Space Station Freedom. With all the downsizing that had taken place, that would be an especially difficult criteria to meet.

The next step in the evolution of the Space Station program was promised on November 1st when NASA would report on the Russian proposal to improve their *Mir* space station and have U.S. and international elements joined to it. An anonymous NASA source was reported as saying: "If in November the international partners and the President and the Congress agree to bring Russia in, we can then, at very little cost – at virtually no cost – rearrange the launch sequence to go with the Russian option." Apparently the NASA source had not been listening to Senator Mikulski, and did not grasp the implications of such a major change, to say it could be done at "virtually no cost." In spite of the confusion, the major players in Congress seemed resigned to the fact that Russia would be a sizable participant, although what its full contribution would be was still unknown.[30]

Some of the unknowns were being analyzed at JSC. Engineers at JSC warned that safety could be compromised and costs increased unless complicated technical and managerial questions were resolved. It was pointed out that the *Mir* station was based on 20-year-old designs. Large differences

---

30. Quotes in the aboce three paragraphs are taken from two articles: "Space Station Backers Are Worried Over Uncertainty on Russia's Role," Kathy Sawyer, *The Washington Post*, September 8, 1993; and "Senate Appropriations mark fences Station funds pending final design," *Aerospace Daily*, September 9, 1993.

existed in the engineering cultures between the two countries and these differences were reflected in the design philosophies for space systems. Mission Operation Director Eugene Kranz and astronaut David Leetsma sent a two page letter to NASA Headquarters followed by an eight page memo containing a long list of problems that the U.S. and Russian negotiators needed to resolve before "binding agreements are made." The memo further noted that little was known of the systems onboard *Mir* and that it was not clear if the power and data management systems could be integrated. Another problem was how to attach the U.S. and international partners' modules to *Mir*, and for that matter to even dock the shuttle. And of course, if you were on the JSC staff, the paramount question was which mission control center would operate the combined *Mir*-U.S. station, Kaliningrad, or JSC? Kranz and Leetsma also warned that all these unknowns could lead to "significant cost impacts in the future."[31] Kranz' and Leetsma's warnings were ignored in the rush to move ahead with the new partnership.

## NEW SPACE STATION IMPLEMENTATION PLAN

On September 7th, Goldin sent the President the Alpha Station Program Implementation Plan reflecting the Russian participation. Some of the highlights: First element launch would occur in September 1998, eleven months later than originally estimated for Option A-1. PHC would be reached in September 2003, three years later than originally estimated in Option A-1. In the cost estimate that followed two weeks later Goldin assured OSTP that the program could proceed under the $2.1 billion annual ceiling with these schedule slips. The US / Russian agreement required modifications to the shuttle for docking to *Mir* and these mods would carry over to the Space Station. The Alpha design assumed that the Soyuz spacecraft would be used as the ACRV and the Salyut FGB tug would replace Bus-1. The orbital inclination would be 51.6 degrees.

In the budget forecasts, it was estimated that science and utilization for Alpha would cost $1.5 billion through FY 1998, 11% of the total costs, and from FY 1998 until PHC would be increased to 15%. Two precursor Spacelab microgravity and life sciences missions were included in the budget, as well as a number of additional orbital experiments that would utilize the Space Station, Spacelab or the shuttle payload bay and mid-deck area. The cost estimate discussed a program reserve in terms of percentages of certain elements including the Russian elements. However, in the accompanying table a specific reserve number was not shown and there was no detail on which to decide if the reserve was reasonable. Low-balling program reserve had caused problems in the past and would hinder the program in the future. NASA would purchase hardware and services from Russia in Phases One and Two (1993-1997) based on a firm, fixed-price contract at a funding level of approximately $100 million in each year during the two phases. An MOU would be negotiated for Phases Two and Three (Phase Three ran from 1997 to PHC) that would list Russian and U.S. contributions and all reimbursable products and services. This seemed to indicate that there was an overlap during Phase Two, leaving the door open for Russia to receive more than $100 million per year during this phase.

So there it was, a new Space Station design with the Russians appearing to be equal partners. Four of the first seven assembly flights would carry Russian and U.S. elements, two would be dedicated to only Russian elements and one would carry just the U.S. Node-1. Very little was said about the original international partners in the Implementation Plan. Canada's RMS would be carried on assembly flight 12, the JEM on flight 19 and the ESA module on flight 21. The U.S. lab would be on flight 8 and the hab module on flight 24.[32] At this point, with all the permutations that had occurred in the last six months, it would be surprising if you could get two NASA engineers to sit down and have both agree on what Space Station Alpha really was. More importantly, could Congress and its staff understand how the program had evolved and agree to support the program?

---

31. Most of this background was extracted from a *Space News* story, Officials Fear Russian Deal Premature, Andrew Lawler, September 6, 1993.
32. Information on Space Station Alpha implementation and costs was extracted from three documents: Alpha Station Program Implementation Plan forwarded to OSTP on September 7, 1993; Cost Report for Space Station Alpha, an enclosure to a letter addressed to The Honorable John H. Gibbons, Assistant to the President for Science and Technology, Executive Office of the President, Washington, DC 20500, signed by Daniel S. Goldin, September 20, 1993; and Alpha Station Addendum to Program Implementation Plan, November 1, 1993.

Meanwhile, Bryan O'Connor and his transition team had been working behind the scenes to devise a new approach to Space Station management. His desire was to: "Build one team and eliminate the distinction and duplication between multiple levels of management and oversight, and keep one management chain: civil service directing the contractor (and) have a direct reporting path to the Program Manager." He wanted to matrix NASA resources to be able to respond in well-defined areas to tasks or facilities management. To achieve his goal he adopted Boeing's successful approach to program management and established multidisciplinary Integrated Product Teams (IPTs). He planned to organize the IPTs into a tiered hierarchy and "bring together the players responsible for a given product." (Eventually, almost 100 IPTs were formed.) He would do all this staying within the ceiling of 1,000 civil servants. He also planned to develop a work breakdown structure that would match the IPT structure. Meetings had been held with the unions to describe the approach and he planned the move to the host center on October 4th.[33]

O'Connor expected that the Boeing contract would be approved in time for them to begin acting as the prime contractor before the end of September. In advance of finalizing their award, Boeing was preparing a 180-day program execution plan that included novating the other contracts (in simple terms, transferring contract reporting and management from NASA to Boeing), and conducting an SRR in 90 days and an SDR in 180 days. The NASA transition team was cleaning up all the loose ends of the terminated SSF program to realign them with the new configuration. One might assume it was the design outlined in the Program Implementation Plan, but this would prove to be a wrong assumption. All the changes had not yet been made.

### Congress Passes FY 1994 Budget –
### Questions and Concerns Rampant

At the end of October 1993, one month into the new fiscal year, Congress passed the NASA appropriation. As forecast during the earlier budget debates, the total NASA budget was $14.549 billion, with $2.1 billion for Space Station Alpha. But the appropriation came with warnings and in a general atmosphere of great discontent. Senator Mikulski and Representative Louis Stokes (D-OH), chairs of the Senate and House appropriations committees with NASA oversight, sent the President a letter warning that Congress might chop $500 million from NASA's budget next year if a number of issues weren't resolved. These issues revolved about how the NASA budget was divided between manned space flight, aeronautics and science – the debate that had raged in and outside NASA for decades. As NASA's budget is included in legislation dealing with veteran affairs and other agencies, such as NSF and HUD, they pointed out that there was pressure to increase funding for programs such as veteran's medical care and public housing. Representative George Brown, chairman of NASA's authorization committee, was quoted as saying in reference to NASA's budget: "I fear that the federal government may be retreating from its historic commitment to invest in science and technology."[34] If NASA's budget were reduced, as some had threatened, the fixed costs of the Space Station program and shuttle flights would consume over 30% of NASA's budget and result in the cancellation or modification of important science and aeronautics programs. These warnings were not to be ignored by the Clinton Administration, just four months from submitting their FY 1995 budget. Would there really be a $3 billion savings by including the Russians in the Space Station as Goldin was now touting, or were the savings just wishful thinking?

Including the Russians in the Space Station had quickly brought severe headaches to the Administration. A front page story in *Space News* headlined: "Skeptics Question White House Station Plan," referring to members of Congress.[35] Both advocates and opponents had questions. Opponents, led by Senator Bumpers, were still trying to come up with language for an amendment to NASA's authorization that would convince a majority of the members to terminate the program. They had almost succeeded a few months earlier, missing by one vote. By including the Russians the

---

33. Space Station Management Approach and Transition Status, September 20, 1993, Bryan D. O'Connor, Director, Space Station Transition.
34. Mikulski, Stokes Warn NASA Funds Will Shrink, Andrew Lawler, *Space News*, November 8-14, 1993.
35. Skeptics Question White House Station Plan, Andrew Lawler, *Space News*, November 8-14, 1993.

Administration had given faltering supporters a reason to not want to do business with "former communists" and agree to cancel the program. Congressman Brown, an advocate, said too many questions remained about the cost of including Russia. Undoubtedly in the back of his mind were the provisions in the U.S. / Russian agreement to pay Russia for becoming a partner. Remember the recommendation of the Working Group to purchase $2.5 billion of hardware and services. Mikulski and Stokes had the same concerns. Goldin, in a meeting with the lawmakers and Vice President Gore, was reported to have said that the Russian help "will allow us to spend significantly less than the $19.4 billion the current design is estimated to cost through completion." That number, $19.4 billion, was the estimate supplied to OSTP in September and was based on Russian participation. Where the new savings would come from was not explained. Goldin was said to have promised more detailed costs in the months ahead. He had just two months until the FY 1995 budget defense would begin.

More of the details of the consequences of involving the Russians began to slowly filter out. Four scheduled Spacelab missions between 1995 and 1997 would be canceled in order to schedule docking missions with *Mir*. Spacelab had flown in the shuttle payload bay several times since its first flight in 1983 with astronauts conducting microgravity and life sciences experiments, some of which were forerunners to more extensive Space Station experiments. The number of shuttle flights scheduled to dock with *Mir* had suddenly increased from the planned five to ten, approximately one every three months. Robert L. "Hoot" Gibson, Chief of the Astronaut Office at JSC said that the increase in the number of flights was unexpected. "We got a call from Goldin while he was in Russia saying that we have just signed up for another nine dockings. The reason for the call was, hurry up and come up with the reasons we will be going out there. This is not the way I recall you generally do a project. You come up with the requirements and then you say, 'How many flights does it take?'"[36] It was becoming more and more obvious that planning for incorporating Russia in the program was being done on the run. Gibson, perhaps for voicing his concerns, was soon replaced in the Astronaut Office by Robert D. Cabana.

Two shuttles, *Atlantis* and *Columbia*, would be modified to carry a docking mechanism so that they could dock with *Mir* and to carry a pressurized module for equipment and experiments. Payloads that had been scheduled for these two orbiters would be delayed and shifted to *Endeavor* and *Discovery*. Missions that had been scheduled to carry a radar experiment for Earth mapping were indefinitely postponed. These changes did not win the approval of the scientific community, especially those who had invested time and resources to meet the original schedules.

Were the disruptions and added costs to the Space Station program worth this new alignment with Russia? Arguments were made on both sides. The Clinton Administration took the long view. In Clinton's first speech to the United Nations, in September, he made the case for using space cooperation as an important ingredient toward achieving improvement in foreign relations. In particular, the Administration used cooperation with Russia as a means to restrain weapons sales and the transfer of nuclear and missile technology to countries that were not friendly to the U.S. The argument went like this: "If you can keep the Russian weapons and missile engineers and scientists busy working on the Space Station, they won't have time or need to build and sell these items to other countries." Not a bad argument; however, the Space Station program could provide useful employment to only a fraction of those formerly working on the ambitious Soviet space and missile programs. Marcia Smith, at a meeting of Women in Aerospace, said, "If the primary political consideration is to build a space station, then you want stability and probably don't want to do it with the Russians."[37] If partnering with the Russians was primarily a political strategy to convince them to cooperate on arms control, it would not be successful. In 1993 there were already documented instances of Russia selling weapons, and nuclear and space technology that could be converted to military uses, to regimes of uncertain reliability.[38] China, Iran, North Korea, Iraq, and other countries benefited from these sales.

---

36. Goldin: More Shuttle Missions to *Mir*, Liz Tucci, *Space News*, October 25-31, 1993.
37. As quoted in *Space News*, October 4-10, 1993, from a Women in Aerospace, Marcia Smith, Specialist in Science and Technology Policy, Congressional Research Service, Library of Congress, September 27, 1993.
38. An example of these sales to China that took place in the early 1990s was revealed in a *Space News* story: Bush, Clinton Blink At Secret Purchase Of Zenit Engines, Theresa Foley, *Space News*, November 1-7, 1993.

Would the real Space Station program please stand up? Was it Option A-1 "plus," announced a few months earlier, or was it an undefined smorgasbord? Negotiations continued in November and December between the Administration and critics and supporters of having Russia join the program. Both sides of the political aisle were involved on both sides of the issue. The international partners were due in Washington the first week of December to be briefed on the latest developments and would undoubtedly express their concerns. Vice President Gore was scheduled to travel to Russia in early December to meet with Russian Prime Minister Chernomyrdin and move the cooperation along. But cooperation to do what?

House Science and Technology Chairman George Brown, just back from a trip to the Kazakhstan launch complex, raised a number of questions on the future of a cooperative venture. In an interview published in *Aerospace Daily* he said, "I don't think we can continue with the Space Station program with a declining NASA budget . . . I honestly think we have undertaken something beyond the scope of what we fully understand in terms of its consequences." He asked, "Are we fully staffed in the White House or in NASA to undertake this massive cooperation?" He answered his question: "I don't think we are." He went on to say that the facilities at Baikonur "raised some serious concerns . . . the surrounding infrastructure is in poor shape and really to be brought up to necessary standards, or U.S. standards . . . would probably require considerable investment."[39] These comments came from a somewhat reluctant supporter of having Russia as a partner in the Space Station program, and did not fully address a larger problem. The Baikonur facility was in the Republic of Kazakhstan (a former Soviet Republic), but was operated by Russia. Friction between the two countries was growing as to who should benefit from the location. The Kazakhs wanted more control.

Critics were even more skeptical. James Oberg, a space consultant and former NASA engineer, in a *Washington Times* article said, "A naive and overeager headlong rush into this celestial alliance may sow the seeds of eventual disillusionment and failure." Then, moving on to other issues that were coming to light as the alliance drew more careful scrutiny he reported: "Americans have made inquiries about fires that have occurred aboard Soviet space stations. Russian officials have announced that this subject is not to be discussed."[40] As more details of the Soviet / Russian space programs became known, Oberg's concerns were reinforced.[41]

### COOPERATIVE AGREEMENT WITH RUSSIA SIGNED – CRITICS DENOUNCE TERMS

Russia and the U.S. signed the bilateral agreement to cooperate on the Space Station program in Moscow on December 16, 1993, expanding the agreement signed by Bush and Yeltsin in 1992. Prime Minister Chernomyrdin announced that by signing he was accepting an invitation to join the International Space Station tendered by the U.S., Canada, Japan, and European Space Agency member states. Vice President Gore, who signed for the U.S. side, told the assembled reporters, "These are significant achievements today . . . As a result of this Space Station agreement the United States will save $2 billion and two years on the launch scheduled to get a Space Station that will be larger, with more energy, better performance, in a better orbit."[42] Somehow, the savings that Goldin had been guaranteeing members of Congress just one month earlier, $3 to $3.5 billion, had been cut almost in half. Why the savings were reduced wasn't explained, but $2 billion must have sounded like a nice round number.

More details of what the agreement would entail were contained in a protocol signed by Goldin and Russian Space Agency Director General Yuri Koptev. The program now had two names: the International Space Station and "Alpha with Russia." The latter name would soon be dropped. The original plan to give Russia $100 million per year for four years for unspecified services had now been extended to ten years. The Soyuz would be used as the ACRV and the Salyut FGB "space tug" would

---

39. Brown says he won't back Station unless NASA budget cuts are revised, *Aerospace Daily*, December 15, 1993.
40. The strange state of U.S. / Russian cooperation in space, by James Oberg, *Washington Times*, December 16, 1993.
41. For a discussion of the history of U.S. / Russian space programs see "Star-Crossed Orbits" by James Oberg, McGraw-Hill, 2002, 355 pps.
42. Russia joins Station effort, will get $1 billion over life of project, *Aerospace Daily*, December 17, 1993.

provide required reboost and station attitude control during the early years. The first element would be launched in 1997, one year earlier than Goldin first announced. Russia would provide twelve assembly launches and six to eight resupply flights per year using Zenit and Proton boosters. There would be ten shuttle-*Mir* dockings, with U.S. astronauts staying aboard *Mir* for a cumulative time of two years in three and six month periods. There would be a joint program to develop a solar dynamic power system that the Russians had originally said that they would provide. The first Russian cosmonaut to be launched on a shuttle was scheduled for February 1994. A U.S astronaut would be part of the crew on a Russian launch one year later. A specific program of technological and scientific research to be conducted on *Mir* would be developed by a joint working group.[43] NASA planned to issue a firm, fixed-price contract to cover the work, but with options to expand the contract if needed.

Samuel W. Keller, a former senior NASA manager who had been closely following the U.S. / Russian agreement, expressed his concerns. He believed that the agreement to pay the Russians $100 million per year over ten years for services and hardware would turn out to be too low and the U.S. would probably end up adding another $1-$2 billion before the program reached PHC. He believed that the Russians would be able to deliver the elements that they promised, but the announced schedule to launch their first element that would comprise part of the "core station" in 1997 was optimistic and would be delayed. His skepticism was based in part on the fact that Russian launches the Europeans had paid for, including a weather satellite and *Mars 94*, were already months behind schedule. Based on his contacts on the "Hill," he said that Chairman Brown and his staff were still skeptical that the joint program would be successful. Not a rosy picture. Perhaps showing his displeasure for his unceremonious transfer on Goldin's orders a few months earlier, Keller said that he did not think that Goldin and his senior staff of Abbey, O'Connor and Shepherd had the management know-how to make the partnership work.[44]

Congressman James Sensenbrenner (R-WI), ranking Republican on the House Science, Space and Technology Subcommittee, who had been a staunch supporter of the original Space Station plans, traveled to Russia in early January 1994. In an interview he said, "I have returned from this trip more concerned about the details of the Space Station and how it will impact our budget . . . and whether the cooperative venture is merely a venture that will allow foreign governments to get us to pay for things they can't pay for themselves." After asking Yuri Koptev, who just three weeks earlier was the protocol cosigner with Goldin, how the $400 million would be spent: "He said that $309 million would be going for rent . . . for American participation on *Mir-1*. Significantly, he said he couldn't build *Mir-2* for the remaining $91 million, which I think is an open invitation that he is going to need more money to build *Mir-2*." Stopping in Paris on the trip, he also talked to ESA officials and described what he found. "As a result of the way Dan Goldin canceled the Freedom design last March, western Europeans are sitting back and waiting to see what happens, rather than participating in the (new) design. Their checkbooks are closed."[45] Sensenbrenner told Goldin at a subcommittee hearing, "It is utterly foolish to propose having Russian space hardware in the critical path to building the international space station."[46] Based on Chairman Brown and Sensenbrenner's statements, NASA and the Administration were facing a united opposition to the direction the program was heading.

Operating under the restrictions imposed by the omnibus Reconciliation Act of 1993, the review of the Clinton Administration's FY 1995 budget request by Congress began in February 1994. This Act, passed with the support of the Clinton Administration, would cap all discretionary spending. NASA's

---

43. NASA News Release 93-222, NASA and Russian Space Agency Sign Agreement For Additional Space Shuttle / *Mir* Missions, December 16, 1993.
44. Based on a interview, 12/21/93, with Keller. Keller was forced by Goldin during his "management purge" to find another position. He was added to the staff of the East-West Space Science Center at the University of Maryland, but continued to draw his NASA salary.
45. Sensenbrenner returns from Russia with added concerns over Station deal, *Aerospace Daily*, January 18, 1994. You may have noticed that there are differences in how the $400 million would be split, $305 versus $309 million, and $95 versus $91 million. Why there were these differences could not be determined. The final split was $305 and $95 million. *Mir-2* would become the critical Russian Service Module of which you will hear much, a 3,500-cubic-foot cylinder weighing 46,300 pounds that would serve as a control center, living quarters and power source for subsequent Space Station operations.
46. Congressman Warns of Station Opposition, Andrew Lawler, *Space News*, February 28-March 6, 1994.

programs fell under the definition of discretionary spending. In keeping with the terms of the Reconciliation Act, the Administration's request for NASA was $14.3 billion, a reduction of some $250 million from the FY 1994 appropriation, and said to have $650 million less "in buying power." Space Station's piece was $2.121 billion. That total included $181 million distributed in the Science, Aeronautics and Technology budgets for a variety of Space Station projects, including life sciences and microgravity research. In addition, two separate line items had $32.5 million for Space Station facilities and $150.1 million for U.S. / Russian cooperative activities (including the $100 million negotiated in Moscow) a grand total of over $2.3 billion.

The funding relationship that was adopted with the original international partners and explained to Senator Lautenberg in 1991 was now officially out the window; the U.S. would pay a partner to join the program. Space shuttle costs were carried in a separate line item, but for the first time Space Shuttle and Space Station, along with operations and related costs, were lumped under one heading, Human Space Flight, representing 40% of NASA's budget. Continuing the push to reduce overhead costs, the FY 1995 budget would shave the civil service work force by more than 800. There would be no forced reductions, it would be handled by retirements and other "non-coercive measures."[47] Other programs that would have reductions were not identified but the Space Station had been singled out earlier as a target. With NASA facing a force reduction, would manpower shortages that plagued the Space Station since its inception continue to hinder the redesigned program?

### Management Changes – Budget and Program Debates

Defending the Space Station request in February 1994 before the several Congressional committees, in addition to Administrator Goldin was his new, handpicked Space Station Deputy Associate Administrator, Wilbur C. Trafton. Randy Brinkley, appointed as the new program manager in Houston, would also be called on to appear at Congressional hearings. Trafton, a retired Navy captain and president of a small Virginia company before coming to NASA, had limited experience dealing with Congress. He would be facing his Congressional trial by fire on a high visibility program.[48]

Trafton had already gotten off to a rocky start while addressing the Washington Space Business Roundtable in January. During his speech he said that the funds for the $400 million Russian contract would come out of the Space Station budget. It would be divided into two parts, $305 million would pay for Russian expenses on the ten shuttle flights to *Mir* and the five missions with U.S. astronauts onboard Soyuz. The remaining $95 million would be used to support launch and construction of the U.S. / Russian elements of the Space Station. When questioned later about using Space Station funds to support the Russians he said, "I misspoke. I knew that's ($100 million in FY 1995) not in the $2.1 billion cap. The money will come from elsewhere and I honestly cannot tell you where."[49] "Where" was specifically identified in NASA's budget request. It was clear he would have to do a lot of homework to prepare for the sharp Washington knives that would be thrown at him in the next months.

At the same Space Business Roundtable meeting, Trafton indicated that NASA had a fallback plan in case Russia did not deliver on its promises. He said that "essentially" the program would go back to the Space Station Alpha plan. "Our contingency plan includes looking at what point we are in the program when they might withdraw. We feel like we have this issue covered; we do not anticipate problems."[50] NASA would have the Lockheed Bus-1 available if the FGB tug had a problem; and if the solar dynamic power system wasn't available, a forth SSF-type photovoltaic system would be added to the complex. This was a somewhat surprising admission coming just one month after the big celebration and ringing endorsements of the partnership by all parties in Moscow. His statement was

---

47. NASA News Release, Background Material, NASA FY 1995 Budget Briefing, Hold for release at 2:00 PM, EST, Monday, February 7, 1994. Remarks Prepared for Delivery: The FY 1995 Budget: NASA Steps Up to the Challenge, Daniel S. Goldin, Administrator National Aeronautics and Space Administration, Press Conference, Washington, DC, February 7, 1994. Details in text extracted from this and above reference.
48. New Station chief sees Hill as biggest hurdle for "Alpha with Russia," *Aerospace Daily*, January 27, 1994.
49. Who Pays Russia $400 Million? Liz Tucci, *Space News*, February 6, 1994.
50. Ibid. 48.

seen as an attempt to disarm Congressional critics and others who questioned the addition of Russia to the program.

"Dear John" letters were received by all the members of the SSAC at the beginning of February. In the letters, Bill Raney thanked the members for their "enduring support of the Space Station program." He wrote that NASA management had decided to change the structure and charter of the committee "to better reflect the revised Space Station program." If interested, send your resumes into the Space Station Program Office and maybe you would be invited to serve on the new committee. It seemed obvious that in the eyes of senior NASA management the SSAC represented the old "bad" Space Station management and, thanks but no thanks, we will call you if we need you. A standing advisory committee was never reestablished, and when advice was requested a committee was formed under the auspices of the National Academy of Sciences. Chaired by Jack Kerrebrock and later by Tom Young, this committee was convened only sporadically through 1997 and 1998.[51]

The Space Station Science and Applications Advisory Committee continued to function. At their mid-February 1994 meeting they made twelve recommendations that were sent to the new space Station managers. Among them was a request to define the roles and responsibilities of the SSPO Science Adviser, including the nature of the interaction between the Chief Scientist for Space Station and the SSPO Science Advisor (a deficiency noted earlier). Other recommendations asked that: a vibro-acoustic control plan be developed, that SSPO move expeditiously to approve an external contamination monitoring plan, include two 20-inch science quality windows, and negotiate agreements with the Russian partners on sharing and access.[52] Several of these requests were carried over from those they had made for Space Station Freedom.

While the debates on the FY 1995 budget were just heating up, the date for NASA to tell Congress what the final Space Station configuration would be in order to "unlock" the $946 million in the FY 1994 budget was fast approaching. To meet the Congressional deadline, NASA held a System Design Review (SDR) on March 23rd, one week before the report was due. As was the practice for earlier design reviews, contractors and all the international partners would participate, including Russia.[53] Actually, the SDR was a step backward from the CDR held the previous spring for Space Station Freedom, after which the go ahead was given to start building flight hardware. With the inclusion of the Russians, all of the SSF elements, including the original international partner's contributions, had been modified or redesigned to some degree. Randy Brinkley, the JSC host center Program Manager, said that after the SDR, "we move from concepts to hardware implementation. This is by far the most important technical milestone in the program since last year's redesign."[54] The SSF flight hardware trusses, scheduled to be on the first SSF assembly flight, were under construction before the redesign. They would no longer be needed until late in the redesigned program and fewer would be required. Brinkley claimed that "By using about 75 percent of the hardware planned for Space Station Freedom, NASA has been able to maintain its investment to date while redesigning the system to be less expensive and more capable."[55] This was an interesting claim in view of the many new interfaces needed between redesigned SSF elements and systems that would have to be retested and reverified, such as ECLSS, data management and power systems. And adding to the ground verification

---

51. Based on interview with Tom Young October 21, 2004.
52. Space Station Science and Applications Advisory Subcommittee Recommendations, Meeting February 15-17, 1994, Presentation to the NASA Advisory Council, April 13, 1994. The twelve recommendations covered the following issues: (1) Support of the Space Station Program Office, (2) Roles and Responsibilities, (3) Microgravity Environment, (4) International Standard Payload Racks, (5) EXPRESS Program, (6) Command and Data Handling System, (7) Attached Payload Accommodations, (8) External Contamination Environment, (9) Science Quality Windows, (10) Exchange of ISSA Assets and Capabilities with Russians, (11) Russian Design Requirements, and (12) Russian Participation on SSSAAS / IFSUSS.
53. NASA News Release: 94-45, Space Station Program Marks Major Milestone, March 17, 1994.
54. Holding a System Design Review (SDR) baffled many old-time NASA managers. What was a SDR and what purpose did it serve? In reviewing NASA regulations setting forth policies for program reviews (NHB 7120.5, Management of Major Programs and Projects, Chapter 2 and Appendix B), unlike PDRs and CDRs, there were no policies for a SDR. It appeared to be a made-up review mechanism to deflect Congressional criticism of the redesigned program and invented to satisfy the requirement to "unfreeze" FY 1994 funds.
55. Ibid. 53.

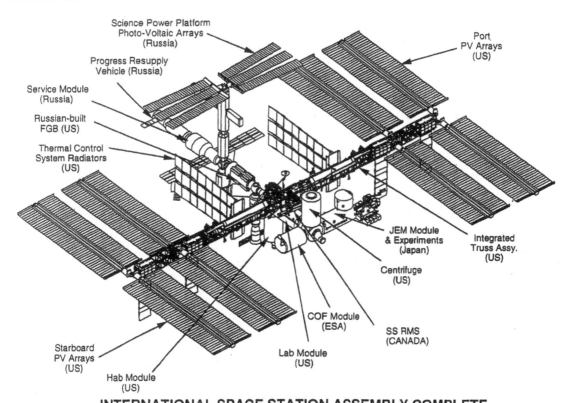

**INTERNATIONAL SPACE STATION ASSEMBLY COMPLETE**

Figure 27. Redesigned International Space Station identifying U.S., Russian, Canadian, Japanese, and ESA elements at assembly complete.

complexity, they would have to mate with completely new Russian elements being built and launched in Russia. Considering the difficulties that SSF experienced during its CDR and the thousands of RIDs that had to be worked off for SSF, his claim seemed unrealistic.

Because of the new configuration and the addition of Russian elements, the assembly sequence that would now use shuttle and Russian launches had to be overhauled. Some of the hardware and systems would first be tested on *Mir* before being incorporated in the redesigned Station, another non-trivial wrinkle that complicated the assembly sequence. The Russian tug would be the first element launched putting it on the "critical path" for assembly of all the other elements. Until it was on orbit and checked out, the node and U.S. lab module, the next two payloads, could not be launched. If there was a problem mating these two elements with the tug, for whatever reasons, the program would come to a standstill until the problem(s) was resolved. When Randy Brinkley and his team tried to put all the pieces together in preparation for the SDR they were confronted by new problems created by the $2.1 billion flat funding cap that Congress had imposed. There would not be sufficient funds in the first year or two to accomplish all that was needed. As a result, the assembly schedule would have to slip at least a few months until the deficiencies were resolved in the next fiscal years.

## SYSTEM DESIGN REVIEW COMPLETED – CBO FORECAST

The SDR was completed on March 24th. The new baseline configuration consisted of the U.S. lab and hab modules and integrated truss; the Russian science power platform, service module, FGB tug and Soyuz ACRV; the ESA laboratory module; the Japanese JEM and exposed facility; and the Canadian RMS. The Station would operate at an altitude of approximately 240 nautical miles in a 51.6 degree orbit. When PHC was achieved, it would accommodate a crew of six. The solar dynamic power system now was listed as an option and a forth U.S. solar array module was baselined. The Ariane V launcher was also in the technical baseline as an alternative launch vehicle for the ESA lab; however, whether or not it would be used was still under study.[56]

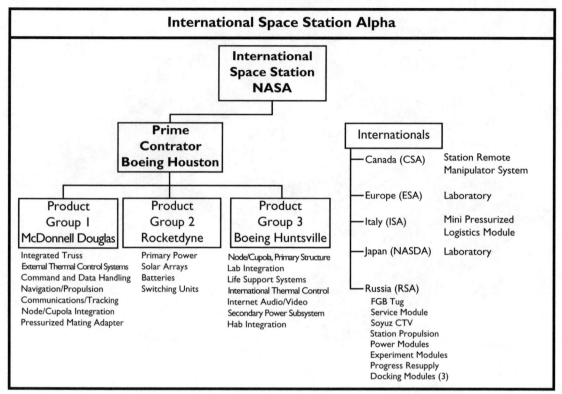

Figure 28. International Space Station Alpha management implementation structure.

The SDR confirmed that the FGB tug would be the first element launched, with a projected date of November 1997. Following the launch of the tug, the major elements in the assembly sequence went as follows: the Russian docking compartment, U.S. Node-1, Russian service module, joint Russian / American airlock, Russian universal docking module, Russian science power platform-1 (SPP), Russian SPP-2, and Russian SPP solar array. After a logistics flight, the tenth flight would carry the U.S. lab, then the Canadian RMS, the Soyuz ACRV, and the first U.S. truss would be on the fourteenth flight. After eleven more assembly and logistics flights the JEM would be launched and seven flights later the ESA *Columbus* module (with a new acronym APM) would be mated to the Space Station.

To arrive at assembly complete in June 2002 would require forty-one assembly and logistics flights – twenty U.S., twelve Russian, two ESA, and seven "utilization" flights (launcher unidentified). Human tended operations would begin at the tenth flight, when the U.S. lab arrived. How all the pieces would come together and be managed was briefly discussed (Figures 27 & 28). Two weeks after the SDR a "block update" was signed off incorporating all the revisions made during the SDR. After the block update, the baseline configuration would be frozen and any changes tightly controlled.[57]

Other results from the SDR were a reduction in EVA hours during assembly and the disclosure that the Station would have a ten-year operational lifetime versus SSF's twenty or more year projection. Ground control, or mission operations, was indicated as building "on the interfaces for the shuttle and Freedom programs." At the close of the SDR it was reported that "only 17 issues remain open." The next major milestone, the CDR, was scheduled for April 1995.[58] These latter items, described by

---

56. SDR discussion based in part on two reports: NASA News Release 94-53, March 24, 1994, and The International Space Station: Engineering the Future, Status Summary, March 1994. There was a major difference in the number of flights that would be needed to achieve assembly complete in the material distributed after the SDR. The NASA News release and the NASA press conference held on March 24th at JSC both stated that 29 flights were required, 16 U.S. and 13 Russian. The assembly sequence contained in the next reference, issued immediately after the SDR, listed 41 flights. Before the SDR NASA was estimating that 90 flights might be required.
57. Ibid.
58. Ibid.

Trafton and Brinkley in a press conference on March 24th, raised many questions. First, the operational life time of ten years cut in half the original plans for the Space Station. The cost-benefit of such a large expenditure, now advertised to cost $17.4 billion from FY 1994 to FY 2002, plus the already expended amount of some $11 billion, would be hard to defend in Congress for a facility that would operate only ten years. What additional funds would be required to operate the Station for the five years after 2002 had not been divulged. At a minimum one would expect it to be $6.5 billion, based on the earlier NASA estimate of $1.3 billion per year for annual operations costs not including shuttle launch costs.

Perhaps most surprising was the low number of open issues. For example, ground control of the Station, as well as who would be designated to command the crews representing many nations, were becoming points of contention with the Russians. Trafton explained the low number of open issues by saying that the designs were very mature and a high percentage of SSF elements were still part of the redesign. But if one read the report closely, and used history as a guide (PDRs and CDRs from earlier designs), one could expect that the CDR in 1995 would find many problems that should have been identified and resolved during the SDR. At least one group felt the SDR was successful. NASA reported that the "Vest Committee" met in Houston at the end of the SDR and, based on the briefings they received, agreed that the program had made significant technical and managerial progress.

Another example of where the SDR may have given an overly optimistic review was the electrical power system. Prototypes of the flight units, known as the Cooperative Solar Array, were designed to combine flight tested Russian structures and mechanisms with U.S. solar array modules. The U.S. arrays would be sent to Russia in May where they would be attached to Russian designed structures. After assembly they would be shipped back to KSC for an October shuttle launch and rendezvous with *Mir*. The new array would be attached to *Mir* during an EVA and tested for some months before it was removed and brought back for a final inspection. All this would be done in two years and then, assuming all went as planned, the Space Station solar array flight hardware would be built to final specifications that would include any changes that might result from the test. Of special interest during the test would be a measure of the power degradation of the U.S. photovoltaic cells that would be on the solar arrays planned for the Space Station.

Anyone who has dealt with space solar arrays, usually stowed for launch folded like a accordion or rolled up like a window shade, knows how often problems are encountered when they are finally deployed in space. There was a long history of problems encountered by Skylab, several commercial satellites and unmanned probes when attempting to deploy their solar arrays. Although the basic design was inherited from SSF, there would be a number of changes to incorporate the Russian components. Yet there was no mention of any concern or "open items" for the power system. There is an interesting side note to the Cooperative Solar Array test. When the Russian Progress supply vehicle collided with the *Mir* in June 1997, rupturing the hull of the *Spectr* module, it bounced off and struck and damaged the Russian solar array shutting down power to *Mir*. Fortunately, the NASA / Russian Cooperative Array attached to *Mir* for testing was not damaged and was used to repower the station.[59]

The Congressional Budget Office (CBO) issued a report the same day that NASA completed the SDR. Titled "Reinventing NASA,"[60] CBO gave a gloomy forecast of NASA's future when forced to live with decreasing budgets, as appeared almost certain. The most quoted line from the report was: "Cheaper, Faster, Better is unlikely to yield major cost savings at NASA." CBO proposed three alternative strategies that might work for NASA within a constrained fiscal environment: (1) A program that emphasized piloted space flight at a sustained budget of $14.3 billion annually; (2) A program that emphasized robotic space science at a budget of $11 billion a year and included piloted spacecraft only for scientific purposes; and (3) A program budgeted at $7 billion annually that eliminated piloted space flight and emphasized robotic space science and developing new technology

---

59. Based on a interview, October 1, 2004, with Dr. John Dunning, LeRC engineer, who continued collaborating with JSC on the solar array design after WP #4 responsibilities were transferred to JSC as a result of the redesign management changes.
60. Reinventing NASA, Congressional Budget Office, March 24, 1994.

for both private industry and public missions. CBO acknowledged that the three "are neither the only options nor necessarily the best ones for NASA as it attempts to adjust to the lower budgets."

CBO did not have too many kind words for the Space Station program. Among other observations it said: "the new cooperation with Russia . . . carries both risks and rewards." Further, CBO stated "Whether the current estimates of costs hold up will not be known until late summer 1994, when it was expected that the final contract with the prime contractor, Boeing, would be negotiated. Integrating U.S. and Russian hardware, computer software and operating procedures could prove difficult" – warnings that had been heard several times before, most recently from Kranz and Leetsma. The Report also pointed out NASA's recent track record for staying within projected costs, saying it had "a strong tendency to underestimate the costs of its projects." Thus, before Congress decided on what the NASA FY 1995 budget would be, its budget oversight arm was adding fuel to the fire for those members who would cripple or try to terminate the Space Station program.

Goldin's response to the CBO report was immediate. "Any of the three alternatives put forth in the CBO report would destroy the essential balance between human space flight, space science and leading-edge aeronautics. They also fail to factor in the tremendous termination and transition costs associated with shutting down a major portion of the space program, not to mention the potential for enormous economic dislocation."[61]

Commenting on the report, Chairman Brown said: "One option Congress may need to consider before the year is out is whether it is wise in the long run to terminate the space station program and focus our resources more sharply on a narrower set of space goals. The CBO report we are releasing today paints very clearly the issues to be weighed in this debate, and the real risks we as a Nation run if we decide we can 'have it all' – space station and a robust space science program."[62] NASA oversight committees were now saturated with reports and advice; the upcoming hearings would decide the future of the program.

---

61. NASA News Editors Note: N94-27, March 24, 1994.
62. SS&T News, Committee on Science, Space and Technology, U.S. House of Representatives, 2320 Rayburn House Office Building, Washington, D.C. 20515, March 23, 1994.

## — CHAPTER 7 —
## THE INTERNATIONAL SPACE STATION (ALPHA)
### (APRIL 1994 - DECEMBER 1996)

Officially, as of April 1994, the new name for the program was the International Space Station. However, some, including members of Congress and the prime contractor, Boeing, called it the International Space Station Alpha, and it appeared as such in Boeing documents. A trivial difference but, perhaps, a reflection of the mindset of Boeing after Trafton announced the fallback plan. If the Russians were unable to deliver on their agreement, NASA would proceed with the earlier Option A-1, Alpha design. The terms of Boeing's contract were still being negotiated with a major disagreement over the price. Boeing was reported to be asking for $800 million more than NASA thought the job required. Boeing was clearly hedging its commitments, because of the many new requirements that had been added, since they were selected as the prime contractor, as a result of the complexities added by Russian participation. With all the other contractor teams now reporting to Boeing there was further uncertainty in what the costs would be until it had the opportunity to renegotiate the novated contracts. NASA's contract negotiations with Boeing would drag on for many months.

After the SDR was completed, underlying problems surfaced that were the result of bringing Russia in as a partner. The first to draw attention was the once-a-day shuttle launch window of five minutes, the result of the requirement to reach the 51.6 degree orbit inclination. A normal launch window was on the order of two hours (some were longer, some shorter) which allowed KSC engineers time to study any problems that might arise, solve the problems, and then proceed with the launch. John Young, NASA's most experienced astronaut, said, "To have a five-minute launch window, you have to change the way we operate in a lot of ways. The vehicle is capable of doing it. The people are capable of doing it." A review of 111 past countdowns of all types of launches, not just shuttles, showed that launches that were achieved within five minutes of scheduled liftoff occurred only thirty-one percent of the time. Since shuttle safety concerns were higher than for unmanned launches, they might be delayed more often than for a unmanned vehicle. Some of the current constraints that might be relaxed to increase the probability of staying within the five minute window were those dealing with rain, lightning, and wind conditions at launch. Launch managers at KSC and JSC might not be willing to reduce the current standards.

Another concern when dealing with such a short launch window is that if a shuttle launch were to be delayed for any reason, it could cost approximately $500,000 to reschedule, with most of the cost coming if there was a need to refuel the huge external tank.[1] NASA began to study the problem under the direction of Deputy AA for Shuttle, Bryan O'Connor. The cost impact of changes that would come out of this study, plus other required design modifications to permit the shuttle to routinely fly to a 51.6 degree orbit, was unknown, but some estimates were as high as $550 million.[2] Whatever the number would turn out to be, these costs had not been included in the Space Station runout costs and would have to be absorbed in some other program.

Communication blackouts were another worry that needed to be analyzed. With the higher inclination orbit, TDRS coverage resulted in a "zone of exclusion" of five minutes per orbit. With sixteen orbits per day, it meant that each day for eighty minutes it would not be possible to communicate or download experiment data. Not necessarily a major concern unless "Murphy's law" should strike while the shuttle was in an exclusion zone and a problem requiring "immediate" attention occurred. NASA did not like to leave such concerns uncovered.

---

1. Tight Shuttle Timing Raises Safety Issues, Liz Tucci, *Space News*, March 28-April 3, 1994.
2. You Should Have Gone Before We Launched, *Space News*, March 28-April 3, 1994. The "should have gone before we launched" refers to the possibility that, jokingly, one of the ways shuttle weight could be reduced in order to carry more payload, was to remove the shuttle head.

## INTERNATIONAL PARTNERS EXPRESS CONCERN WITH PROGRAM CHANGES

At the U.S. Space Foundation's 10th National Space Symposium in Colorado Springs in early April the international partners revealed some of their frustrations. All were facing tight budgets, just like the U.S. program. Jean-Marie Luton, Director General of the European Space Agency, participating on a Symposium panel, discussed the difficulties ESA faced to stay in the program in view of "an economic climate in Europe which has not been seen since the '30s. It is well understood that a program of this size and importance needs to be constantly scrutinized and debated. But there comes a time when you have to decide whether you want the project or not. I think we have now reached this point." He went on to say that the deliberations leading up to the next ESA ministers meeting "could be made much easier if there are no more doubts about the objectives and the political will to carry out a program that we see in the political debate in the United States, the leader in this effort." At the same Symposium, Masato Yamano, President of the National Space Development Agency of Japan (NASDA), said what was needed was "a robust and responsible forum in which to exchange information and coordinate plans."[3] This last comment indicated that, despite NASA's claim that the international partners were full participants in the redesign effort, some felt that their views were not completely accommodated.

And all was not proceeding smoothly with NASA's new Russian partner. With Russia now playing a major role, the Houston program office had not had the opportunity to establish a permanent liaison office in Russia. As a result, negotiations on any issue required teams to travel back and forth between the countries. An example of how this deficiency slowed negotiations was the disagreement over the amount the U.S. would pay Russia for their partnership. Russia had submitted a proposal, five weeks late, that was $250 million above the previously agreed $400 million. NASA, working toward a June 15th deadline to complete the contract to satisfy Congress, was behind schedule and under pressure to accept the additional amount. Another issue that would require close cooperation to resolve, discussed earlier, was back on the table. Which mission control center would control Space Station operations? Its final resolution would drag on for many months, as you will read.

At a hearing of the House Space, Science and Technology Subcommittee on April 13, these and other issues were debated. Responding to the Subcommittee's request, NASA submitted answers to 100 specific questions on Russia's role. One of the questions dealt with where mission control would be located. Before the U.S. lab was launched on flight ten, seven of the nine launches would be Russian hardware forming the "core station." Logically, at least during this phase, it would make sense to control the Station from Russia. To complicate this issue even further, before PHC the Russians wanted a crew of cosmonauts to be continuously on board for maintenance. In its response NASA wrote: "RSA feels strongly that they need a crew on the Station at the earliest opportunity to tend their equipment and perform maintenance tasks. The U.S. position is that the crew should be made up of astronauts of all the international partners. If Russian hardware requires early human presence, then we should put an international crew on board and use a Soyuz or a modified Soyuz to perform the ACRV function."

In response to a related question, NASA said that while the Russians had agreed in principle that the Station would be operated as a single integrated vehicle, it had not agreed to NASA controlling it from the JSC Mission Control Center. Russia wanted to control their elements from their mission control center outside Moscow.[4] These were not minor points of disagreement. If Russia insisted that only they would have the expertise to control their elements there was only one solution, Russia's mission control must be in command. And if there would be a high maintenance workload, clearly cosmonauts would be better trained and in a better position to understand what was needed. Issues such as these would have to be ironed out quickly or the disagreements would fester and jeopardize the relationship that was coming under increased scrutiny and criticism in Congress.

## SDR RESULTS IN NEW COST ESTIMATES – CONGRESSIONAL REACTION

Ever-changing budget numbers and other modifications in the program, some described above, began to catch up with NASA management. Congressional frustration over what it was hearing from NASA

---

3. *Aerospace Daily*, Station budget fight overshadows partners' concerns, April 7, 1994.
4. Some of this discussion was extracted from *Aerospace Daily*, Houston Station office doesn't have Russian liaison as issues stack up, April 15, 1994.

management took center stage again. At the April 13th hearing before the House Science, Space and Technology Subcommittee, Goldin told the members that the estimated cost of the International Space Station (ISS) through FY 2002 had grown to $17.9 billion, $500 million more than he had estimated in February. He also said that the $17.9 billion level would include a reserve of $3.7 billion. Neither of these numbers were in the totals that Trafton used when briefing the NASA Advisory Council (NAC) the next day. To say there was confusion at NASA's top management would be an understatement. Goldin testified that NASA also had $980 million in unresolved costs. Responding to this new information, Ralph Hall (D-TX) told Goldin that Congress would not tolerate such large budget problems and said, "I cannot impress upon you the need to resolve these overruns as quickly as possible."[5] However, simple addition of all the known differences and new demands that were being tallied behind the scenes as a result of the new design actually placed the potential overrun at more than $1.5 billion, but this number was not given to the Subcommittee.

In view of the Space Station budget controversies, Chairman Brown, one of the program's strongest supporters, told the Administration that he was preparing legislation to terminate the Space Station program and redistribute the funds freed up to other NASA programs. His rationale for this move was a fear that if the program were canceled, as seemed possible, without this legislation, the money saved would be removed from NASA's budget and given to various social programs. Whether or not he would have to take such drastic action would be known by May, when the committee allocations would be known.[6] The budget battles were becoming a war of attrition. NASA's inability to correctly estimate Space Station costs placed the program in the greatest danger of cancellation it had faced since its inception in 1984.

Two weeks after the SDR, Trafton briefed the NAC on the status of the program. Many questions were asked resulting in interesting answers, or no answer. A few of the major subjects covered during the briefing: the baseline design had not yet been frozen but some hardware was being built; the SSPO Science Advisor position had not been filled; command and control would be done from Houston with backup support from mission control in Kaliningrad; thirty-four flights would be needed to complete assembly (compared to the 29 announced after the SDR and the 41 listed in the SDR assembly sequence table); if for some reason Russia backed out of the program, it would only mean a one-year delay to complete assembly (summer 2003 versus June 2002) based on the Alpha design; the original international partners, plus Italy, planned to spend a total of $8.95 billion to complete their contributions; and the $17.4 billion estimate ($500 million lower than the number Goldin had just given to the Subcommittee) to complete the U.S. portion now included a $2.3 billion reserve ($1.4 billion less than the amount Goldin provided Congress one day earlier). There was also a $194 million carryover from FY 1993.

NAC Chairman Bradford W. Parkinson of Stanford University asked several questions. In response to his question "What two things worry you most?" Trafton responded, "The Russian request that was $600 million higher than the agreed to $400 million" (this number was $350 million higher than that reported to Congress), and "the schedule to launch Node-1 by December 1997." One of the reasons for this latter concern was that KSC now estimated that they would need six months to process the Node after delivery. Parkinson asked how much slack was in the schedule but Trafton could not provide an answer. The short, ten-year operating lifetime was discussed and when asked how the Station would be deorbited, Trafton did not have an answer. He did indicate that the program was exploring how to extend the Station's lifetime. Trafton's answers or no response left many NAC members with an uneasy feeling about the program's future.[7] What Trafton's briefing revealed was NASA senior management's disarray and lack of control over critical aspects of the Space Station program. These deficiencies would afflict the program up to first element launch, even when new managers were appointed.

During a hearing on April 20th before the Hall subcommittee, State Department representative James Collins, who was testifying in favor of the Space Station, became the person to jump on for the

---

5. Goldin Tells Congress Station Costs in Flux, Liz Tucci, *Space News*, April 18-24, 1994.
6. Brown Trumpets Charge Against Station, Andrew Lawler, *Space News*, April 11-17, 1994.
7. The discussion of the NAC meeting is based on notes taken by the author and Trafton's presentation: The International Space Station, Status Summary Briefing for NASA Advisory Council, Wilbur Trafton, Director Space Station Program, April 14, 1994.

Administration's decision to bring Russia into the program. Representative Sensenbrenner, showing his increasing dissatisfaction, called Russian participation "The most goofed-up foreign policy . . . that I have ever seen." He added, "Unless this deal is renegotiated (removing Russia from the critical path) so that we can make this an American space station, the time has come for the Congress to cut our losses . . . and vote against it when it comes up in June."[8]

NASA, attempting to convince Congress before the June committee votes that it had a handle on Space Station costs, made public new cost control measures. Daniel C. Tam, newly appointed to lead SSPO's Business Management Office, during an interview with Liz Tucci of *Space News*, described some of the measures that had been instituted.[9] Among the steps being taken were fixed-price contracts to reduce costs for standard subcomponents, closer monitoring of fees based on satisfactory performance, encouraging contractors to lower overhead rates, and reducing the number of contract changes. The latter practice, commonly required for many government projects, is perhaps the biggest reason for cost overruns, as it allows contractors to add on costs of doubtful affiliation to the project or balloon the cost of the changes. Another effort to save money would lower quality specifications for some hardware instead of requiring the usual NASA practice of demanding the highest possible certification levels for all parts. This latter position could not have sat well with the JSC engineers. That was Tam's good news. The bad news was that canceling Space Station Freedom came with contract termination costs that he estimated could be as much as $100 million, probably an under-estimation.

Funding problems were not confined to the U.S. and ESA efforts. Canada informed NASA in March that some Canadian political decision makers were pushing to withdraw from the program. By April, high-level discussions between the U.S. and Canada had arrived at a compromise that would reduce the costs to Canada for some of its contribution, with the U.S. absorbing the difference. The final terms of the agreement were still being worked out with a resolution expected in May. How much this new relationship would cost the U.S. would not be known until the agreement was signed.[10] The timing of this change did not improve the climate in Congress and gave opponents one more example of how fractured Space Station planning had become, with new revelations of change almost every day.

How did all of this dissembling and disorder in the Space Station program play in Congress? At the end of May Senator Mikulski, on the eve of the final House and Senate budget debates, published a long opinion piece in *Space News*.[11] In it she briefly traced the volatile history of the program and wrote, "The mood of the Congress with respect to space is tenuous. With respect to space station, it is downright cranky." Referring to the appropriations for which her subcommittee had oversight, she said, "Congress faces a $900 million shortfall between what is requested (by the Administration) and what is available . . . I don't know if the space station will make it this year . . . But if the station is killed in the House, because of fiscal pressures and other program demands in the VA-HUD bill, I do not think it will be restored by the Senate . . . Space policy should no longer be driven by the domination of space station. Or else every year the space program is just one vote from being put out of business [referring to the latest attempt by Senator Bumpers to cancel the program that failed by one vote] . . . This nation must establish a consensus on space policy and not lurch from one year to the next, or bet the farm on a single program." Adding many other thoughtful observations, Senator Mikulski clearly outlined NASA's problems. Although a strong supporter for Space Station, and NASA programs in general, she challenged the Clinton Administration: Decide where your priorities are, request the required funds, and make the case to the Congress. Together, the Administration and Congress would have to decide what the future held. Was anyone listening to the Senator's concerns?

Perhaps some were listening. In an interview published in *Aerospace America* magazine, W. Bowman Cutter, Deputy Assistant to the President for Economic Policy, answered several questions on how the Administration viewed NASA and the Space Station. "We are obviously aware that the space station is

---

8. Space station plan has a rough time, Sean Holton, *The Orlando Sentinel*, April 21, 1994.
9. Cost Control Measures Aimed at Saving Money On International Station, Liz Tucci, *Space News*, April 11-17, 1994.
10. Canadian Station Role Revised, Andrew Lawler, *Space News*, April 25-May 1, 1994.
11. Hawks and Doves vs. Space Station, Sen. Barbara Mikulski, adapted from a May 18 speech, *Space News*, May 30-June 12, 1994.

a matter of great contention and disagreement . . . Some think the station is basically a pork barrel issue . . . We concluded that it really is an important piece of infrastructure (a laboratory in space in which you will do certain scientific and technical things) . . . and you basically buy it all of a piece." In answer to the question "Will NASA have to predict what returns it expects from a program, and then have to deliver them by a specific date or risk losing its funding for that program?" Cutter replied, "I have watched the space station program evolve for a very long time. No matter how often you said to NASA that you are building a station that can't be bought, their reaction was always that there is more money available, and they had very little institutional interest in fundamental rethink of the problem . . . it is going to have to come to the realization that budget resources are finite, and it is going to have to find a way to do a better, faster job with the same or fewer resources."[12] Not exactly a ringing endorsement of the Space Station. In other parts of the interview he emphasized that NASA had a long way to go changing the way it did business as part of the Clinton Administration's push to "reinvent" government.

## RUSSIAN AGREEMENT MODIFIED – PROGRAM AND BUDGET CONCERNS CONTINUE

On June 23rd one of the problems hanging over the program was partially resolved. An Interim Agreement and a contract were signed between NASA and the Russian Aviation and Space Agency (RAKA). The Interim Agreement governed Russian participation in the Space Station Program until a Intergovernmental Agreement (IGA) and a NASA-RAKA Memorandum of Understanding (MOU) were signed. Negotiations on the IGA and MOU were scheduled to begin later in the summer and no date was forecast for when they would be completed. The Interim Agreement, however, contained important management covenants. It established a NASA / RAKA Program Coordination Committee to review design and development activities. RAKA became a member of the Space Station Control Board that controlled requirements, configuration and interfaces through assembly and initial operational verification. RAKA was also included on the Multilateral Coordination Board that would coordinate the operation and utilization activities. And finally, the Interim Agreement provided for the establishment of technical liaison offices in Houston and Moscow, the lack of which up to this point had severely hampered development of good working relations between the two countries.

The contract expanded the 1992 Human Space Flight Agreement and specified what the U.S. would receive in return for its $400 million over four years. For the moment, Russia was willing to live with this number. U.S. astronauts would spend up to twenty-one months on board *Mir* conducting the first long-duration experiments for U.S. experimenters. The shuttle would dock with the *Mir* station an additional nine times, bringing equipment, experiments and new crews. Solar thermal electric power generation was back in as a joint project and joint EVAs would be conducted. A new claim on the $400 million was the agreement that the U.S. would help fund initial development of the FGB tug.[13] NASA would pay RAKA $25 million for this work and then Lockheed Missiles and Space Company would negotiate a contract with Khrunichev Enterprise of Moscow to buy the tug with NASA funds.[14] Where all these monies would come from was not identified.

All during the month of June 1994 Congress continued to argue over if, or how, the Space Station should continue. On June 9th, the House Appropriations VA, HUD and Independent Agencies subcommittee sent forward a markup for NASA of $14 billion, reducing the Administration's request for human space flight by $127 million. The markup included $2.1 billion for Space Station. Other parts of the Human Space Flight line item would be reduced by the $127 million at the agency's discretion. As the first step in the budget process, Space Station proponents, most notably Chairman Brown, were encouraged that the $2.1 billion meant that the program could be saved as well as other science projects. The Subcommittee recommendation went on to the House Appropriations Committee and on the 22nd the $14 billion survived the second step. A full House vote, step three, was scheduled on June 28th and the knives would be sharper during the floor debates as the opponents were

---

12. Conversations with W. Bowman Cutter, Interviewed by Johan Benson, *Aerospace America*, May 1994.
13. NASA News Release 94-101, NASA and Russian Space Agency Sign Space Station Interim Agreement and $400 Million Contract, June 23, 1994.
14. Russia, U.S. Reach $400 million Agreement, Andrew Lawler, *Space News*, July 3, 1994.

determined to terminate the program. Even if it survived in the House, it would then have to face a compromise in a conference committee with whatever total the Senate thought appropriate, step four. Word was that the Senate was considering $13.7 billion as the most that NASA could expect, some $600 million below the Administration's request.

Before these votes were cast, new problems surfaced. Trafton announced that schedules were slipping. The U.S. lab might not be launched until December 1998, six months later than the date planned in April. Other elements might be delayed five to seven months. The explanation was – to reduce technical risk and costs. Yet as part of the new plan he stated that testing would be reduced or eliminated on a wide range of hardware, and the number of spares reduced. How this would reduce risk was not explained. Reducing hardware had the potential to impact science and technology experiments as the ORUs were on the list of items to be reduced. Qualification and acceptance tests would be done at the factory and testing that normally took place at KSC to assure, among other concerns, that there was no damage in transit would be eliminated.

Some NASA managers said, off the record, that if the plan was approved the program would need additional funding in future years to fix problems due to fewer spares and reduced testing. Dick Kohrs, now an industry executive, said that when he was in charge his managers had considered reducing testing to save money but he had rejected the idea as too risky. "If you have done everything right it might not be a problem. But my experience was that hardware shipped by a contractor is never 100 percent ready for flight."[15] Other cost cutting recommendations were being considered and a final plan would have to be adopted soon so the program could stay within the $2.1 billion ceiling, assuming that was what NASA would receive for FY 1995. A lower number would require more deferments and severe choices to be made.

Once again GAO issued a Space Station analysis at a critical juncture. The study, requested by Senator William Cohen (R-ME) a Space Station opponent, said that Russian participation would not save the $2 billion claimed by the Administration and NASA, but rather would add $1.4 billion and in the long run "result in little or no net savings." The GAO study said Russian participation added $258 million for shuttle fleet modifications, $400 million for the recently signed contract, and $746 million for two additional shuttle flights. Senator Cohen described the program as a "financial black hole." For NASA, the alternative to the "black hole" created with Russian participation was positive. Goldin said the GAO report proved it was a "win-win" and, "The GAO report is a validation of what we have been saying for months – Russian participation in the Space Station is a good deal for the American taxpayer."[16] Which of these diametrically opposed interpretations would be believed by the Congress would soon be known.

The "win-win" side of the black hole won. In early July the House agreed to a $2.1 billion appropriation followed soon after by a similar Senate appropriation. The compromise bill emerged from the conference committee on August 18th and was quickly passed by both chambers, providing Space Station with $2.113 billion in FY 1995. Critics would have to wait until next year to mount their attacks again. The course now seemed open to continue with the new partner and Station configuration, and maintain the schedule to launch the first element in November 1997. Meanwhile, the nine or ten U.S. / Russian missions to *Mir* would fill in some of the blanks.

However, at a meeting at JSC on July 12th, the Space Station assembly sequence changed again. In order to increase power to 13.5 kW, the launch of the U.S. photovoltaic power array would be moved up, replacing a utilization flight. As a result there would be a fourteen month gap in utilization flights that threatened the integrity of many of the first human-tended, life sciences and microgravity experiments. Their planned "shelf life" on board the Station could not tolerate such a long period before being returned to Earth. SSPO scrambled to find a solution. By the end of August there was another change: shuttle flights to *Mir* had been reduced to seven. The shuttle *Atlantis* would be the only orbiter configured to dock with *Mir*. Providing a docking mechanism for a second shuttle was

---

15. New Plan Poses Higher Risks For Station, Andrew Lawler, *Space News*, June 27-July 3, 1994.
16. This text extracted from two stories that ran at that time: High Hopes in the House, Andrew Lawler, *Space News*, June 27-July 3, 1994, and an AP story, GAO: Russia will not save much money on space station in *The Washington Post*, June 25, 1994.

estimated to cost $28 million and that money was not in the budget. The first flight to *Mir* with astronaut Norman Thagard on board was now scheduled for March 1995 aboard a Russian Soyuz. The first shuttle docking was tentatively set for May 1995, with the other six missions scheduled every five to six months. The primary goals announced for the missions were to test command, control and communication procedures and prototype Space Station hardware. If all went as scheduled, U.S. astronauts would log up to two years on *Mir*.

As summer turned to autumn, more changes and schedule slips were announced. Ten months after Boeing and NASA had signed a letter contract they still had not reached agreement on a final contract, and Boeing was in the same position with its subcontractors. Disagreements still existed on what the fee would be and the total cost. And, if NASA made any changes in the baseline design or assembly sequence, NASA would have to find the money to cover these additional costs. Considering the recent history of the program, that requirement had to be a troubling thought. A new problem entered into the discussions. Pamela McInerney, Trafton's spokesperson said, "We're providing Boeing with reassurance that should the program be canceled, they are not left holding the bag." Negotiations with the international partners had also dragged on and probably would not be completed until some time in 1995. An industry official was quoted as saying, "Everything is sliding. The program is simply too big and too complex."[17]

### RUSSIAN FUNDING TRANSFERS BECOME A PROBLEM

Financial problems in Russia that would effect the Space Station partnership began to surface. Paychecks were not being received by Russian workers and many were on unpaid leave. U.S. officials were quoted as saying that the persistent budget crisis and government infighting could jeopardize Russia continuing as a Space Station partner. In October, Congressman Sensenbrenner was authorized by the Committee on Science, Space and Technology to visit the European partners and Russia and report back his findings. His trip, covering five days, took him to Paris and then on to Moscow, reprising his January trip. His cover letter to the report made the following observations: "(1) While significant progress must be made before consideration of the Fiscal Year 1996 NASA budget, I have much greater confidence in the overall space partnership today than I did last January . . . I have confidence that negotiations and technical discussions can be concluded prior to consideration of the next NASA budget request. (2) In terms of partnership arrangements, however, three areas continue to cause some concern. First . . . the European Space Agency has yet to formally make political and financial commitments to the technical program . . . Second, Russia and the United States have not yet agreed on a Memorandum of Understanding that would clearly outline Russia's partnership role in the program, . . . Third, the U.S. / Russian MOU must be agreed to before Russia will sign the multilateral Intergovernmental Agreement of all apace partners. To the extent technical progress was evident, work on Phase 1 of the U.S. / Russian cooperative agreement appears on track. The March 3, 1995 launch date has been kept for U.S. astronaut / physician Dr. Norman Thagard, despite a potential three-month delay of the U.S.-sponsored Russian *Spektr* module." In the final paragraph of the letter he wrote: "it was neither appropriate nor possible for the CODEL (Congressional delegation) to determine the extent of U.S. dependence on Russian elements at the time of the visit."[18] This last statement was very puzzling in view of Sensenbrenner's often expressed complaint that Russia was on the critical path.

The delegation did not visit Baikonur where, on Sensenbrenner's January visit, he expressed concern about the decrepit condition of the city and poor living conditions of its workers. Thus there was no

---

17. NASA, Industry Struggle To Meet Station Schedule, Ben Iannottta & Andrew Lawler, *Space News*, September 19-25, 1994.
18. The purpose of the trip was to "update Appendix B of the Chairman's Report, 'Oversight Visit; Baikonur Cosmodrome,' March 23, 1994, (House Report 103-451), which enumerated a number of concerns with the proposed program plan and associated agreements between NASA and the Russian Space Agency (RSA) to develop and operate the international space station, in partnership with our long-standing space partners in Western Europe, Canada, and Japan." The report was titled, Update Investigation of U.S.-Russian Space Cooperation, October 17-21, 1994. Report of The Hon. F. James Sensenbrenner, Jr., Ranking Republican Member, Subcommittee on Space, Committee on Science, Space and Technology, U.S. House of Representatives, Washington, DC 20515. The date of the cover letter was December 1, 1994.

discussion of the overall decaying situation of the work force, which was general knowledge to all of the recent U.S. visitors and a potential impact on delivering the Russian elements on schedule. Perhaps attending the several lunches and dinners hosted by their Russian counterparts dulled the senses and blunted the pens of the U.S. delegation. As anyone who visited the former Soviet Union or Russia on official business remembers, lunches and dinners were accompanied by many friendly toasts, even during the "cold war" era.

Soon after Sensenbrenner returned with his generally upbeat assessment of the Russian effort (although his report described several additional concerns not listed above), negotiations with the Russians on the price of the FGB tug broke down. The Russians were asking for $245 million to supply the critical path first element to be launched. Recall that Lockheed, as a Boeing subcontractor, was responsible to deliver this element teaming with Khrunichev Enterprises. Larry Winslow, Boeing's Space Station manager, told reporters at the Houston Space Exploration '94 Conference, "The Khrunichev offer is way, way out of the ballpark." This was in spite of the fact that NASA had agreed to pay $25 million up front to help refine the design for Space Station application. Of note, according to the report of Sensenbrenner's October visit, he never had a chance to look at the FGB as he had for other Russian elements being fabricated; he was just briefed on its status. Why his delegation received this different treatment was not explained in his report, but the subsequent revelation of problems with the FGB might be the answer.

## ELECTIONS BRING CHANGES TO CONGRESS – PARTNERS ADD CHANGES

Before the Sensenbrenner report was officially released, Washington's political world turned upside down. After forty-five years (1949-1994) of Democrats controlling the House of Representatives, the November 1994 elections returned the Republicans as the majority party. Sensenbrenner would become the Chairman of the Committee on Science, Space and Technology when the 104th Congress was sworn in, replacing longtime chairman George Brown. All the other House committees and subcommittees would also change leadership. A similar change took place in the Senate, although not as earthshaking, since the Republicans had controlled the Senate during Reagan's first term. The implications for NASA, Goldin and the Space Station program would be revealed in the spring of 1995 when Clinton submitted his first budget to a Congress not controlled by his party. Building relations and educating new members on both sides of the aisle would be a top priority. Both Space Station supporters and opponents had been replaced by 86 new House members and 11 new Senators. Overshadowing the forthcoming debates was the Republican Party's 1994 platform "Contract with America," which promised to strengthen national defense while at the same time reducing the size of government and balancing the budget. This would be difficult to achieve without reducing the amount spent in "discretionary" programs.

After the elections more bad news surfaced. At the ESA Council meeting at the end of October, France notified its partners that it would not share the funding, as promised, to develop a data management system that would be installed on a Russian Space Station module. Two of ESA's partners, Germany and Italy, scrambled to take up the slack, but a solution was not arrived at, leaving the ESA contribution some $40 million short. France's change of heart was reflected in other contributions it had agreed to make. Some thought there was a possibility that France would withdraw support in building the *Columbus* module, ESA's flagship contribution to the Space Station. An official at CNES, the French space agency, was quoted as saying, "It is clear that if we could leave the space station gracefully, we would do that. Our preferred position is that our contribution level drop to zero. But this is difficult to do because we want to be good Europeans." CNES' Director General, Jean-Daniel Levi, explaining the French position said, "ESA was asking for commitments that go beyond the original three-year period (1993-1995) that we had agreed to."[19] Sensenbrenner's report had hinted at these difficulties when he listed as a concern the lack of a signed MOU with ESA to nail down partner contributions. A month later his concern turned into a potential major problem that would not be resolved for many months.

---

19. ESA Funding Insufficient To Meet Station Goals. Peter B. de Selding, *Space News*, November 14-20, 1994.

Added to the above problem, there was a delay in delivery of experiment equipment for U.S. astronaut Thagard's first stay on *Mir*. Thagard planned to conduct a number of tests to determine how productive astronauts could be during long-term exposure to weightlessness, and compare that information with Russian data that was being shared for the first time. The equipment, almost one ton of various devices such as treadmills, designed to assist him in his tests, was to have been launched on a Russian Spektr before he arrived. But because of problems in clearing Russian customs, the equipment was delayed too long to make the Spektr launch.

As a result, Thagard would be launched without the equipment already at *Mir*. NASA would have to devise other types of tests to take the place of those planned, otherwise Thagard's stay would be unproductive. The alternative, waiting for the equipment to pass customs and be flight certified, estimated to take three or more months, was unacceptable and would have disrupted experiments scheduled after Thagard's flight. Passed off as a minor misunderstanding, Goldin explained, "There's nothing wrong with Russian customs, but we don't know how to work with them." A work-around was devised; 300 pounds of new equipment was assembled for launch on a Russian supply ship to *Mir* in February, one month before Thagard was scheduled to join the *Mir* crew.[20]

## PROGRESS, BUT OLD PROBLEMS PERSIST

At the end of November 1994, Boeing published a new assembly schedule.[21] Six of the first ten elements launched would be on Russian launch vehicles. Four, ending with the U.S. launch of the lab module, would be on the shuttle. A total of 44 launches, with first element launch in November 1997, would be needed to complete assembly, ending with the outfitting of the U.S. hab module. The ESA *Columbus* module, now called the Attached Pressurized Module (APM), would be launched on an Ariane rocket. The difference in the required number of launches between what NASA originally advertised, thirty-four, and the Boeing prime contractor number was that NASA only counted assembly flights carrying individual elements while in reality an additional ten supply flights were also required. Thirty-four certainly sounded better than forty-four. Another nineteen flights of Russian supply vehicles would be necessary to transfer fuel to the FGB for Station reboost and were not included in the Boeing schedule. EVA time to complete assembly was now estimated to be 500 hours versus the estimate of 434 in early 1994. EVA hours for cosmonauts were not included and were expected to run several hundred hours. Total weight of the ISS on orbit would be approximately 440 tons.

At assembly complete, the ISS would measure 355 feet by 244 feet. It would consist of eight modules providing 46,200 cubic feet of pressurized volume in which the crews would live and work. Atmospheric pressure would be maintained at 14.7 psi. Outside the pressurized modules, a number of experiment payloads would be attached to the trusses to be serviced during EVAs. A total of 110 kW of power generated by the solar arrays would be available for experiments and housekeeping, with approximately 46 kW reserved for experiments. If the schedule held, the U.S. lab module would be launched at the end of 1998, the JEM in 2000, and the ESA APM in early 2001. The Canadian RMS would be attached at the end of 1998. On flight thirty-six a centrifuge module would be attached to Node-2. On flight forty in 2002, the U.S. hab module would arrive. As soon as the second ACRV was attached on flight forty-three, the ISS could accommodate a crew of six. Italian modules, carrying supplies or attached as working space, would be launched after assembly was completed.

A new Congress would have to be dealt with in 1995 and then an even more difficult problem would have to be addressed. An edict from the White House was delivered on January 12th to come up with a strategy to cut $5 billion from NASA's budgets over the next five years. When carried out it meant that NASA's annual budget would be approximately $13.2 billion by FY 2000.[22] If one looked at the budget reductions strictly from a Space Station perspective, this was double trouble. The probability of being

---

20. U.S. / Russian Space Mission Stumbles, With Delivery Delays at Customs, William J. Broad, *The New York Times International*, November 29, 1994.
21. International Space Station Alpha, Reference Guide, November 30, 1994, Prepared by: Boeing Missiles & Space Division, Defense & Space Group.
22. Goldin Pledges To Fulfill Vision Fighting Budget Cutbacks, Criticism and Low Morale, NASA Chief Forges Ahead, Ben Iannotta, *Space News*, April 3, 1995.

able to stay on schedule in the next few years within a $2 billion level budget, while inflation reduced buying power, seemed slim. On top of that, with the overall reduced agency budget, it would give critics more ammunition to cancel the program in order to continue other popular programs that also would be effected by the lower total budget (by 2006 the contract value would grow to $13.3 billion).

Although there were still many loose ends to be pulled together in 1995, as the new year started some progress could be shown. NASA and Boeing finally agreed in January on a $5.63 billion contract. Under the terms of the contract Boeing would complete assembly by 2002 and then support the program for one additional year. There was the potential for Boeing to receive fees of up to $500 million if the program stayed on schedule. What remained to be done? Boeing would now have to negotiate with its subcontractors to determine their payments within the remaining $5.13 billion, not an enviable task as it would mean lower contract values for the subcontractors.

The Administration's budget request for the Space Station in FY 1996 was $2.115 billion, maintaining the promised level of funding. Negotiations on the price of the FGB were continuing, with some narrowing in the difference of opinion on the price, but as the months dragged on it was becoming more critical to get an agreement on the first element to be launched or the schedule would slip again.

ESA continued to debate what their contribution would be in light of budget problems. In the latest discussions two elements were under consideration as potential contributions. A crew rescue vehicle would be canceled and the development of a automated transfer vehicle would be postponed for fifteen months. The APM launch on an Ariane would be delayed six months until late 2001.[23] There was also disagreement on how to split the annual maintenance costs among the partners. A meeting that would resolve these issues was not scheduled until October.

But smooth sailing was never in the cards very long for the program. After rendezvousing with *Mir* (but not docking, only *Atlantis* was configured for docking), *Discovery*'s February mission included conducting EVAs in space suits that had been redesigned for use with the Space Station. The modifications included improving mobility and new insulation in the gloves to allow astronauts to grip Space Station elements in shadowed areas where temperatures might reach minus 125 degrees F. Mobility improvements were needed so the astronauts could position pieces of equipment with masses of many tons. While moving a 2,800 pound satellite in *Discovery*'s cargo bay, astronauts Bernard A. Harris and Michael C. Foale found that their hands became too cold to continue the test. As a result the EVA was terminated without accomplishing all the objectives.[24] Improving space suit gloves had been a technology challenge since Apollo days, so it was back to the drawing boards to devise a solution.

EVA time for Space Station assembly would not go away as a problem, reminiscent of the EVA uproar that occurred in 1990. Ignoring the problem of designing a space suit that would permit the astronauts to more easily carry out their assigned tasks, the estimate of required EVA assembly time began to grow again. The new estimates were 648 hours for U.S. astronauts and 240 hours for cosmonauts. In addition, another 171 hours would be needed each year to maintain the Station.

NASA attempted to defuse the concern over the growth in EVA hours. First, NASA EVA experience on tasks such as repairing the Hubble telescope showed that complicated procedures could be done with a minimum of problems. Although always carefully planned and trained for, NASA said EVA was becoming routine. To improve training and suit design, JSC had consolidated the shuttle and Station EVA offices. Flight rules were being revised to assure the safety of EVAs, including limiting the maximum length of time, number of EVAs per astronaut per flight, and amount of rest time between EVAs. Rules were also being drawn up for EVAs originating at the Space Station when a shuttle would not be docked. Use of robotics to reduce EVA hours continued to be studied and the missions to *Mir* would provide opportunities to test techniques and equipment before being needed at the Space Station.[25] But in the last analysis, providing a space suit that would live up to the

---

23. ESA Makes Cuts, Delays to Space Station Pledge, Peter B. de Selding, *Space News*, February 6-12, 1995.
24. For those interested in reviewing all the shuttle flights including crew names, dates, mission objectives and results, this can be done by accessing the NASA website: www.nasa.gov. Using "Find it @NASA" search for STS Missions. Click on "Documents – STS Mission Summaries". Select a mission when the list appears. Good luck! It worked for me.
25. Space walk Estimate To Build Station Soars, William Harwood, *Space News*, February 20-26, 1995.

requirements was the key ingredient. A need to cut costs in FY 1990 had eliminated the development of a space suit designed specifically for Space Station.

Economic conditions in Russia continued to imperil the program. Yuri Koptev told the Russian Duma in February that without more financial help the shuttle-*Mir* program was threatened with disruption or cancellation.[26] Inevitably, with a deeper understanding of what the new partnership entailed, problems were bound to arise. However, with each new problem, regardless of the origin or what needed to be fixed, the U.S. would receive a bill from Russia or the original agreed upon cost would increase. Russia needed cash to continue and the only place they could find it was in the U.S.

James Oberg, with close connections in both Russia and Houston, drew a dismal picture of what the Russian partnership brought to the table. Based on his accounts, U.S. officials were seeing incompetence, corruption and greed. Although tens of millions of dollars had been sent to RAKA, very little had been passed on to Kazakhstan. Visitors could not see any improvements in conditions for the engineers and their families living in Leninsk, the city outside the launch site at Baikonur. Oberg reported that a European astronaut who spent 30 days on *Mir* searched for some of the equipment sent up earlier for his use and never found it.[27]

Shuttle docking with *Mir*, scheduled for June, had become a larger than anticipated problem. Modifications were needed for both the shuttle and *Mir* for the docking. Vladimir A. Solovyov, Director of the *Mir* mission control center, was reported as saying combining U.S. and Russian work habits had been very difficult. The realities of the shuttle-*Mir* program were proving to be more complicated than first thought.[28]

As predictable as a satellite orbiting overhead, Space Station Congressional opponents renewed their attacks in May 1995. Led by representatives Richard Zimmer (R-NJ) and Tim Roemer (D-IN), five Republicans and thirteen Democrats cosponsored a bill to cancel the program and authorize $500 million for termination costs. Their strategy was to enlist the support of the many new House members who were elected in 1994 with the promise to balance the budget. However, in the early going, none of the 86 new members on either side of the aisle had signed on. New House Speaker Newt Gingrich (R-GA) had endorsed the Space Station in years past and reaffirmed his support in a February letter sent to European space leaders.[29] All indications were that the critics would not succeed in killing the program during the FY 1996 budget debates.

Throughout the remainder of 1995 the program made real progress. After the *Discovery* mission, four more shuttle flights that included Space Station related activities were successfully launched. In June *Atlantis* docked for the first time with *Mir* and transferred crew, leaving U.S. astronaut Bonnie Dunbar, and bringing home Thagard and two Russian cosmonauts. *Endeavor*'s mission in September was similar to *Discovery*'s – practice assembly techniques and test modified space suits. This time the astronauts did not describe any discomfort while holding cold objects. In October *Columbia*'s primary mission objective to conduct microgravity experiments as precursors to Space Station work was successful. And in November *Atlantis* returned to *Mir*, docking and changing out crews. *Endeavor* was on the launch pad at the end of the year awaiting a January 1996 launch to continue evaluating Space Station assembly techniques.

Negotiations with the Russians at the end of the year finally resolved the price of the FGB. NASA agreed to pay $190 million more for this critical element. When added to the first $25 million transferred to assist in the design, it came to a total of $215 million, $30 million below the original Russian request. Congress, now under Republican rule, focused most of its disagreements on domestic social programs. The early organized opposition to the Space Station fizzled out, and the Space Station FY 1996 budget sailed through the House and Senate and the program actually ended up with more

---

26. Russian Space Officials Decry Their Small Budgets, Peter B. de Selding & Anton Zhigulsky, *Space News*, February 27-March 3, 1995.
27. Grim Russian Realities imperil joint space effort. Incompetence, corruption assailed, Adnan Berry, *The Washington Times*, February 26, 1995.
28. Shuttle-*Mir* Dockings Pose Logistics Problems, Peter B. de Selding, *Space News*, February 27-March 5, 1995.
29. Station Foes Renew Attack, Ben Iannotta, *Space News*, May 29-June 4, 1995.

than the original request, $2.244 billion. Boeing successfully negotiated contracts with the subcontractors and, for the moment, their $5.613 billion budget appeared to satisfy all the participants. In retrospect, the last six months of 1995 probably raised the fewest controversies of any previous six-month period.

But as the year wound down new problems, or the old problems dressed in a different disguise, arose with the Russians. Their efforts were running on fumes, and rubles were not being made available to refill the tank. Blaming political and economic difficulties, NASA was told by their Russian partners they had bitten off more than they could chew. They suggested rather than shut down the ten year old *Mir* in 1997 (as currently planned), use it as the core of a new international space station during the first few years of construction thus reducing their required contributions at the beginning of assembly. The U.S. team meeting with the Russians agreed to take the proposal under consideration. However, Goldin rejected the idea and was quoted as saying, "That's the one non-negotiable item [proceeding with the currently planned ISS]." The U.S. counterproposal was to allow the Russians to delay all of their contributions except for those elements critical to the completion of the ISS. If they agreed to this change, Goldin said, "This saves them a barrel of cash."[30] Saving a barrel of cash might not be enough, without additional concessions and the U.S. absorbing more of the load.

Russia's proposal to continue operating *Mir* beyond the agreed to deorbit date in 1997 and incorporate it into the ISS was still on the table as Goldin and Koptev prepared to meet at the end of January 1996. Sensenbrenner had gone back to Russia with another Congressional delegation at the beginning of January to warn them that changes to the program would not be permitted. On returning, Sensenbrenner, in a letter to First Deputy Prime Minister Oleg Soskovets, expressed concerns about the slip in production of critical Russian elements and wrote "concrete action by the Government of Russia is required to ensure the ISS is built on time . . . Failing such actions to put the Service Module back on schedule, the U.S. Congress will be forced to conclude that ISS must be built without the benefit of our historic cooperation." [31]

Bryan O'Connor explained in an interview that NASA was attempting to identify options to add shuttle flights to *Mir* and reduce some of the Russian costs. "We're asking ourselves: Is there anything we can do to help? How much benefit do we get from helping? Is the problem worth it?"[32] A number of nagging problems, seemingly swept under the carpet during the earlier negotiations, were now front and center as the Russians sought ways to obtain additional funding from the U.S. For instance, Soyuz, planned as the ACRV, could not accommodate about half the U.S. astronauts, they were too big. It would have to be modified and that took money. To stay docked to the Space Station as the ACRV for a year, without losing any reliability, required other modifications. And a new request: Russia was now asking the U.S. to pay for logistic flights to *Mir* carrying food, clothing, and other consumables for the Russian crews, an estimated six to eight metric tons. O'Connor estimated that the incremental cost of adding a shuttle flight beyond the seven planned was approximately $50 million, plus the shuttle launch itself of about $500 million.[33] With the expected budget pinch and Congressional concerns about the Russian partnership, adding another shuttle mission and paying to supply cosmonauts seemed unlikely.

Negotiations with the Russians at the end of January resulted in more program changes. They included: extending *Mir*'s life until 2000 (this would allow Russia to continue to schedule experiments and be paid from hard currency countries for the use of *Mir*); delivering 6,000 kilograms of supplies on shuttle flights, and delivering the Russian science power platform on a shuttle, thus replacing three Zenit launches. Attempting to put the best spin on these changes, Trafton said this last modification would cause a small "ripple effect." The JEM launch would be delayed another five months, delivery of the

---

30. NASA Renegotiating With Partner in Space, Russia Lacks Funds to Keep Mir and Build New Station, Kathy Sawyer, *The Washington Post*, December 26, 1995.
31. Letter from the U.S. House of Representatives Committee on Science addressed to The Honorable Oleg Soskovets, First Deputy Prime Minister, Government of the Russian Federation, The White House, Moscow, Russia, signed by Jerry Lewis, Chairman Appropriations Subcommittee on VA, HUD, and Independent Agencies and F. James Sensenbrenner, Jr. Chairman Science Subcommittee on Space and Aeronautics, dated 8 March, 1996.
32. Shuttles Shuffled, *Aviation Week & Space Technology*, January 29, 1996.
33. Ibid.

centrifuge eight months, and a logistics flight with supplies or crew change-out, scheduled on the fortieth assembly launch, would be delayed until after assembly was completed. In return for these concessions, Russia agreed to foot the bill to modify Soyuz to fill the ACRV role.[34] In addition to these accommodations, Russia asked for more funds. Trafton said: "They wanted us to give them $100 million a year for the privilege of going to *Mir*. We rejected that." However, Goldin was quoted as saying that the new agreement would not require the U.S. to pay Russia "big bucks . . . we are not going to subsidize . . . *Mir* . . . But if there is a specific task that we need that we should pay for, we will."

Although stating that with these changes (total cost not revealed) NASA would still be able to adhere to the $2.1 billion spending cap, the door was open to provide Russia additional funds if they were needed to keep the partnership alive. The only question was, what would be sacrificed? The agreement added two shuttle flights to *Mir* through 1998. Using O'Connor's taximeter math, this would add a minimum of $100 million, plus whatever might be charged against the shuttle launch as unique to the new schedule. To meet the schedule two shuttles would be needed. *Discovery* would have to be modified like *Atlantis* for docking, another new cost added to the program. Counting the January launch of *Endeavor*, which would conduct Space Station-related experiments, five shuttle launches to support the Space Station were scheduled in 1996, three would be docking missions and one other would test new engines intended to improve performance and reliability.

### O'CONNOR RESIGNS – MORE MANAGEMENT CHANGES

At the end of February, Bryan O'Connor abruptly resigned as the Deputy Associate Administrator for Space Shuttle. His resignation was the result of a disagreement with Goldin on the management restructuring that combined shuttle and Space Station management at JSC. He was replaced on a temporary basis by another former astronaut, Stephen P. Oswald. It was reported that when O'Connor was told about the management change he appealed to Goldin to exclude the shuttle from the restructuring. Goldin did not agree with O'Connor's request and went ahead with the change.

In a public forum, O'Connor explained his reasons for objecting to the change. His primary concern was shifting the shuttle management office from Headquarters back to JSC. This would place the office under the direction of George Abbey, the new JSC Center Director who had moved back to Houston after assignments in Washington. One of the reasons for moving the shuttle office back to JSC was to save money. O'Connor said, "Even if all twenty-five of them (staff working in his office) went away that's not a big saving for the taxpayer compared to the risk we're putting on the program . . . Now the boss says we have a new NASA . . . We're not going to have disagreements between the centers as in the past. I disagree. In time the center-to-center rivalries will show up again."

He went on to recall that in the months before the *Challenger* accident, engineers at MSFC knew there were problems with the solid rocket boosters but they hesitated to reveal the problems with their rivals at JSC. Responding to the *Challenger* Accident Investigation Board's recommendations to rectify this internal communication gap, NASA had increased the Headquarters shuttle office's authority to referee between the centers. As head of that office O'Connor was the chairman of the Flight Readiness Review (FRR) and had the final sign-off before any manned mission launch. By making a senior JSC manager chairman of the FRR it removed the impartial check and balance to approving a launch. It just took seven years, almost to the day, for a shuttle tragedy to be repeated. Overruling O'Connor's concerns was even more surprising when placed in the context of the NASA Aerospace Safety Advisory Panel's 1995 annual report. It warned that NASA cuts in space shuttle personnel might impact safety and called for a team approach to restructuring the shuttle program. The team had just been reduced in size. (One could conjecture that it was this management move, made in 1996, that was a major reason for the communication problem that contributed to NASA management's inability to focus on the problem that contributed to the *Columbia* accident.)

NASA's Space Station budget request for FY 1997 was $2.149 billion. Hearings in the Senate went relatively smoothly except when the questioning turned to whether or not Russia would meet the schedule amidst rumors that construction had bogged down. Goldin told the Senate Commerce Space

---

34. U.S., Russia Reach Deal, Ben Iannotta, *Space News*, February 5-11, 1996.

Subcommittee on March 26th, "I would say if it [delays in construction] goes beyond a month or so we begin to lose an ability to hold a schedule for the Service Module." He said he was "cautiously optimistic" that Russia would deliver the module on time. If that did not occur, NASA would be forced to replace additional Russian hardware in addition to that already committed, estimated by Chairman Sensenbrenner to cost more than $300 million.[35]

When the House began its committee hearings in March NASA ran into all sorts of trouble. Chairman Sensenbrenner, of the newly named Science Subcommittee on Space and Aeronautics, had learned that the Russian government was late again in transferring funds needed to build the FGB module. Khrunichev State Research and Production Space Center, responsible for supplying the FGB, had received its first $10 million payment on time but had not received the next payments and was unable to pay its subcontractors. With the Service Module also in trouble that meant that two of the three first elements scheduled for assembly could be delayed. At a hearing on March 28th, Representative Roemer accused NASA of allowing Russia to manipulate the Space Station program. Goldin reiterated the position he had taken two days earlier with the Senate saying "other appropriate actions" were under way, and plans were being refined at JSC.[36]

The plans that Goldin referred to underway at JSC were different than the contingency plan described by Trafton a year earlier. Instead of falling back to Option A-1 (Space Station Alpha) as Trafton discussed, JSC and MSFC were studying the design of a new module that would substitute for the two Russian modules. Originally, as you may recall, the contingency plan, if the Russians did not supply the FGB, was to use Lockheed Bus-1 designed for DOD applications, but no substitute was identified for the Service Module. Scott Croomes, in the MSFC Space Station integration office, described the new 45-foot-long module that might be the replacement as being just in a conceptual design phase. It would provide a "stable platform from which we could continue to build the rest of the Station without the Russians being there."[37]

With such a late start and urgency to complete the module, the cost was uncertain but could be as much as $1 billion. Delays of nine months or more would probably occur in FEL, and the Station assembly schedule would have to be juggled to place the U.S. lab module on orbit as early as possible, since the new module would not provide living accommodations. If the Russians were to completely pull out, NASA might go back to a lower-inclination orbit for the Space Station. With these latest developments, it was not a very pretty picture for Space Station management to contemplate only nineteen months before the first scheduled assembly flight.

Because of the possibility of the late delivery of the first components, schedule concerns compelled NASA to modify the plans for processing Space Station components at KSC – the old integration and verification problem that the SSAC had been studying for Space Station Freedom. At the end of April, Randy Brinkley, the JSC Program Manager, announced that John "Tip" Talone and a team at KSC would provide government oversight for Space Station elements from the factory to the launch pad. He described Talone's job thusly: "He's responsible for management of overall ground processing, for manufacturing through successful launch, for reviewing the transfer of work from manufacturing sites to KSC."[38] In his new role, although a KSC employee, Talone would report to Brinkley at JSC. Brinkley was quick to say that this assignment did not reduce Boeing's contractual responsibilities, but clearly NASA felt that an extra overview was needed. Remember Kohrs' warning that he had never experienced the delivery of a piece of flight hardware from a contractor that did not have some deficiency. Historically, KSC always had a role in checkout and verification of flight hardware once it arrived at KSC and before it was stowed on whatever spacecraft was being prepared for launch. But Talone's team would also work "outside the fence" at the various manufacturing sites. His reach,

---

35. Goldin gives Russia six weeks to get Station on Schedule, *Aerospace Daily*, March 27, 1996.
36. Hearings, Backup Plans Indicate U.S. Concerns, Ben Ianotta, *Space News*, April 1-7, 1996.
37. NASA Engineers Design an Alternative In Case Russia Defaults on Agreement, William Harwood, *Space News*, April 1-7, 1996.
38. NASA Assembles Shuttle Flight Team, William Harwood, *Space News*, April 30-May5, 1996.

## Cost Runout Concerns – GAO Issues New Report – Changes Anger Experimenters

Throughout the summer of 1996, problems in controlling program costs continued to surface. GAO released another report which estimated that NASA would overrun the $17.9 billion budget between $60 million and $400 million by 2002. NASA disputed those claims. Vice President Gore and Russian Prime Minister Viktor Chernomyrdin signed a new cooperation agreement in July and apparently the FGB module was back on schedule. However, the Service Module was estimated to be six to eight months behind schedule. (Recall that Goldin had said in February that if the Service Module got a month or so behind schedule NASA would have to take action. That statement was not acted on.) The GAO report continued to question the ability of Russia to meet all of its obligations and whether the program's reserve would be sufficient to take up the slack if Russia could not deliver all its promised contributions. There were reports that Boeing was overrunning its contract, with numbers ranging from $170 to $450 million.[40]

Budget problems were seen as threatening funding for the science experiments being developed for the Space Station. Overruns had forced NASA to look for new sources of money within its capped budget to pay for unexpected costs, such as improving the structural integrity of Node-1, which had failed initial pressure tests. Andrew M. Allen, the Headquarters deputy program director, said that design and hardware changes needed to improve performance could exhaust the program reserves set aside through 1998. One potential source was the $280 million a year in the program's budget to design and build science facilities that would be carried in the laboratory module, nodes and other modules where the crew would live and work. These facilities included the centrifuge, furnace, and a bioscience laboratory that PIs were counting on to conduct their experiments. If Congress agreed to the shift in funds to meet the new demands, the impact would be a postponement of several shuttle flights and a delay of many months in sending experiments to the Station, currently scheduled to be launched beginning in 1999.[41]

NASA officials tried to calm the anger of the PIs when they heard of these late changes, but at the end of August the decision was made to raid the science budget. There was no alternative other than to slip the schedule for assembly, and that was rejected. The Space Station taximeter was clicking away at almost $7 million a day, so any delay was costly. JSC had lobbied to delete science entirely for the first two years but Goldin rejected that course. As a compromise, science payloads would be severely restricted during the first two years, but some experiments would be carried out. Searching for other ways to keep science on schedule, NASA asked ESA and Japan to consider building the centrifuge and the module that would house it. Together, these two pieces were estimated to cost $500 million, the most expensive facility effected by the funding shift. Neither partner responded immediately to the request.[42]

The funding ax cut most deeply into the bioscience community and NASA asked it to come up with new plans that would take into account schedule delays. Martin J. Fettman, a professor of veterinary and biological sciences at Colorado State University and former NASA Payload Specialist who chaired the NASA Space Station's biology advisory panel said: "It's ludicrous. They want us to reprioritize the science we've been working on for ten years, and they want it done right away." He asked whether "the time has come for us to solicit the participation of professional scientific societies . . . in lobbying for greater protection of scientific interests."[43] A revolt in the science community, whose participation

---

39. Based in part on a interview with John Talone, October 15, 2004. Talone credits George Abbey as the person most responsible for insisting that integration and verification testing be strengthened by the addition of his team and instituting regular status reviews.
40. Cost overrun data came from two articles: GAO Says Troubles Could Overwhelm Station, Anne Eisele, *Space News*, July 29-August 4, 1996, and *Aviation Week & Space Technology*, August 26, 1996.
41. Construction Costs May Bite Into Science, Andrew Lawler, *Science*, Vol. 273, 9 August 1996; Cost Increases Add To Station Woes, Joseph Anselmo, *Aviation Week & Space Technology*, August 26, 1996.
42. NASA Scales Back Science on Station, Andrew Lawler, *Science*, Vol. 273, 6 September 1996.
43. NASA Slices Biology on Station, Andrew Lawler, *Science*, Vol. 274, 25 October 1996.

was the raison d'être for the Space Station, was the last problem that NASA wanted to face, but recent management decisions were taking it down that path. Congress was sure to hear from their constituents in the bioscience community.

Russian and U.S. officials held a incremental design review in Houston at the end of September to try and resolve the many problems created on both sides by financial constraints. Key to the Russians staying on schedule was the release of funds by the Russian government. They asked the U.S. to apply pressure for their release. At this meeting they said the Service Module was not as far behind schedule (three months) as the U.S. had been led to believe, and if Khrunichev and their subcontractor, Energia, received the funds owed they believed they could catch up by adding shifts. As a result of the meeting the assembly sequence was modified. The centrifuge facility and ESA's *Columbus* module would be slipped until after the assembly of U.S. elements was completed in June 2002. Another issue resolved was the designation of William Shepherd as commander of the first manned Station mission scheduled in May 1998 and Anatoly Solovyev as module manager. Summing up the meeting, Randy Brinkley said the review "identified a number of issues all of which are resolvable. We did not find any issues that were considered show stoppers."[44] The question of who would build the centrifuge was not resolved. NASA had terminated the contract to build the 2.5 meter centrifuge with a U.S. company and was negotiating with Japan to build the facility. However, the negotiations were not expected to be completed until April 1997.[45] Eventually, after further negotiations, a new MOU was signed and Japan agreed to provide the centrifuge.

To try and reduce costs associated with shuttle launches, NASA made a bold move. At the end of September 1996 the agency signed a contract with United Space Alliance (USA), a joint venture between Lockheed Martin and Boeing, that consolidated various shuttle contracts under a single entity. Under the contract, USA became involved in astronaut and flight controller training, flight software development, shuttle payload integration and vehicle processing, launch, and recovery operations. USA also provided training and operations planning for the ISS. This was a major break with past NASA traditions. Although NASA oversight continued, some at NASA were concerned that without more direct interaction with the many contractors involved in all the operations now consolidated under USA, problems would arise.

The November 1996 elections maintained the status quo. Republicans kept their majorities in both the House and Senate. Some of the committee chairmanships would change but most would have familiar faces. Although late in reaching final approval, the Space Station FY 1997 appropriation passed by Congress, $2.149 billion, exceeded the Administration's request. Congress had given NASA a chance; it would have a few months before the FY 1998 budget hearings to show that the Space Station would stay on schedule in spite of all the problems with its Russian partner and the latest forecast overruns with its prime contractor, Boeing.

But other problems continued to pile up. *Discovery*'s mission at the end of November to test Space Station assembly techniques, the last scheduled launch in 1996, ran into an irresolvable failure. Astronauts Tamara E. "Tammy" Jernigan and Thomas D. Jones were forced to cancel the two EVAs they had trained for to demonstrate the ability to carry out specific Station assembly tasks. They could not open the hatch to permit entry to the payload bay. Ground teams labored over the problem but were never able to duplicate whatever jammed the hatch handle. Before the mission launched, astronaut F. Story Musgrave, who was one of the crew on the mission, said that NASA was not putting enough emphasis on EVA training. He said, "We are going to have to ramp up resources and capabilities to handle increased Station EVA requirements – and we are not there yet." His complaint was prophetic, although a possible latch problem was not the focus of his concern. Astronaut Jerry Ross, Branch Chief for EVA and Robotics in the astronaut office and scheduled to perform the first Space Station assembly EVAs in 1998, said that problems such as encountered with the hatch left open the possibility "of some surprises on orbit."[46] EVA would continue to grow as a concern in 1997.

---

44. Russian Budget Remains Station's Biggest Hurdle, Anne Eisele, *Space News*, September 30-October 6, 1996.
45. NASA Spins Centrifuge Work to Japan's NASDA, *Space News*, October 7-13, 1996.
46. Hatch Problem Focuses Attention on EVAs, Craig Covault, *Aviation Week & Space Technology*, December 9, 1996.

## Inability of Russia to Deliver Their Elements Continues – Spies in the Woodwork?

But the biggest problem at the end of 1996 was the continuing delay in the availability of the Russian Service Module that would provide power and working quarters during the first assembly flights. Since the incremental design review at the end of September, the projected delay for the Russian module went from three months back to eight. Instead of the scheduled April 1998 launch, the earliest launch had become December 1998. And that date continued to be contingent on the release of the needed funds to complete the fabrication – actions that seemed beyond the ability of the Russians to resolve.[47] NASA began to study the options that were available – shuffle the schedules or substitute a U.S. spacecraft to replace Russian modules.

Refurbishing the Naval Research Laboratory's spare Clementine spacecraft became an attractive alternative. Clementine had been designed for launch on the shuttle and, with modifications, could perform as an interim control module in place of the FGB. Trafton estimated that modifying Clementine would cost less than $100 million, and those funds could come from program reserves.[48] A replacement for the Service Module would be harder to find. Surprises and problems at the end of each year had become standard fare for the program.

With so much depending on Russia for the Space Station's success, one would have thought that, after the collapse of the Soviet Union, the old distrust would have faded in the new Russian partnership. Not true. At the end of December, FBI Director Louis B. Freeh gave disquieting testimony to a Senate subcommittee. He told the committee that Russian espionage activities to obtain information on U.S. military and space technology had continued unabated. He said the U.S. is "No. 1 on the list of the SVRR (the Russian agency that replaced the KGB). We have to ensure that our defenses are better than they ever have been and that our countermeasures work." To do this, strict guidelines containing a set of regulations had been established on how to handle day-to-day cooperation with the Russians at JSC. James Van Laak, Space Station Phase 1 Deputy Director said, "Equally important is that some of the work going on is proprietary to individual companies and universities. Those activities have to be treated every bit as sensitively as the general issue of technology transfer."[49] One could imagine, counterespionage being what it is, that the same concerns held true for the Russians. That may have been one explanation as to why the Russians wanted cosmonauts always onboard once the Russian elements were on orbit. This was just one more concern to add to the overall problem of technology transfer that complicates programs with international partners.

---

47. NASA Prepares To Drop Russian Station Module, Anne Eisele, *Space News*, December 9-15, 1996.
48. Station Officials Search for Design Solutions, Anne Eisele, *Space News*, January 13-19, 1997.
49. NASA Maintaining Vigilance Against Any Potential Russian Espionage, *Aviation Week & Space Technology*, January 6, 1997.

## — CHAPTER 8 —
## ASSEMBLY BEGINS – A NEW STAR IN THE EVENING SKY
### (JANUARY 1997 - JANUARY 1999)

In January 1997, as a result of the November 1996 elections, the chairmanships of the Congressional committees with NASA oversight changed. On the Senate side, William H. Frist (R-TN) replaced Conrad Burns as chairman of the Commerce Committee Subcommittee on Science, Technology and Space. Frist, first elected in 1994, did not have a long track record of positions on Space Station issues, so where the Subcommittee would go under his leadership was an open question. In the House, Sensenbrenner moved up to chair the Science Committee and Congressman Dana Rohrabacher (R-CA) became the chairman of the House Science, Space and Aeronautics Authorization Subcommittee. Continuing the Clinton Administration's policy to reduce NASA's budget over the next five years, the total FY 1998 request was $13.5 billion, some $300 million below FY 1997. However, the Space Station request came in at $2.121 billion, $320 million above the previous year. What Rohrabacher's Space Station position would be when hearings began was a major concern, as he had not been an enthusiastic supporter. It was believed he would push NASA to commercialize aspects of the Space Station, but how he would propose to do this was not clear.

The first hearings before Rohrabacher's Subcommittee took place on February 12th. Described once again in the trade press as a contentious hearing, Rohrabacher made his position clear. "This fatally flawed approach to working with the Russians has made the whole Space Station project flawed." He told Goldin and Presidential science advisor John H. Gibbons that the administration made a mistake bringing the Russians in as partners, rather than as contractors from whom NASA would buy the hardware and services needed. Goldin replied he was "cautiously optimistic" that the Russians would meet the schedules. At an impromptu press conference during a committee recess Goldin discussed the actions NASA had taken to speed up delivery of the Service Module (SM). A review would be held in March with the SM subcontractors and $20 million had already been transferred from the shuttle / *Mir* program to Khrunichev to pay the subcontractors. In addition, alternatives to the SM were being studied as noted in the last chapter. Russia was preparing a second FGB which, with modifications, could perform SM functions. The Naval Research Laboratory's "interim control module" and the shuttle also could be used to substitute for some of the SM functions.[1] Goldin's "cautious optimism" would soon be tested.

November and December 1997 were the projected magical dates when the results of fourteen years of effort by tens of thousands of people would culminate in the launch of the first Space Station elements. When joined in orbit the first elements would signal the beginning of a new era in space exploration. But as 1997 began, a faltering program, beset by many uncertainties, signaled that the start of the new era would be delayed. Phase I shuttle missions to *Mir* were continuing as scheduled, with *Atlantis* launched on January 12th to change-out crews, resupply *Mir*, and bring back experiments and trash. However, the problems that surfaced at the end of 1996 threatened to derail the assembly schedule. The estimate that the Russian SM would be at least six months behind schedule could mean a potential six-month or more slip of the first element launch of the Russian FGB. Without the SM attached to the FGB and Node-1 (the second element to be launched) to provide power, reboost and living space, the launch of the first two elements would have to be postponed. They could not remain on orbit independently until a delayed SM could be launched. None of the SM options that Goldin discussed would be ready in time to meet the current assembly schedule.

Ignoring the uncertainties of whether or when the Service Module would be ready, Boeing and its subcontractor McDonnell Douglas began a three-shift effort to get Node-1 ready for a December launch. Node-1, the first U.S. element to be launched, would be mated to the FGB and would be a critical link to the other elements that would follow, including the Service Module. It had six hatches to which

---

1. Key House member blasts NASA / Russia deals as 'fatally flawed', *Aerospace Daily*, February 13, 1997.

elements would be attached, including two pressurized mating adapters, and an environmental control system that would allow the astronauts to work inside. Each element that would be mated would have its electrical connections, data links and other housekeeping functions flowing through the Node, a complicated piece of plumbing and wiring if nothing else. The earlier decision to increase processing of Space Station elements at KSC versus at the manufacturing site would begin with Node-1. If all went as planned, Node-1 was scheduled to be installed in *Endeavor*'s payload bay in November 1997.²

## RUSSIAN SERVICE MODULE CONTINUES TO SLIP – ASSEMBLY SCHEDULE CHANGED

At a meeting in early February, Russian Prime Minister Chernomyrdin assured Vice President Gore that the U.S. equivalent of $100 million would be given to RKA by the end of the month for Space Station-related work, and an additional $250 million by the end of the year. Soon after the February 12th hearing, Sensenbrenner returned to Russia for the third time, accompanied by Rohrabacher and other committee members. His purpose was to determine if the Russians would live up to their commitments and transfer the funds Chernomyrdin promised so that the SM could meet the schedule. Sensenbrenner did not receive the assurances he was looking for. The Prime Minister's commitment given to Gore was an empty promise. He did not authorize cash payments, only a loan guarantee of some $260 million. Stopping in Paris after his visit to Russia, Sensenbrenner held a press conference and provided his conclusions on what he had learned. He said Russia would be given until the end of March to provide the funds. "We're getting close to the point where the back of the camel is going to break. If Russia does not fulfill its promises here, its government should know that it will effect its relations with the rest of the world on matters far beyond space."³

Before Sensenbrenner could unpack from his trip, at a press conference in Moscow Yuri Koptev made a announcement describing a Russian decision. Russia would delay the launch of the FGB seven months so that it would have enough time to complete the SM. This new schedule, not yet coordinated with NASA, would launch the FGB in June 1998, followed a month later by Node-1, and the SM in November or December 1998. This schedule was dependent, of course, on the Russian government sending the required funding to the Khrunichev space production center. A NASA spokesman, commenting on Koptev's press conference, said that all the options were still being studied. Any decisions on a schedule change would be made in conjunction with Congress and the White House.⁴ Koptev's unilateral declaration did not sit well with his U.S. partner, reinforcing the concern that Russia was still on the critical path and in a position to make or break the schedules.

Schedule slips were not the only changes under way. At the beginning of March NASA and ESA announced a new agreement. In exchange for launching the ESA laboratory on a shuttle, versus on an Ariane booster, ESA would provide two nodes and several pieces of "advanced technology" laboratory equipment. The ESA laboratory, now called the *Columbus* Orbital Facility in addition to the designation APM, would be carried on a shuttle flight, at no cost to ESA, sometime in late 2002 or early 2003. The ESA nodes would be an outgrowth of ESA's previous agreement with Italy to develop mini-pressurized logistics modules. Node-2 would be delivered at the end of 1999 and Node-3 would follow two years later. For ESA this agreement killed two birds with one stone. ESA and Italy had been in a dispute over $120 million in payments owed to the Italian Space Agency that they should have received in return for contributions to ESA programs. Italy would be paid for the nodes, and the "advanced technology" equipment would be "mainly various types of freezers" to be delivered between August 2001 and August 2002.⁵

---

2. U.S. Module Engineers Try To Beat the Clock, William Harwood, *Space News*, January 27-February 2, 1997.
3. Two articles describe Sensenbrenner's trip both in *Space News*, February 24 and March 2, 1997: Sensenbrenner Departs Russia With New Doubts, Peter B. de Selding, and NASA Seeks Solutions to Russian Delays, Peter B. de Selding and Anne Eisele.
4. Koptiev [Koptev] says Russia will slip first Station launch seven months, Article: 32450, *Aerospace Daily*, February 25, 1997.
5. Extracted from: ESA Press Release, ESA Trades Space Station Hardware for Shuttle Launch Of European Laboratory, March 7, 1997, No. 07-97, and Station Negotiations, Peter B. de Selding, *Space News*, February 17-23, 1997.

Would any experimenters still be interested in using the ESA freezers? At the same time that ESA was announcing the new agreement, NASA was telling the NRC's Space Studies Board (SSB) that not much science would be done on the ISS until after the turn of the century. Raiding the science budget to make up for overruns in building the ISS, mentioned earlier, had been quantified – $50 million had been used from the FY 1996 budget, $177 million from FY 1997, and $235 million would be taken out of FY 1998. Steven Isakowitz, a White House budget official, was quoted as saying, "It was the only thing we could do." The budget cap had struck again. Members of the SSB were understandably disturbed. Martin Glicksman, a board member, said, "These communities (experimenters) are going to dissipate." Simon Ostrach, another board member, complained, "We have built up a world-class community, and now we have to sit for six or seven years and not do anything." Mark Uhran at NASA Headquarters who had oversight for science payloads, said that using funds designated for science facilities (to solve other problems) leaves "a very, very thin program." He said that when the program budget authority was transferred to JSC, his office lost control of the funds.[6]

By the beginning of April nothing had changed relative to the Russian funding delays. Former astronaut Thomas Stafford, returning from a fact-finding trip, told Goldin that Russia had still not transferred the needed funds for the Service Module. Russian contractors were now at least eight months behind schedule, but because of the $20 million that NASA had diverted to the SM, some work was continuing. Stafford wrote to Goldin: "I believe that the Russian government has a credible process in place that, if implemented as planned, will adequately fund Service Module requirements in 1997."[7] This was the same analysis heard many times before – the Russians talked a good talk, but did not deliver. NASA continued to put a smiling face on the problem.

As a result of the certain delay in the availability of the SM, NASA took the only course open, telling Congress it would change the assembly schedule and slip the first element launch by eleven months. At $7 million a day to keep the program running, that would add over $2 billion to the program's runout costs. At a hearing before the House Science, Space and Technology Subcommittee on April 9th, Trafton attempted to explain how the program would work its way out of the dilemma in which it found itself. He indicated that if the Russians did not send $140 million to the Service Module contractors by the end of May, believed to be very unlikely, the SM would not be available until October 1999, an additional ten-month delay. Of the options previously discussed, a decision had been made to develop the interim control module (ICM) to backup the SM. NASA proposed transferring $200 million from the shuttle account and $100 million from other accounts to fund ICM development at the Naval Research Laboratory.

Subcommittee members were not impressed. Sensenbrenner listed eight times that the Congress had been promised by NASA and Russia that the funding for the SM would be resolved. He said, "The program is falling apart around us because of the Russians." Chairman Rohrabacher charged that the U.S. taxpayer was being forced to foot the bill for a Russian government run by "nincompoops" and "dunderheads," and suggested that any additional funding be taken out of U.S. aid programs with Russia.[8]

The House Science Committee took the next step. It unanimously approved a bipartisan amendment to the NASA appropriation that barred NASA from spending U.S. funds on those Space Station elements that Russia had the responsibility to build. The amendment required NASA to develop contingency plans to replace all Russian-built hardware. It also required the Administration to decide by August 1st whether or not U.S. hardware would replace Russian elements. Sensenbrenner said he came close to including in the amendment language terminating Russian participation. In what could only pass as typical bureaucratic buck passing, Russian style, it was reported that President Yeltsin blamed Koptev for not telling him about the problem. Once again, as a result of public exposure that it had defaulted on meeting its Space Station commitments, Russia said it would transfer $260 million to their contractors.[9]

---

6. Science Slides to Bottom of Schedule, Andrew Lawler, *Science*, Vol. 275, 14 March 1997.
7. Russia's Check Still Undelivered, Anne Eisele, *Space News*, March 31-April 6, 1997.
8. From an article in *Aviation Week & Space Technology*, Russian Problems Force 11-Month Station Delay, Joseph C. Anselmo, April 14, 1997.
9. This story extracted from three press reports, Russian Problems Force 11-Month Delay, Joseph Anselmo, *Aviation Week & Space Technology*, April 14, 1997; NASA, Partners Revising Station Assembly Plan, Anne Eisele, *Space News*, April 14-20, 1997; Station Funding Promised; Congress Fires Warning Shot, Nicolay Novichkov and Joseph Anselmo, *Aviation Week & Space Technology*, April 21, 1997.

Despite all the uncertainties in the main Space Station program, Phase I of the joint operations on board *Mir* continued. A fire on *Mir* in late February, while U.S. astronaut Jerry Linenger was on board, had raised serious concerns about safety conditions on the aging Russian space station. Repairs had alleviated some of these concerns but they did not completely go away. Experiments were being conducted but had been restricted at times due to problems with *Mir*'s life support systems, power supply, and communications. It seemed that most of Linenger's time, along with that of his Russian crew mates, was taken up repairing failing systems, although Linenger was reported to be conducting additional science experiments not on his official timeline. A five-hour EVA was conducted on April 29th by Linenger and Russian cosmonaut Vasily Tsibliev to install and retrieve several experiments, the first time that the two countries had participated in a joint EVA.[10] After 122 days aboard *Mir*, Linenger returned to Earth on *Atlantis*. This Phase could be considered at least partially successful, and certainly the U.S. and Russia were gaining experience working together, at times under difficult conditions. Hopefully, this would payoff once the Space Station was on orbit.

Hearings in the House and Senate the first week of May put the Space Station back in the budget debate cross hairs. In the Senate, Appropriations Subcommittee Chairman Christopher Bond (R-MO) asked Goldin at what point would NASA have to ask to raise the program funding cap. Using his favorite expression, Goldin's response was, "I am cautiously optimistic that our reserves of just under $1 billion will take us through 2002 . . . But it will be rough." At the House hearings the same concerns were voiced, and NASA gave the same answers in response to questions on the delay in Russia sending funds to complete their contributions. Goldin's cautious optimism would once again be proven wrong. Chairman Sensenbrenner told a *Space News* interviewer "Placing Russia on the critical path has already been a $700 million mistake." How many times had he expressed that point of view in the last four years? Sensenbrenner said he told the Administration "that the honest way to deal with the issue of Russian failure to perform is by seeking a supplemental appropriation, rather than transferring funding from other programs."[11] Members of Congress would talk a tough game, but in the end it was business as usual. The hard decisions were never made.

While this was playing out in Washington, NASA received word that another installment of funds had been received by the Russian contractors. The big question was: would the funds continue to flow? Back at home NASA had transferred $20 million to the Naval Research Laboratory (NRL) to begin modifying the ICM. NRL was telling NASA that to continue the work it would need more money by May 15th or it would have to break up its team. In addition to the NRL request, more unexpected bills might be slipped under the door. Reports were that Boeing was behind schedule in some of its work and overruns were expected.

Following a meeting of the Space Station Control Board at KSC that included U.S., Russian, ESA, Japanese, and Canadian principals, a new assembly sequence was announced on May 15th. (The new schedule was formally accepted by the heads of the five space agencies at a meeting in Japan on May 31.) The first element, the U.S.-financed FGB, would be launched on a Proton booster on June 20, 1998. It would be followed in early July by Node-1 and two mating adapters carried on a shuttle. In December 1998 the Russian SM, the third element, would also be launched on a Proton rocket. In case the SM was not ready, a shuttle would be ready in December to carry the ICM as the replacement. If the SM was on schedule, the ICM shuttle flight would be changed to a logistics flight. The fifth shuttle launch, in January 1999, would be a logistics flight carrying equipment for the crew who would be on the next launch. The sixth launch, in January 1999, would be a Soyuz carrying the first crew of William Shepherd and cosmonauts Sergei Krikalev and Yuri Gidzenko. They would stay on orbit until June, using the SM as their living and working quarters. The Soyuz would remain attached as the ACRV.

---

10. Two stories in *Aviation Week & Space Technology* contributed to the Phase I text: NASA Increases *Mir* Station Safety Vigilance, Craig Covault, April 14, 1997, and Mir Status Improves As Cooperation Evolves, Craig Covault, April 28, 1997. For further discussion on *Mir* safety concerns, see James Oberg, Star-Crossed Orbits, McGraw-Hill, 2002.

11. Two news stories contributed to this discussion: NASA Needs Station Money by Mid-Month, Anne Eisele, *Space News*, May 5-11, 1997 and Struggle To Find Station Money Continues, Anne Eisele, *Space News*, May 12-18, 1997.

In June a replacement crew, this time consisting of two Americans and one Russian (individuals not identified), would be launched on a shuttle flight that would also carry the U.S. lab element, and Shepherd's crew would return in the shuttle. The third crew, two cosmonauts and one American, would be launched on a Soyuz in late November or early December 1999. In between some of the flights listed above, other assembly and logistics flights would be bringing the next elements or resupplying the core station.[12] A total of 81 shuttle, Soyuz and Progress supply vehicles would be needed to achieve assembly complete in 2003; 48 would be Russian and the other 33 would be U.S. shuttle launches. Using Ariane and the Japanese H-2 Transfer Vehicle for logistics was still possible but not essential to making the Space Station operational on this new schedule.

At the same May meeting an agreement was reached on the composition and command of the crews, the latter a previously sticky issue. On the first five crews there would be a total of eight Russians and seven Americans. While Shepherd would be the first commander, subsequent commanders would be chosen on the basis of experience and other factors. This meant that Russian cosmonauts could be selected as commanders regardless of crew makeup. Crew members from the other international partners would not be included in the first five crews. Other important decisions made at this meeting included guidelines on which language would be used during crew training, and the language that would be imprinted on equipment labels. Both English and Russian would be used, depending on location of the training sites and equipment supplier. Astronauts and cosmonauts would be required to be bilingual, at least to a working level.

### EVA TIME AND MIR CREW SAFETY REMAIN AS PROBLEMS

New estimates of EVA time for assembly and maintenance through assembly complete were released in June. The last numbers estimated in February 1995 were: 819 hours for U.S. astronauts and 240 hours for cosmonauts. The new numbers were: 1,105 hours for the U.S. and 414 for the Russians, an increase of 460 hours. Cosmonaut EVAs would concentrate on assembling and maintaining their elements and U.S. astronauts would do the same for the U.S. elements.[13] The increase in the amount of EVA was a troubling reminder that the full impact of the redesign was not completely understood. The number of EVA hours was still much less than the estimates made for Space Station Freedom just before it was canceled but, like Freedom, every time a new study was completed EVA hours increased. One of the rationales for the redesign was the belief that there would not be a need for so much EVA during assembly. Marcia Smith, commenting on the new numbers, said, "The question is, what is driving the substantial increase in the number of EVA hours, and whether that signals some fundamental problem with the design of the Space Station. Even if it doesn't, it raises the question of whether or not they can complete the EVAs in the allotted time and if not, whether this will cause more schedule delays. And, of course, there is always the risk to crew safety when they are out on EVA." This latter concern mentioned by Smith was the most difficult to address. The required U.S. EVA time would be almost 400 hours more than the total combined EVA experience of NASA, including that accomplished on the lunar surface during Apollo.

*Mir* maintenance was the next red flag raised as the U.S. and Russia continued Phase I of their joint Space Station cooperation. On June 25th, a Progress resupply module collided with *Mir*. It punctured and depressurized the *Spektr* module that housed many experiments and cut electric power to *Mir* by 50 percent. (The power failure was discussed in the previous chapter.) This forced the crew to seal off *Spektr* and remain in the intact sections. Repairs were needed and another Progress was launched in July carrying the repair hardware and other supplies. Once the Progress was docked to *Mir* the two Russian cosmonauts, Vasily Tsibliev and Alexander Lazutkin, donned their EVA suits and used the repair kits to reconnect the electric cables that ran from the solar arrays, through the damaged *Spektr*, to the core modules. This was a tricky repair as part of *Mir* would be depressurized in order to reach inside *Spektr* while the cosmonauts remained in the small ball-shaped node that connected *Spektr* to the

---

12. Two stories contributed to the details of the new launch schedule: New Agreement Reached On Station Crew Selection, William Harwood, *Space News*, May 19-25, 1997; and *Mir* Resupplied As Station Modified, Craig Covault, *Aviation Week & Space Technology*, May 19, 1997.
13. Some data on EVA was extracted from a *Space News* story: Spacewalks To Increase, William Harwood, *Space News*, June 16-22, 1997.

core station.  U.S. astronaut Michael Foale did not participate in the repairs as he did not have a EVA suit.  Instead, in case there were any problems, he stayed in the Soyuz capsule while *Mir* was depressurized.[14]  (The repairs were made successfully on August 22nd.)

Reports of the above problem and several others prompted Chairman Sensenbrenner in July 1997 to request the NASA Inspector General (IG) to analyze: (1) the suitability of the Russian *Mir* Space Station for habitation by U.S. astronauts, (2) research productivity on board *Mir*, and (3) cost effectiveness of continued NASA involvement in the *Mir* space station program.  Roberta L. Gross, NASA IG, responded to Sensenbrenner in August with a nine page letter and ten Appendices.[15]

Based on interviews with astronauts who had been on board *Mir* and other NASA and contractor employees, her letter was a overwhelming indictment of how NASA was ignoring the problems encountered while working with the Russians.  Citing instance after instance of the difficulties faced in living on board *Mir* – the lack of training, impediments to conducting useful science experiments, poor communications, power failures and many other problems– Gross wrote, "various sources have voiced their concern about the objectivity and/or adequacy of NASA's risk / benefit assessment process in the face of stated national policy to maintain the Russian / American partnership."  Or in other words, NASA would say everything is going great regardless of the risks and danger in order to maintain the partnership.  On the positive side, the IG report included statements by three astronauts who had crewed on *Mir* who "praised the cosmonauts for their inventiveness and ability to cope with the constant maintenance requirements."  Astronaut Frank L. Culbertson, NASA's Shuttle / *Mir* Program Manager, said that coordination had "significantly improved."  The NASA party line was that Phase I was always intended to be primarily a learning experience of how to work with the Russians, and certainly that had occurred.  The question was, were the lessons learned being applied to future *Mir* missions and were they transferable to Space Station operations?

Gross proposed four alternatives to Sensenbrenner: (1) Maintain the status quo.  If the situation on *Mir* degrades further, emergency evacuation on the Soyuz remained an option. (2) Remove astronauts from *Mir* but provide support to the Russian space station. (3) Monitor the Russian's ability to improve conditions on the *Mir* and, if safety conditions improved, return an American astronaut(s) to continue participating in *Mir* activities. (4) Terminate American involvement in Phase 1 activity and proceed to Phase 2. (The four alternatives above are abbreviated versions of those in the letter.)  You did not have to read between the lines to understand how concerned the IG was after conducting her interviews.  She ended the four options writing that 80% of the Phase I funding had already been expended, so the monetary impact of any of the options offered "appears minimal."  The letter seemed to be an invitation to Chairman Sensenbrenner to have Congress make the call to continue or terminate Phase I, but no action was taken.  The Appendices to IG Gross' report contained many recommendations on how to

---

14. Information for the *Mir* repairs extracted from a story in *Aviation Week & Space Technology*: Russia and U.S. Chart Daring *Mir* Salvage, Craig Covault, July 7, 1997.
15. Letter addressed to: The Honorable F. James Sensenbrenner, U.S. House of Representatives, Chairman, Committee on Science, Suite 2320 Rayburn House Office Building, Washington, DC 20515.  The letter, dated August 29, 1997, was signed by Roberta L. Gross, Inspector General.  Many comments in the Appendices described the problems the crews encountered while on board *Mir* and include: Appendix B – "NASA management has accepted a different standard for human safety for the Phase I Shuttle / *Mir* Program than it has been willing to accept for either the Shuttle or the International Space Station."; C – A table listing the number of failures on board *Mir*, a total of 91; G – Merbold debrief – He recommended that NASA adopt Russian philosophy of maintenance and repair, less bureaucratic, rely on the crew to perform as needed (he estimated they spent 2 hours/day on maintenance) but thought U.S. Space Station equipment would not require as much maintenance.; H – Astronaut Linenger thought the oxygen canister fire on *Mir* burned uncontrollably until all the chemicals were consumed.  The Russians said they controlled the fire.  The JSC team investigating the fire could not confirm the Russian statement.  Gross' report indicated a general low quality in Russian reporting and documentation of incidents on *Mir*.  All of these concerns led in Appendix I to a detailed Lessons Learned and Recommendations for 14 categories of activities that included operations, mission control, training, and communications.  Appendix J is an interesting AP story filed from Moscow (8/13/97) about how the Russian cosmonauts were paid with an average salary of $3,000 (US) per month (very high by Russian standards).  They could earn bonuses for such things as performing EVAs and making repairs.  A rather entrepreneurial approach considering their societal background.

avoid repeating the problems on Space Station that were encountered while working with the Russians during the *Mir* missions.

Inspector General Gross' report to the House Science Committee covering problems encountered by U.S. astronauts while onboard *Mir* resulted in a number of changes to Space Station design and operating procedures. It was announced that methods were being developed to exchange technical data between Russian and U.S. mission controllers, to permit simultaneous decisions to be made when needed. Stowage and inventory control techniques would be developed to avoid the long delays encountered on *Mir* in finding equipment and setting up experiments. Increasing the number of experiment racks was also being studied, as well as increasing the number of body restraints to steady the astronauts at the work stations.[16]

*Mir* experience also showed the need for more pre-mission science planning and easier communication with the PIs in the event that the prescribed experiment procedures required changes. (If the PIs, astronauts, and mission specialists were not occasionally surprised by some of the results, one might question the need to do an experiment.) Beginning with Apollo experiments, direct communications between astronauts and PIs through mission control were not allowed. Mission control's position at that time was: "The astronauts were smart, well trained and knew what they were doing; don't bother them with a lot of questions and advice." Space Station experiments, including some that would be conducted over periods of months, needed a different approach.

On July 22, Senator Dale Bumpers took the occasion of the debate on NASA's FY 1998 budget to try for the sixth time to terminate the Space Station. Bumpers cited the GAO April 1996 report that the program would cost $100 billion and called it a "boondoggle." To show that the scientific community opposed the Space Station, he quoted from a number of physicists, including Harvard's Nicolaas Bloembergen who said that "microgravity is of micro importance." Also, using the GAO report as his source, Bumpers said the program overruns had more than tripled and listed the social programs that would benefit if the funds saved were transferred to those programs. He reminded the Senate that it was his amendments that killed the Clinch River Breeder Reactor and the Super Collider. At the beginning of his one-hour and twenty-minute narration he introduced an amendment to NASA's budget that would cancel the program except for termination costs. The vote went against his amendment, 69 to 31 (a rare 100 member vote). He received six fewer votes than his similar attempt in 1996.[17] The House had already passed an appropriations bill to give NASA $13.65 billion, $148 million more than the Administration's request. It included $2.121 billion for Space Station. With the Senate version at $13.5 billion, the House bill would have to go to conference to keep the Space Station moving ahead.

In July a decision was announced to modify two more shuttles, *Discovery* and *Endeavor*, to allow them to rendezvous and dock with the Space Station and *Mir*. The remaining shuttle, *Columbia*, was too heavy, even with modifications, to achieve a 51.6 degree orbit with a useful payload. This was the second change as to which shuttles would be modified. In 1993 NASA believed it would require two shuttles, *Atlantis* and *Endeavor*. In 1994, NASA decided to have just *Atlantis* modified for these flights. At the time, the reason given to modify only one rather than two was the extra cost, estimated to be $28 million. New cost figures were not given to modify both *Discovery* and *Endeavor*, but assuming the 1994 estimate was accurate, and with a little inflation factored in, this would add another $60 million to the program. Funds for the modifications would be taken from the Office of Space Flight's budget line. In addition to a standardized docking mechanism, the biggest change would be made to the airlock, repositioning it from the crew cabin in the mid-deck to the cargo bay. The airlock would be large enough to accommodate two space suited astronauts. The modifications would be done at KSC and the shuttle Palmdale, California, facility and were scheduled to be completed by October 1998.[18]

---

16. Ibid.
17. Material for Senator Bumpers amendment and speech comes from two sources: Congressional Record, Department of Veterans affairs and Housing and Urban Development and Independent Agencies Appropriations Act, 1998 (Senate-July 22, 1997) and, Space Station Condemnation, Mary McGrory, *The Washington Post*, August 17, 1997.
18. NASA Starts Shuttle Facelifts for Station Duty, Anne Eisele, *Space News*, July 21-27, 1997.

Cost overruns by Boeing became a matter of record on September 18th. At a hearing before the Senate Commerce, Science and Transportation subcommittee, Goldin told Chairman Frist that he expected the Boeing contract would overrun by $600 million through assembly complete. NASA and Boeing managers had been unable to control the cost growth resulting from the many design changes. The program needed to immediately transfer $330 million to allow Boeing to pay for already incurred expenses. It also needed $100 million for a new Russian Program Assistance account that had been set up in case Russia failed to deliver its contributions. Goldin asked that he be given "the flexibility to handle most of this within the agency's [1998] budget" and said he would take the funds from mission support and uncosted carryovers. Replying to a question from Senator Jay Rockefeller (D-WV), he said he did not think the transfers would adversely affect NASA science programs. If the $2.1 billion cap would not be lifted, the alternative was to slip the schedule and delay assembly.[19] It would be several weeks before Goldin received an answer to his request.

Brazil proposed becoming a Space Station partner at the beginning of October 1997. In exchange for conducting research on Space Station, Brazil would make five contributions: a window observational research facility, a technology experiment facility, a logistics site to house experiments outside the Station, an unpressurized logistics carrier, and an EXPRESS pallet to deliver hardware to the Station. Terms of the agreement would be negotiated by President Clinton and Goldin during a trip to Brazil the second week of October. When the agreement was signed, Brazil joined fourteen other nations supporting the U.S. and the International Space Station.[20]

### MIR DECOMMISSIONING CREATES NEW PROBLEMS – ADDED COSTS

At the end of October there was a little good news, followed by several problems in November. The good news was that the FY 1998 Space Station appropriation was approved at $2.441 billion, $320 million above the request. Despite the increase, NASA immediately came back to Congress for a supplemental appropriation of another $200 million to pay for cost overruns. At a November 5th hearing of the House Science, Space and Aeronautics subcommittee, Trafton argued his case saying that without the $200 million the launch schedule would have to slip again. "We can try to find more money within the program, but I don't think it is there. We can transfer funds from other accounts; or we can ask Congress for more money." But, he warned, slipping the schedule would in the long run add costs, an argument that NASA always trotted out when it came hat in hand for more money for any program, and undoubtedly true. Chairman Rohrabacher told Trafton that NASA would have to eat the $200 million somewhere in the existing appropriation and suggested that the EOS program would be a good source, but not to take it from shuttle, Space Station or the reusable launch vehicle accounts.[21] Seven weeks after Goldin appeared before the Senate Subcommittee requesting authority to reprogram funds to take care of overruns from his "favorite place," mission support, Trafton seemed to be telling a House subcommittee that this would not be a good idea, just give us more money. Trafton's request bore out White House staffer W. Bowman Cutter's complaint in 1994 that NASA's "reaction was always that there is more money available" when the Space Station got in trouble.

Another problem that emerged in November was how to decommission *Mir*. Tentatively scheduled for sometime in 1999 or 2000, analyses showed that four Progress spacecraft would be required to supply the fuel to slow *Mir* for a safe burn-up over the South Pacific Ocean, the preferred location. During the same period that *Mir* deorbit supply flights were needed, six or seven Progresses also would be required to resupply Space Station elements that would be in orbit. James Van Laak said, "Both sides realize the plans today call for a requirement for more Progresses than it looks like [the Russians] can launch." NASA began to study alternatives that included using the ICM attached to the SM to provide station keeping power if the Progresses were not available to refuel the SM.[22] This would require modifications to the SM to permit the ICM to dock. Decommissioning *Mir* would continue to be a point of contention

---

19. Two press stories covered the hearing: Station Costs Pinch Other Programs, Andrew Lawler, *Science*, Vol. 277, 26 September 1997, and Station Cost Soars 20% Above Cap, Anne Eisele, *Space News*, September 22-28, 1997.
20. Brazil Will Become Participant In NASA Barter Deal, Anne Eisele, *Space News*, October 13-19, 1997.
21. NASA Told To Cover Costs, Anne Eisele, *Space News*, November 10-16, 1997.
22. Spacecraft Shortage Bedevils *Mir*, Station, William Harwood, *Space News*, November 10-16, 1997.

between the U.S. and Russia. And the problems did not end there. Reports out of Russia in November were that the contractors working on the Russian elements, in particular the SM, which was furthest behind schedule, were owed 1.5 to 1.8 trillion rubles, or some $250 to $300 million dollars.

One proposal to reduce projected Space Station runout costs was to commercialize the use of some of the facilities. Congressman Rohrabacher was an early proponent of commercializing the Space Station. He was quoted in a *Aviation Week & Space Technology* story as saying, "We've had some problems getting people at NASA to really take this seriously. Basically, we've got to make sure we force it upon them." In November an industry consortium made a different offer. It proposed to operate the Space Station. Consisting of Daimler-Benz Aerospace, Boeing, Alenia Aerospazio, Mitsubishi Heavy Industries, and Spar Aerospace, the consortium offered end-to-end operations service. Sami Gazey of Daimler-Benz said, "Station operations need a new approach. They cannot be conducted in the traditional way if cost effectiveness and optimal performance are to be achieved. We need an industrial approach."[23] The consortium proposal sounded like a bold move, but the consortium soon dissolved without detailing how it would operate and NASA never had to seriously consider the offer.

Back in May, representatives from ESA, Japan and Canada had urged NASA to structure the program to allow them to sell or barter some of their rights to use Station facilities to countries or companies not in the program. Ian Pryke, ESA's Washington office manager, said the individual partners "can certainly commercialize their portion of the (Space Station) utilization, but once you start getting into the operations arena, that concerns all of the partners."[24] With the continued delays, cost overruns, and other problems, the members of the industry consortium must have been relieved that their proposal was not acted on. NASA claimed to be interested in commercializing the Space Station; however, any approach would have certainly required heavy subsidization in some form by the government, especially for mission operations.

### ASSEMBLY SEQUENCE CHANGES – SCHEDULE DELAYS CONTINUE

By the end of 1997, additional assembly sequence changes were required to accommodate potential delays, the third revision since the redesign in 1994, and the second following the changes announced in May 1997. The first two elements to be launched, the FGB and Node-1, remained the same. However, the availability of the SM, scheduled as the third launch, was still in doubt. The payloads for the fourth and fifth launches were dependent on whether or not the SM was on orbit. If it was, the fourth launch would be a Soyuz with fuel for the SM and the fifth would be a shuttle carrying the Spacehab and spare parts. If the SM had not been launched, the third flight would be the shuttle with the ICM to dock with the FGB. It would carry 11,000 pounds of propellant, enough to keep the FGB in orbit for a year. The next launches would be delayed until the SM was on orbit. Then the former fourth launch, the Soyuz with fuel for the SM, would become the fifth flight. It was the only way to plan in order to assure that the first two elements launched, the FGB and Node-1 with two mating adapters, would be able to maintain their orbital altitude; that was the job the ICM would do if needed. After the fifth launch, and assuming the core Station was functioning, the sequence would continue. The sixth launch would be *Atlantis* carrying the Z-1 truss, Ku-band antenna system, control moment gyros (CMGs) and pressurized mating adapter-3. This would be an important set of equipment. CMGs provided non-propulsive attitude control and Ku-band communication would support the first science experiments. After the seventh launch, a Soyuz carrying the first three-man crew, experiments would begin. The eighth and ninth launches would be Soyuz refueling flights and the tenth launch would be a shuttle carrying another truss and solar arrays. The eleventh launch would bring the U.S. lab module with five integrated science racks and mark the beginning of extensive man-tended experimentation. Five more logistics and equipment flights would complete Phase 2 in August 1999 when the Canadian RMS was attached on the fourteenth launch.

Phase 3 (post-PHC) would require at least thirty-eight more shuttle and Russian launches to complete assembly in late 2003 or early 2004. At assembly complete the JEM would be mated to Node-1, (eight launches later than the sequence released in 1994), and the ESA *Columbus* Orbital Facility, similarly delayed, would be attached to the Italian-supplied Node-2.[25]

---

23. Group Calls for Private Station Management, Space Log, *Space News*, November 3-9, 1997.
24. Space Station Partners Make Contingency Plans, Anne Eisele, *Space News*, May 19-25, 1997.

NASA began to study in 1997 a new inflatable module called TransHab. It became a candidate to replace the original U.S. hab module being built by Boeing and scheduled to be one of the last launches in 2003. There were two reasons for this possible substitution. First, although the hab module was approaching completion, it was overrunning costs. Second, NASA believed an inflatable module could be the best technology to use as living quarters for astronauts traveling to Mars and living on the Moon. It would have a large volume, and be lightweight and durable. By attaching it to the Space Station NASA would have the opportunity to evaluate it over a long period of time to determine if it would be suitable for Moon and Mars missions.

As the 1998 new year began, the launch of the first Space Station element, the FGB, seemed less and less likely to hold to its June date. Besides the continuing problem with the SM, the NASA shuttle manifest was contributing to the delays. Problems preparing the X-ray telescope (AXAF) were delaying its Shuttle launch. As a result, all of the shuttle schedules were being rearranged. Until AXAF's checkout and integration were completed, with forecasts that it would be delayed at least three months, all the other shuttle launch schedules would be affected in order to accommodate the high-priority telescope launch. Thus, Node-1 would be delayed, which meant that the FGB would also have to be delayed. Instead of a July launch, Node-1 would not be ready until September or later, pushing the FGB to August at the earliest.

With launch dates continuing to slip, the delays allowed "Tip" Talone and his staff at KSC to go beyond their original charter and develop a plan to conduct end-to-end testing before launch of the primary U.S. Space Station elements. By maintaining the "pre-slip" delivery schedule of U.S. elements to KSC, they would now be at KSC at the same time instead of arriving one by one after each of the preceding elements had been launched or were on the launch pad. Of particular concern was the cooling system plumbing that would transfer fluids from element to element, and the electrical cables carrying power and computer connections between elements. Prior to setting up the new test program, these connections would have been made for the first time on orbit. The original plan was to test each element at KSC as an individual spacecraft. Now KSC would be able to test the first elements as an integrated Station. Talone described the Multiple Element Integrated Testing (MEIT) program as "the best and final insurance policy we've got that what we put up there is going to work."[26] A series of four tests were planned that would link Node-1, the Z-1 Truss, Pressurized Mating Adapter-1, the P6 Photovoltaic module, the U.S. lab module, a logistics module, and a shuttle simulator. Except for the FGB and SM, all the initial Space Station core elements would now be tested as a functionally complete assembly before launch. The concerns expressed by the SSAC Test and Verification Panel, four years earlier, were being resolved. Sometimes schedule slips worked to NASA's advantage.

## FY 1999 Budget Hearings – Runout Costs Debated

President Clinton submitted his FY 1999 budget request to Congress at the end of January. NASA's share was $13.465 billion, which included $2.27 billion for Space Station, slightly more than the FY 1998 request but below the final FY 1998 appropriation. It included a new $5 million line item to start development of the crew return vehicle (CRV – formerly called CERV), proposed as a six-person vehicle to replace the three-person Russian Soyuz. Congressional hearings on the budget promised to be a repeat of the past year's discord. Among the issues that would be debated were NASA's latest runout cost estimates, $3.9 billion above the agreed to cap, and the siphoning of $462 million from the science budgets to pay for some overruns. At the first hearings in February, Goldin and Associate Administrator for Space Flight Joseph Rothenberg, who had replaced Trafton in January, requested that NASA be allowed to transfer $173 million from the science budgets and $27 million from Aeronautics and Technology within the FY 1998 appropriation. Committee members did not seem disposed to support the request, and approval, if it came, would have to await further debate as the hearings continued.

---

25. Two stories appearing in *Aviation Week & Space Technology*, December 8, 1997, written by Craig Covault provided the background for the assembly sequence: Station Sequence Combines Complex Hardware, and Mars Initiative Leads Station Course Change.
26. NASA Decides To Connect, Test Station on the Ground, William Harwood, *Space News*, January 5-11, 1998.

Soliciting interest in commercializing the Space Station received a boost at the end of February 1998. A free, interactive teleconference, "Open For Business,"[27] was sponsored by NASA and televised by PBS. Participants were able to obtain information about ongoing scientific and commercial research in the fields of biotechnology, materials processing and agriculture. Panelists representing NASA, universities and industry, including a venture capital firm, made presentations and answered questions from the participants about the opportunities to conduct additional experiments. Despite this overture, NASA received few new inquiries.

Hearings on the FY 1999 budget continued into April. Resolution of NASA's money problems with Russia were complicated by the disclosure that Russian President Yeltsin had fired Prime Minister Chernomyrdin, Vice President Gore's and NASA's senior Russian contact and supporter on the Space Station program. Yeltsin's nominee to fill the position, Sergei Kiriyenko, was quickly approved by the Duma. His position on Space Station matters was unknown. At the last Gore-Chernomyrdin meeting in March before his firing, Chernomyrdin once again had assured Gore that delinquent payments of some $20 million would be sent to the Russian contractors, but one month later they were still "in the mail." How the Russians would respond to the new leadership cast the entire program in doubt. If Kiriyenko did not immediately address the funding problems, Russia's contributions could be in further trouble.

Backing up a few months, cost runout uncertainties during the final hearings on the Space Station FY 1998 budget forced Goldin to request the NAC in September 1997 to establish a cost control task force to conduct an independent "analysis of the management, operational, and programmatic factors that effect cost growth."[28] Congress required that this review be conducted by "a qualified independent third party." The NAC responded immediately, and in October the Cost Assessment and Validation (CAV) Task Force was chartered. One month after starting their review, Trafton (prior to Rothenberg's appointment) requested the CAV to address the House and Senate Appropriations Committees' specific requirement to provide answers to questions dealing with how the prime contractor costs would be controlled.[29] These answers were needed before NASA could secure the release of $851 million in the FY 1998 budget that had been "fenced" (could not be spent). That became the CAV's charter.

Chaired by JMR Associates President and former TRW executive Jay Chabrow, the CAV took on a busy schedule over the next seven months, including visits to production facilities in Russia and ESA. The comprehensive report released at the end of April 1998 contained a number of findings and conclusions that cast serious doubts on all of NASA's runout projections.[30] The CAV concluded that, although the program had not described any "show stoppers, . . . The program size, complexity and ambitious schedule goals were beyond that which could be reasonably achieved within the $2.1 billion annual cap [within] a $17.4 billion cap." To overcome this shortage the CAV believed "additional annual funding of between $130 million and $250 million will be required." If the analyses were correct, it would mean that the mandated program ceiling would be $7.3 billion short. The Report recommended that "realistic major milestone dates should be established as the basis for development of the program plan, and internally defined target dates should be used for execution. If necessary, program content should be eliminated or deferred to fit within funding constraints." It was not the first time that NASA had heard this recommendation. The CAV also recommended that NASA develop a cost and schedule risk evaluation and mitigation strategy in the event that Russia did not deliver on its promises.

In evaluating the Space Station spares policy, the CAV concluded that in order to stay within spending constraints the program had shortchanged the spares needed "to support the current schedule . . . Not

---

27. 5th Annual International Space Station Teleconference, Open For Business, Produced by NASA and WHRO Center for Public Telecommunications, Presented by the PBS Adult Learning Satellite Service, February 26, 1998. Six years after the broadcast, in response to questions from the author, PBS was unable to find records as to how many participated in the teleconference.
28. NASA letter to Dr. Bradford Parkinson, Chair, NASA Advisory Council, Washington, DC 20546, dated September 17, 1997 and signed by Daniel S. Goldin.
29. NASA letter to Mr. Jay Chabrow, President, JMR Associates Inc., 8841 Cortile Drice, Las Vegas, NV 89134-6142, dated November 6, 1997, signed by Wilbur C. Trafton, Associate Administrator for Space Flight.
30. Report of the Cost Assessment and Validation Task Force on the International Space Station to the NASA Advisory Council, April 21, 1998. The report was commonly referred to as the Chabrow Report after its chairman.

procuring adequate spares during the initial production run . . . may lead to quality and consistency issues as well as increased cost." How to deal with the lack of spares had been one of the frequent issues raised by the defunct SSAC. Development of the Crew Return Vehicle also came in for criticism. The CAV believed that the CRV was critical to achieving a permanent human presence on the Space Station. The CRV and X-38 research vehicle programs were closely linked, with the X-38 already ten months behind schedule. CAV recommended combining the programs and increasing the funding profile by $120 million in order to accelerate the CRV's start date.

The Report contained two interesting tables. One (Table 3-1) estimated schedule slippage based on major risk elements such as Russian deliveries, training, software integration, and availability of the new CRV. It calculated a most optimistic slippage of only 10 months and least optimistic of 38 months. The second (Table 3-2) provided an estimate of cost growth associated with the major risk elements; most optimistic $174 million, least optimistic $968 million. The Report's Executive Summary stated in its final conclusion: "Completion of ISS assembly is likely to be delayed from one to three years beyond December 2003."

When the Report was first released, Goldin, at a hearing before the Senate Appropriations Subcommittee, said, "I do not acknowledge or accept the $7.3 billion overrun right now." He promised to give Congress a new estimate at the end of May. It was not clear how Goldin's estimate would differ from the CAV's independent report, the one required by Congress. As a show of progress, Goldin told the Subcommittee that even though all the funds had not been received by the Russian contractors, construction of the SM was continuing, and that 95 percent of SM components had been installed. He said a team led by Rothenberg would travel to Russia at the end of April to assess the status of the Russian programs.[31] Would Goldin try to shoot the messenger or address the problems the CAV identified? Testimony the committees had heard during the FY 1998 and FY 1999 hearings certainly supported the CAV's shortfall estimate. CRV development alone would add hundreds of millions. NASA's request of $5 million in FY 1999 to start a study was too little and too late to have the CRV ready when it would be needed, as the Chabrow Committee rightly concluded.

Congress certainly took the Chabrow report seriously. At the end of April it passed an emergency supplemental appropriation giving NASA permission to transfer $63 million from other accounts to Space Station. Chairman Sensenbrenner told attendees at a Space Transportation Association meeting that: "Congress does not want to write a blank check." He indicated that Congress would likely increase NASA's FY 1999 budget, but NASA and the White House would have to provide a credible cost runout estimate. At a hearing called by Sensenbrenner in May to force the Administration to take the necessary steps, Goldin finally admitted that bringing Russia in as a Space Station partner had backfired. Instead of reducing costs, costs had increased. Goldin said, "We were too naive in expecting [the Russians] to act like we act."[32] The Chabrow report forecast that because NASA would have to pay for Russia's failure to live up to its agreements, costs might grow from $17.4 billion to $24.7 billion. Now NASA and the Clinton White House would have to come up with a plan to keep the program on schedule. New estimates indicated that every month of delay would cost NASA $120 million. At the May hearing Franklin D. Raines, OMB Director, said, "We will look for offsets first from within the $6 billion spent annually on [the] human space flight account, and second from other non-priority areas." Sensenbrenner questioned how NASA would be able to find $7 billion in those accounts. Former Chairman Brown said the Administration should admit it made a mistake and take the money "from a proper source, like foreign aid."[33]

### Phase 1 Completed – New Launch Dates Announced

All the schedule uncertainties came to an end (for the moment) on May 22nd during a video conference of the Space Station partners chaired by Randy Brinkley. The launch of the FGB (named *Zarya*) from

---

31. Based on stories in *Space News*, Goldin Rejects New station Cost Estimate, Anne Eisele, April 27-May 3, 1998; and *Aerospace Daily*, Goldin not sold on report finding Space Station overruns, April 24, 1998.
32. Two stories contributed to this discussion: Cost Overruns May Prompt Budget Hike, Anne Eisele, *Space News*, May 4-10, 1998, and Partnership in Jeopardy, Anne Eisele, Space News, May 11-17, 1998.
33. White House, Congress Debate Station Funding, Joseph Anselmo, *Aviation Week & Space Technology*, May 11, 1998.

Baikonur would slip from August to November 20th. Node-1 (named *Unity*), the first U.S.-built element, would be launched December 3rd from KSC (Figure 29). Launch of AXAF, which had complicated the shuttle launch schedules, would slip to the last week in January 1999. The new schedule, called Revision D, was expected to be formally approved by the end of May. It would have 34 shuttle assembly flights and nine Russian launches (not including logistic flights), a reduction of three Russian flights from the previous schedule to reduce their costs. Gretchen McClain, Deputy Associate Administrator for Space Flight, said launch schedules after the first two launches would be worked out later in the month.[34]

On May 30 and 31, the International Space Station Control Board confirmed the schedules.[35] In addition to agreeing to the launch dates for the first two elements, other changes were announced. An exterior attached "warehouse" for spare parts, supplied by Brazil, would be added to the schedule. When and where it would be attached were not explained. The SM would be launched in April 1999 and the first Station crew in the summer of 1999. The U.S. lab module was scheduled for a October 1999 launch, with the ESA *Columbus* orbiting facility (COF), JEM, and two Russian lab modules to follow. The Canadian RMS was now scheduled for a December 1999 launch. One other decision, when to deorbit *Mir*, was also discussed. RSA would work to deorbit *Mir* "as is safely possible . . . with a goal of . . . July 1999." The press release said: "The agencies' leaders also acknowledged the atmosphere of cooperation, and look forward to the smooth transition to Phases 2 and 3 of the International Space Station." Was the program back on track, although far behind its earlier schedule?

Phase I of the redesigned Space Station program came to an official end on June 12, 1998. The shuttle *Discovery*, the last orbiter to rendezvous and dock with *Mir*, touched down at KSC, returning the seventh and final U.S. astronaut to visit *Mir*, Andrew Thomas, who had completed 140 days on board the Russian space station. During the ten flights that began in June 1995, nine of them docking flights, almost 30 tons of supplies had been transferred by shuttles to and from *Mir*. Frank Culbertson, NASA Shuttle / *Mir* manager said, "It's very difficult to imagine beginning the assembly of the International Space Station, beginning operations, without doing what we've done during the shuttle / *Mir* program." Culbertson would now move on and become ISS Deputy Program Manager at JSC. *Mir* would continue operating, until deorbited, with crews of Russian cosmonauts and French astronauts paying to visit *Mir* to conduct various experiments.[36]

### RUNOUT COST ESTIMATES UNRESOLVED – HARDWARE DELIVERED

Goldin did not keep his promise to give Congress a new estimate of Space Station runout costs by the end of May. Instead, on June 15th, NASA officially agreed that the Chabrow report conclusions were correct. The program was at risk of significant cost growth, but not the $7.3 billion estimated by the CAV. NASA's FY 1999 run-out estimate of $21.3 billion ($3.9 billion above the estimate made in 1993 of $17.4 billion) was $3.4 billion below the Chabrow Report high estimate. The White House told Congress that the near-term budget problems discussed in the Chabrow Report were addressed in the FY 1999 request's future year budget projections and the longer-term problems would be addressed in the FY 2000 submission. However, while waiting for the promised estimates, the House Science Committee recommended that the FY 1999 Space Station request ($2.27 billion) be reduced by $170 million. Appearing before the Science Committee on June 24, Goldin did not specify how NASA would close the funding gap to avoid further schedule delays. But he did say, "We are not giving up at NASA, and will not ask Congress for another penny until we feel additional funds are necessary to get the job done."[37] The FY 1999 budget disagreements would continue, and the Senate had yet to be heard from.

---

34. First Station Flight Set for No. 20, William Harwood, *Space News*, May 25-31, 1998.
35. ESA Press Release: International Space Station Partners Adjust Target Dates For First Launches, Revise Other Station Assembly Launches, June 2, 1998, No. 18-98.
36. Three press stories contributed to this wrap-up of Phase I, Shuttle / *Mir* Sails Into History, Craig Covault, *Aviation Week & Space Technology*, June 15, 1998; Shuttle / *Mir* Outlay Touted as Bargain, William Harwood, *Space News*, June 1-7, 1998; and French View *Mir* Visit as Station Training, Peter de Selding, *Space News*, June 1-7, 1998.
37. White House Balks at Station Cost-Control Decision, Warren Fester, Space News, June 29-July 5, 1998.

Figure 29. STS-88 crew that would attach first U.S. element *Unity* to *Zvezda*. Unity shown on KSC work stand. From left: Pilot Frederick W. Sturckow, Mission Specialist Nancy Jane Currie, Commander Robert D. Cabana, and Mission Specialist James H. Newman. KSC-97PC-944.

On July 17th the Senate acted. It proposed to add $130 million to the Administration's Space Station request and small amounts to other accounts bringing the NASA total to $13.6 billion. But in return the Senate would restrict NASA's ability to shift funds from other accounts by creating a separate Space Station line item. Senator Mikulski, former chairperson of the Senate Appropriations VA, HUD and Independent Agencies subcommittee, and now the ranking minority member said she was "deeply troubled by the increased frequency with which NASA reallocates funding within its appropriation account structure to accommodate these Space Station overruns . . . Ultimately the Space Station must be fiscally responsible and held accountable, like any other NASA program, and its cost overruns shouldn't affect other programs."[38] Her complaint was beginning to sound like a broken record, and repeated over and over again by other members of Senate and House committees. But there was never any resolution of the underlying problems. Program redesigns, mismanagement, contractor overruns, the addition of the Russians, and budget battles, had made it impossible for NASA managers to provide reliable cost estimates. The conference committee would decide which approach or compromise would be accepted.

38. Senate Bill ends Station Raids on Other Programs, Brian Berger, *Space News*, July 27-August 2, 1998.

Budget disagreements did not stop flight hardware from being delivered to KSC. The first of three Italian modules arrived on August 1st. Called multipurpose logistics modules (MPLMs), they were designed to carry experiments and supplies in a pressurized environment to and from the Space Station in the shuttle payload bay. Built by Alenia Aerospazio, they were large enough (14.8 feet in diameter by 21.3 feet in length) that, when docked to the Space Station, two astronauts could work inside. ESA also used the cylindrical shell of the MPLM design as the basic building block for the *Columbus* Orbiting Facility, including the MPLM thermal control system. The Italian Space Agency (ASI) and Alenia, starting from a modest commitment in 1992, had rapidly become a major player. It agreed to deliver, in addition to the MPLMs, two nodes, a thermal control system and cargo carrier for the automated transfer vehicle to be launched on an Ariane, *Columbus* module lab equipment, and two cupola units to allow astronauts better visibility of activities taking place outside the Station.[39] And in contrast to the Russian elements, all this hardware appeared to be on schedule.

Bad news continued to be reported from Russia. By the end of July, only four months before the first element was scheduled for launch, almost no Russian government funding had been transferred to contractors developing the elements that were needed after the FGB. NASA asked the White House to allow it to fund additional backups to the ICM in view of the reports that production of Progress and Soyuz vehicles had stopped. Several alternatives were proposed to supply reboost if the Progress vehicles were not available, including a new propulsion module, and using a Spacehab Inc. docking module that would allow the shuttle to use its aft jets to provide reboost. A new propulsion module was estimated to cost between $300 and $400, million while the Spacehab concept was much cheaper at $35 to $40 million. OMB Director Jacob J. Lew told the House Science Committee that the Administration would wait and see if additional backups were needed, and if they were, they would be addressed in the FY 2000 budget request.[40]

But the persistent Russian funding problems would not go away. NASA examined a number of options that would allow it to transfer additional funds to Russia to purchase needed hardware, such as Soyuz spacecraft, a different problem than developing backups for a shortage of Progress vehicles. There was no backup for Soyuz. The CRV, even if developed on an accelerated schedule, would not be ready for four or more years. If Soyuz was not available for emergency crew return, crews could not stay on the Space Station without a docked shuttle. PHC on the present schedule could only take place if Soyuz was available. During the remainder of August and through September NASA went round and round with the Administration and Congress to find a solution. NASA proposed to pay Russia an immediate $60 million "advance" as part of a $660 million payment over the next four years for needed hardware and services. The $660 million was part of a new $1.2 billion plan to make NASA independent of any further Russian involvement if it failed to resolve its funding problems.[41] The Administration said any additional funds would have to come from reprogramming in the NASA appropriation; Congress was saying it would not allow NASA to gut the agency's science and technology accounts to pay for Space Station overruns – a classic Washington impasse, with NASA caught between its two masters.

Adding to the debate, many close to the program believed that solving all of the Space Station problems might require double the $1.2 billion estimate to make the program independent of Russian involvement. The tug of war was partly resolved on October 15th. It was a complicated deal disguised as a tradeoff between the U.S. and Russia. In return for Russia leasing part of the SM's "space" to the U.S., in which it could conduct experiments and store equipment, NASA would give Russia $60 million from its 1998 budget to complete building the SM. Because of the delays in receiving funds from the Russian government, the SM launch would be delayed three months, until July 1999. NASA described the deal to Sensenbrenner in an October 13th letter and the Chairman responded that, before he would agree, NASA would have to send his committee a plan "at the earliest possible date" to remove Russia from the program. Goldin sent him the 1997 contingency plan and Sensenbrenner

---

39. Station Momentum Builds As Alenia Hands Over Module, Michael Taverna, *Aviation Week & Space Technology*, July 27, 1998.
40. White House Rebuffs NASA Plan To Fix Station, Joseph Anselmo, *Aviation Week & Space Technology*, August 10, 1998.
41. NASA Plans $660 million Station Bailout for Russia, Joseph Anselmo, *Aviation Week & Space Technology*, September 21, 1998.

agreed to the transfer.[42] That was the easy part; still unresolved was the matter of the remaining $600 million. It would be some time before NASA, the Administration and Congress would return to that issue. For the FY 1999 budget, the House-Senate conference split the difference. The total NASA budget was increased $200 million to $13.653 billion, and the Administration's original request for Space Station, $2.27 billion, survived.

Backing up one month, on September 6th, following a review and certification, NASA accepted the Boeing-built *Unity* module. *Unity* had been delivered to KSC in June 1997 to begin its processing. While at KSC, *Unity* was prepared for launch, along with other major pieces of hardware, in the Space Station Processing Facility. *Unity* would be carried on the shuttle *Endeavor* to the Space Station with two mating adapters to which other elements would be joined. *Unity* would then serve as the connection between U.S. and Russian elements.

At the end of October Congress directed NASA to assess the potential for commercial interests to play a role in the ISS and provide "commercial goods and services for the operation, servicing, and augmentation . . . and . . . commercial use of the International Space station."[43] As previously described, for the past three years some in Congress, led by Dana Rohrabacher (R-CA), chairman of the House Science Space and Aeronautics Subcommittee, had been pressuring NASA to open up the Space Station Program to commercial activities. With the passage of the Commercial Space Act of 1998 these advocates now had the mandate to get NASA's attention and conduct a market study of potential commercial interest in the Space Station. NASA was given 180 days from enactment to conduct the study and report back to Congress. If the schedule was kept, the report was due in April 1999.

### First Elements Finally Launched – Budget Battles Continue

November 20, 1998 arrived and the 22-ton *Zarya* departed, more or less on schedule. Less, if you remembered the November 1997 launch date first proposed in 1994, and still further behind schedule if you remember the 1985 projection of a 1991 first element launch. But that could be forgotten in the excitement of finally placing the first Space Station element in orbit. Design changes along the way increased power and fuel on board *Zarya* allowing it to remain on orbit almost 500 days, in case the SM was delayed beyond the new July 1999 launch date. The decision to have the U.S. fund *Zarya* through the Boeing contract had paid huge dividends. Without that arrangement it was likely that this critical element would have met the same fate as the SM. After achieving its orbit, checkout of *Zarya* systems went well except for a few minor problems. One of its antennas did not fully deploy and battery charging was not proceeding as expected. When the shuttle *Endeavor* arrived with the *Unity* node, the crew would be prepared to conduct repairs.

*Unity* also launched on schedule, December 3rd. On December 5th mission specialist Nancy Currie began the delicate job of extracting the 12.8 ton node from the payload bay using the Canadian manipulator arm and then attaching *Unity*'s pressurized mating adapter (PMA) to *Endeavor*'s docking system. After these steps were completed, mission commander Robert Cabana started the maneuvers to rendezvous with *Zarya*. On December 6th Currie captured *Zarya* with the manipulator arm and brought it to within 4 inches of linkup with *Unity*'s mating adapter. At that point Cabana flew the shuttle upward and achieved a hard docking, mating *Zarya* and *Unity*. During the two days of maneuvering, three mission control centers, two at JSC (shuttle and ISS) and the Korolev Center near Moscow, coordinated their activities.

After joining *Zarya* and *Unity*, U.S. astronauts James Newman and Gerald Ross conducted three EVAs over a five day period connecting cables between the spacecraft, installing handrails and insulation blankets, and many other tasks, all successful except for a few pieces of equipment that floated away during handling (a lesson learned, leading to improved tethering of EVA equipment on future flights). During the second EVA the stuck antenna on *Zarya* was pulled free. On December 10th, Cabana and Russian cosmonaut Sergei Krikalev entered *Unity* and turned on the lights.[44] The first two Space

---

42. Details of the funding transfer were derived from two *Space News* stories: Moscow Station Deal Crafted To End Service Module's Woes, Simon Saradzhyan, October 5-11, 1998, and NASA Receives Nod To Give Russia Funds, Amy Wittman, October 19-25, 1998.

43. Commercial Space Act of 1998 (Public Law 105-303) October 28, 1998. Title I – Promotion Of Commercial Space Opportunities, Sec. 101. Commercialization of Space Station.

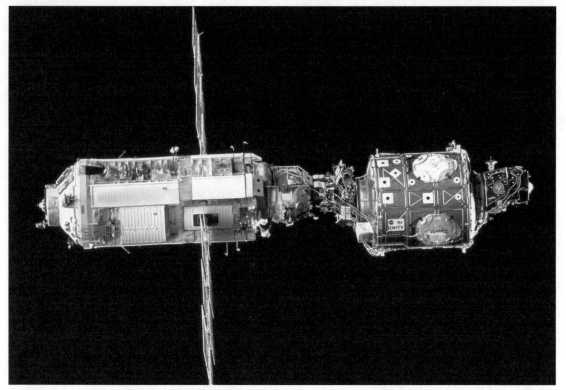

Figure 30. NASA photo of *Zarya / Unity* in orbit. Unity at right. STS088-703-019.

Station elements, one built in Russia, the other in the U.S., were ready to be cast free from *Endeavor* to begin their perilous odyssey – orbiting 240 miles above the Earth as the first components of the International Space Station (Figure 30).

Back on the ground everything was not going as smoothly. The Administration's FY 2000 budget was sent to Congress at the end of January 1999. At first glance, Clinton's FY 2000 NASA budget request of $13.578 billion, which included $2.483 billion for Space Station, seemed generous. But the devil was in the details. Plans were dropped to set aside $110 million a year for the next four years to subsidize Russian efforts. Russian services and hardware would continue to be purchased from funds set aside in the 1999 budget, while they lasted, but extra funds were not requested. Two accounts, one called Russian Program Assurance (the contingency plan to eliminate Russian participation), and the X-38 / CRV program would receive $348 million. A decision was expected in 2000 on the design of a six-person CRV. Funding for research that would be carried out on board the Station was reduced by seventeen percent. Aeronautics and technology funds would be drastically reduced, with one of the casualties being research on a supersonic transport.

These last two reductions angered many in the Congress whose constituents would be affected by the cuts. In a January 22 letter to Goldin, Sensenbrenner wrote: "Past administration promises to maintain a healthy research budget for ISS by paying back science funds 'borrowed' to make up for Russia's failures have not been kept."[45] Concerns were expressed again that when the Space Station was ready to accommodate experimenters, few would take advantage of the facilities. Budget hearings were scheduled in late February 1999 to restart the debates. Although the first elements of the ISS were successfully in orbit, nothing had changed in the Congressional hearing rooms; the same concerns were being raised and the same unsatisfactory answers were coming from NASA. Assembly of the ISS would move ahead, but no one could have completely forecast the next round of problems.

---

44. A step by step account of the mating of *Zarya* and *Unity* is contained in the December 14, 1998 issue of *Aviation Week & Space Technology*; U.S., Russian Modules Linked to Begin Station, Craig Covault.
45. Two *Space News* articles contributed to the FY 2000 budget discussion: Research Grounded by Budget, Brian Berger, February 8, 1999, and NASA Seeks $178 Million Hike for Station in 2000, Brian Berger, February 15, 1999.

## — CHAPTER 9 —
## THE STORY CONTINUES
### (1999 - 2005)

After the successful launch and assembly of the first two Space Station elements at the end of 1998, the Space Station's future was looking brighter. The U.S. / Russian partnership had survived formidable, and at times tense, differences, and cooperation was improving. However, NASA faced a continuing challenge to keep the program on track and relevant. Distracted by domestic and international issues, the Clinton Administration did not place the program near the top of its priority list. Following the change of administrations in 2001, the Bush Administration was immediately confronted with the enormous challenge of dealing with the devastating terrorist attacks of September 2001. Meanwhile, Space Station elements and equipment were arriving at KSC and Baikonur for checkout and launch. The most unsettling event, in terms of moving the Space Station program along, was the loss of space shuttle *Columbia* in February 2003. Returning to flight after the accident would become more difficult and costly than the recovery from the *Challenger* accident.[1] Larger events and growing budget deficits made it increasingly difficult to imagine how President Bush's space vision, announced in 2004, would play out. This chapter will briefly review highlights of each year from 1999 through 2005.

Before the *Columbia* accident placed the ISS assembly schedule in disarray, many other events unfolded, some a replay of past problems, some showing progress. When NASA's FY 1999 budget was last debated by Congress, the Chabrow committee had predicted that ISS runout costs would be $24.7 billion, and NASA had conceded it might be $21.3 billion. Unofficially, some NASA sources were now saying that the Chabrow committee estimates were closer to the truth. Runout costs were more likely to be $23.4 billion, with the possibility they could rise to $25.3 billion if the Russian SM was delayed beyond July 1999.[2] All of the uncertainties of FY 1999 carried over into Congressional consideration of NASA's FY 2000 budget request for $13.578 billion, which included $2.483 billion for the Space Station. Not paying back the funds "borrowed" from previous years' science budgets as promised, in order to keep the Space Station on track, continued to be a sore point in Congress and the science community. Launching the first elements had not succeeded in changing the outlook for the program in this regard.[3]

At the beginning of March 1999, NASA announced that a contract had been signed by the first Space Station paying customer to conduct research on board the ISS. The contract was between the Colorado School of Mines Center for Commercial Applications of Combustion in Space (CCACS) and Guigne Technologies Ltd. CCACS would lease the use of a Guigne-funded "furnace" to conduct research on ceramics and advanced materials. Using mostly NASA funds, CCACS would eventually invest some $12 million in developing the device, and in return planned to perform experiments when it was installed on the Space Station. CCACS had begun its research under a NASA grant and was preparing to conduct preliminary experiments on board the Spacelab to investigate the design of combusters, fire safety and suppression in zero "g" (a hazard of prime concern on manned spacecraft), sensors, and controls.[4] The

---

1. Recommendations made by the 13 member *Columbia* Accident Investigation Board (CAIB) chaired by retired Admiral Harold Gehman were contained in their report released on August 26, 2003. A good summary of the recommendations can be found in the September 1, 2003 issue of *Aviation Week & Space Technology*. Various NASA press releases were issued in 2003, 2004, and 2005 describing NASA's response to the CAIB recommendations and other actions taken by NASA to return to flight.
2. NASA Seeks $178 Million Hike for Station in 2000, Brian Berger, *Space News*, February 15, 1999.
3. Research Grounded by Budget, Brian Berger, *Space News*, February 8, 1999.
4. Based on a interview with Dr. Michael Duke, CCACS director, January 26, 2005. As of the beginning of 2005, CCACS had yet to perform an experiment as the "furnace" (total cost some $40 million) remained at KSC awaiting the return to flight after the *Columbia* accident. The "furnace" was actually an acoustic levitation containment device that allowed high temperature exothermic reactions to take place without interacting with the device walls. CCACS experiments were on board the *Challenger* Spacelab when it crashed, but almost 90% of the experiment results were recovered in real-time during the mission from down-linked telemetry.

CCACS / Guigne contract was the first fruit for Daniel Tam, who had left NASA and then was rehired by Goldin in February to commercialize more of NASA's programs. Tam predicted that within six months NASA would be able to show increased commercialization of certain NASA assets.

An article in *Science* magazine in May 1999 described the ambivalence that continued to exist in the science community to support or reject the program.[5] Supporters and critics were quoted. On the support side were those, such as University of Alabama, Birmingham, crystallographer Lawrence DeLucas, who were looking forward to longer and more frequent experiment periods to advance their disciplines compared to what was available on shuttle and Spacelab missions. To keep experimenters active and happy, NASA had increased grants to continue ground-based experiments, in spite of the "raids" on some of the science budgets. Comments from the critics were more colorful. University of Michigan professor emeritus Thomas Donahue said that in return for his original support NASA science budgets increased, but "I made a pact with the devil . . . The whole program has been a botched mess." Margaret Geller, an astronomer at Harvard-Smithsonian Observatory in Cambridge, Massachusetts, was quoted as saying, "You don't make Faustian bargains. It was obvious to me the Station had nothing to do with science. People were jumping on board to get money." Her observation was undoubtedly true to some extent, but how many university researchers were not looking to the government for support, whether it be from NASA, NSF, DOE or some other agency? Not very many.

NASA submitted an initial response to examing the commercial potential of the Space Station in May 1999.[6] In that report Goldin indicated that it was not possible to identify potential cost savings, revenues, or reimbursements that would accrue without specific commercial offers in hand. The report promised that NASA would continue to study the issue and identify initiatives that it would take to respond to the Congressional directive. In June GAO, in response to a request from Congressman Sensenbrenner, provided their analysis of whether or not commercial activities might reduce the annual cost of operating the Space Station.[7] After reviewing the actions that NASA had taken, and discussions with private companies, GAO "concluded that many businesses are skeptical of the station's commercial usefulness." GAO further stated that: "There is insufficient information at this time to estimate whether commercial activity would eventually reduce the space station's cost of operations." Wrapping up this subject, by the end of 1999 NASA had a number of studies under way, including a National Research Council Task Group, with the objective of developing specific recommendations and eventual legislation.[8]

After the launch of *Zarya* and *Unity*, it would be 19 months (July 2000) before the next element, the Russian Service Module *Zvezda*, joined them in orbit. Between December 1998, the launch of *Unity*, and July 2000, the NASA-Russian partnership endured another series of confrontations centered on when the SM would or would not be ready for launch. By continuously predicting readiness dates that came and went, Russia forced NASA to rationalize the delays to Congress and critics as to why the SM was late, and to downplay the impact. Promised availability dates slipped from July 1999 to September to November, and then into 2000. All during that time, the specter hovered over the program that, lacking an alternative, propulsion on *Zarya* might run out by April 2000, causing the first elements to reenter and burn up.

Ignoring the above difficulties, and continuing to look ahead to the time when the ISS would be fully operational, Mission Control at JSC was preparing to be the primary ground control. Six other ground stations were gearing up to monitor various aspects of the activities taking place during assembly or when the crews were on board. At MSFC the Payload Operations Integration Center would coordinate payload operations, planning and safety. Outside the U.S., in addition to the Russian control center at

---

5. A $100 billion Orbiting Lab Takes Shape. What Will It Do? David Malakoff, *Science*, Vol. 284, 14 May 1999 and attached story by Andrew Lawler.
6. NASA report in response to the Commercial Space Act of 1998 (P.L. 105-303 sec. 101, October 28, 1998), Commercial Development Plan for the ISS, May 14, 1999.
7. GAO, National Security and International Affairs Division, B-282682 to The Honorable F. James Sensenbrenner, Jr., Chairman, The Honorable George E. Brown, Ranking Minority Member, Committee on Science, House of Representatives, Subject: Space Station: Status of Efforts to Determine Commercial Potential, June 30, 1999.
8. Briefing to NASA Headquarters: ISS Commercial Development Program Progress Report, 9/28/99.

Moscow, control centers at Oberfaffenhoffen, Germany, and at Toulouse, France would monitor and control ESA activities. In Japan, the control center at Tsukuba would keep watch over Japanese hardware. The control center at St. Hubert, Quebec would monitor the operation of the Canadian robotic arm.

ESA and its signatory partners had been negotiating with NASA for three years to reduce its payments in return for launching the *Columbus* laboratory and other services. Final agreements were announced in March that resulted in ESA promising to develop an automated cargo supply vehicle, thus greatly reducing its payments. ESA's bartering negotiations with Russia for the same purpose had not been as successful. Although ESA would provide some in-kind help, they still would be forced to give the Russians cash in return for services once their astronauts were included in the crews. The total of the cash payments to both NASA and Russia could not be firmly fixed until the ISS assembly was complete.[9]

In May 1999 an issue that continues to resonate in Congress caught the Space Station in its web. Russia was accused of providing Iran with missile technology. Congressman Benjamin Gilman (R-NY) introduced a bill to withhold payments to the Russian Aviation and Space Agency (RAKA) that were to be paid in return for Russian crew time aboard the Space Station. NASA had already transferred $60 million for these services and was preparing to transfer an additional $100 million. RAKA officials denied that it had transferred missile technology, but apparently there was credible evidence that somewhere in the Russian space industry it had occurred.[10] Five years later, there is no doubt that Russia, along with North Korea, which also received Russian missile technology, supplied Iran with technology that allowed it to upgrade the performance of its long-range missiles. The Iran Nonproliferation Act of 1999, was worded to put pressure on Russia to stop the flow of ballistic missile technology to Iran. It eventually passed unanimously in both the House and Senate in 2000. It contained language that directly affected the Space Station. If the President would not certify that the flow of such technology had stopped, then NASA would not be permitted to pay Russia for any services, including the use of Soyuz for crew rotation and as the crew emergency return vehicle.

A Proton launch failure in October 1999, following an earlier failure in July, eventually caught up with the Space Station assembly schedule. The Proton was the launch vehicle for the Russian SM. Yuri Koptev, in November 1999, said the SM launch could be delayed for months until the investigation determined the cause of the rocket's second stage failure.[11] By the end of December the Russian investigating commission had not been able to determine the cause of the Proton crashes. Also in December, fleet-wide shuttle repairs for various electrical wiring and cable problems were projected to delay NASA assembly and resupply flights. The repairs impacted the overall shuttle flight manifest, including the launch of science payloads.

## 2000

January 2000 saw NASA still scrambling to compensate for the delayed launch of the SM. The latest estimates coming out of Russia were that a launch would be no earlier than July or August, in order to have at least three successful flights of the modified Proton before launching the SM. A delay of this length put in jeopardy the two elements already in orbit, since the FGB was only certified for operation through March. Batteries on board the spacecraft had already begun to fail because of overcharging errors before launch and other upgrades were needed. In order to conserve fuel on the FGB, reboost was curtailed and orbital altitude was beginning to decay at a rate of two miles per week. NASA quickly planned two resupply missions. The launch of *Atlantis*, originally scheduled to take place after the SM was in orbit, was moved up to March or April. *Atlantis*, while docked to *Zarya* and *Unity*, would use its steering jets to increase the Space Station altitude by twenty-three miles. A second supply launch would be scheduled as soon as possible after the SM was mated to *Zarya* and *Unity*.[12]

---

9. Bartering Cuts ESA's Payments for Station, Peter B. de Selding, *Space News*, March 8, 1999.
10. Russian, U.S. Space Officials Decry Iran Bill, Brian Berger, *Space News*, July 26, 1999.
11. Proton Review May Delay Station Module Launch, in *Space News*, November 15, 1999.
12. NASA Plans Two Missions For Vacant Space Station, William Harwood and Simon Saradzhyan, *Space News*, January 17, 2000.

Staffing shortages continued to be a problem for the Space Station and other programs. In the FY 2001 budget request NASA asked for funds to hire an additional 500 new staffers for the human space flight centers JSC, MSFC, and KSC, and a total of 1,850 full and part-time employees. The Aerospace Safety Advisory Panel (ASAP) had warned in its 1999 report: "NASA must continue to address workforce problems aggressively and establish program priorities that ensure a workforce capable of achieving long-term safe and effective operations." It added: "Emphasis should be placed on . . . recruiting younger S&Es [scientists and engineers] who can develop into experienced and skilled future leaders." They were especially concerned that: "Space Shuttle processing workload is sufficiently high that it is unrealistic to depend on the current staff to support higher flight rates and simultaneously develop productivity improvements to compensate for reduced head counts." Congress' tight hold on NASA's personnel budgets had crippled the Space Station in the past and would continue as the personnel ceiling was never increased to meet the obvious shortages that caused the ASAP concerns.

After several weather-related launch delays, *Atlantis*, after a one-year absence, rendezvoused and docked with *Unity* in May. Batteries were replaced and *Zarya* and *Unity* were nudged into a higher orbit. Other repairs made during this mission included securing the U.S. crane mounted outside *Unity*; attaching pieces to the Russian crane, also mounted outside *Unity*; and replacing fans, smoke detectors and other gear that had exceeded their lifetimes. In June ESA committed to the launch of nine ATVs with the Ariane-5 launch vehicle during the period 2003-2013.[13]

All the friction and frustration that had developed between the U.S. and Russia over schedule delays faded with *Zvezda*'s launch on July 12, 2000 and the successful docking with *Zarya* and *Unity*. *Zvezda* measured 43 feet long and 95.5 feet across, including the deployed solar panels. It had three pressurized compartments, two staterooms and 14 windows. With this event behind them the partnership looked forward to a heavy schedule of nine U.S. and three Russian launches over the next fifteen months. *Zvezda*'s successful mating resulted in NASA canceling the ICM contract with the Naval Research Laboratory. In November 2000 Bill Shepherd, and his Expedition 1 Russian crew mates Yuri Gidzenko and Sergei Krikalev, reached the Station for a four-month stay (Figure 31). Arriving on a Soyuz spacecraft that remained docked as their CRV, they became the first sleep-over Space Station occupants.

*Zvezda* became their new home, where they would provide needed maintenance and assist in assembling the next elements. Very little time would be available to conduct research. They were immediately confronted with the problem that noise levels on both *Zarya* and *Zvezda* were too high for comfortable occupation. Background noise created by fans and other equipment had been a problem on *Mir* requiring the crews to sleep with earplugs. This complaint was noted in the report made by NASA IG Gross in 1997 and apparently was an innate Russian technology problem that had not been corrected for *Zarya* or *Zvezda*. Engineers back on Earth went to work to come up with a way to reduce the noise levels that threatened the crews' ability to operate productively during their stay. Skipping ahead eighteen months, despite several attempted "fixes," the problem never was solved for later crews. Whereas the U.S. elements were quiet, astronaut James Voss, who slept in *Zvezda* during his four-month stay, suffered a hearing loss. Crew mate Susan Helms, who slept in *Destiny*, occasionally wore ear plugs to muffle the continuous high-level background noise generated in the attached Russian modules.[14]

*Endeavor*'s primary objective during its December mission was to attach and deploy the 65 kW, 240-foot, solar array it carried to the ISS. Problems were encountered during deployment, causing the array to develop strong undulations that risked damaging the fragile structure. Tiger teams were assembled at JSC and at the manufacturer's site to find a solution. An astronaut team, working in the JSC neutral buoyancy tank, rapidly developed repair procedures that were sent up to the crew members who would attempt to fix the problem during EVAs. When *Endeavor* left, the arrays had been deployed and were providing power. However, when *Endeavor* landed it was discovered that an external tank separation detonator had not fired. The redundant detonator had done its job thus avoiding the catastrophic loss of the orbiter. Until this problem was understood and corrected, the fleet was grounded again.

---

13. International Space Station Gets A Boost From Europe, ESA Press Release No. 37-2000, June 7, 2000.
14. Space Station too noisy for astronauts who live there, AP story printed in *The Jacksonville Times* Union (Jacksonville, Florida), November 19, 2001.

Figure 31. Expedition I crew. From left: Commander Willam M. Sheperd, Sergei K. Krikalev, Yuri P. Gidzenko. STS098-353-0006 For a complete listing of all Expedition crews visit NASA web site http://spaceflight.nasa.gov/station/ Click on expedition press kits. Pick A Kit or crew.

## 2001

January 2001 saw the inauguration of President George W. Bush, bringing in a new team to deal with the continuing ISS cost overruns. The $25 billion cost ceiling imposed by Congress in NASA's FY 2000 authorization, and signed by President Clinton in October 2000, was finally acknowledged by NASA to be $3 billion or more below its needs.[15] The new administration's response was to scale back the program when it submitted its FY 2001 budget, replacing the budget proposed by the former administration. Elements affected included the hab module, CRV, and the centrifuge accommodation module. As a result, crew size would be restricted to three people and the future of the centrifuge, once the centerpiece facility for many scientific investigations, was left in limbo. With a related action, the Bush Administration announced it would stop funding the X-33 and X-34 programs that had already received over $1 billion, thus terminating efforts for a potential shuttle replacement.

Restricting the crew to three persons opened up a new set of problems. Recently completed studies indicated that it would require three crew members just to operate all of the Space Station systems, with little time left over to perform other work. Representative Tim Roemer (D-IN), was quoted as calling the changes "the incredible shrinking science program."[16] In a letter to Joseph Rothenberg, NASA Associate Administrator for Space Flight, Martin Fettman at Colorado State University, who was planning to conduct animal experiments on the Space Station, wrote: "We were at the extreme edge of maintaining a credible science endeavor; we might as well completely discontinue."[17] Once again, a

---

15. Major Cost Growth Looms For Space Station, Brian Berger, Space News, February 19, 2001.
16. Smaller Station Crews Could Reduce Scientific Output, Brian Berger and Peter B. de Selding, *Space News*, March 5, 2001.
17. New Cuts in Space Station Spark Walkout, Andrew Lawler, *Science*, Vol. 291, 23 March 2001.

new administration had imposed a change of direction on the Space Station that clouded its future and reopened the question of how useful the Space Station would be as an experiment facility and laboratory.

In February, the U.S. lab module *Destiny* was docked with the other three elements in orbit. The 28 x 14 foot laboratory, with 4,750 cubic feet of working space, was designed to open up a new dimension in ISS operations. *Destiny* arrived with thirteen computers and five science racks; an additional nineteen racks were scheduled to be added on later flights. MSFC's Payloads Operations Center came on line to coordinate science activities.[18]

With *Destiny* now attached to the Space Station, NASA briefed the NAC on the new schedule (Table 1) and utilization plan (Table 2) that would carry the program through 2006.[19] Mark Uhran, Headquarters Director of Research Integration & Product Development, told the NAC, "We are ready to begin research on board the Space Station." He provided a long list of equipment that would be delivered and experiments that would take place. He reviewed the status of NASA's plans to develop a Non-Government Organization (NGO) to manage Space Station utilization for both domestic and international users. Planning for the NGO had been under way since the end of 1999 in response to Congress' directive to commercialize the ISS. In complying with this direction, Uhran said the plan would be submitted by the end of September.

## GENERAL SCHEDULE

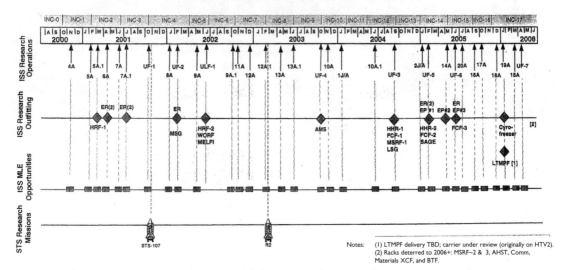

Table 1. Schedule for Space Station assembly 2000-2006.

After fifteen years and one month in orbit, *Mir* was successfully deorbited in March 2001. Elements that did not burn up splashed down safely in the Pacific Ocean. *Mir* crews (104 individuals including seven American astronauts) set many records. Among them: longest space mission, 438 days, Valery Polyakov; longest total time in space, Sergei Avdeyev, 747 days.[20]

---

18. *Destiny* Laboratory Faces ISS Science, Software Challenges, Craig Covault, *Aviation Week & Space Technology*, February 12, 2001.
19. International; Space Station Utilization Planning. Briefing to the NASA Advisory Council, 15 March, 2001. Mark Uhran, Director Research Integration & Product Development / UM.
20. *Mir* records from a AP story, Odyssey for Mir Station Nears Its End, in *The Florida Times Union*, March 11, 2001.

| UTILIZATION CAPABILITIES | NASA | CSA | ESA | NASDA |
|---|---|---|---|---|
| **User Accommodations** | | | | |
| a. NASA Lab Module<br>    Annual Average Rack Locations DRE | 97.7% | 2.3% | | |
| b. NASA CAM<br>    Annual Avera Rack Locations DRE | 97.7% | 2.3% | | |
| c. NASA Truss Payload Accommodations<br>    Annual Average Truss Attach Points AP | 97.7% | 2.3% | | |
| d. ESA COF Module<br>    Annual Average Rack Locations (DRE)<br>    Annual Average COF Attach Points AP | 46.7% | 2.3% | 51.0% | |
| e. Experiment Module<br>    Annual Average Rack Locations (DRE)<br>    Annual Avera a JEM-EF Attach Points AP | 46.7% | 2.3% | | 51.0% |
| **Utilization Resources** | | | | |
| Annual Average Power (kW)<br>Annual Total Crew Timi hours | 76.6% | 2.3% | 8.3% | 12.8% |
| **Rights to Purchase Supporting Services** | | | | |
| a. Space Station Launch and Return Services<br>    Pressurized Up Mass (kg)<br>    Unpressurized Up Mass (kg)<br>    Pressurized Down Mass (kg)<br>    Unpressurfzed Down Mass (kg)<br>    Pressurized Volume (DRE)<br>    Pressurized Down Volume ORE | 76.6% | 2.3% | 8.3% | 12.8% |
| b. Communications data transmission capacity<br>    Annual Average Down Link Mbps | 76.6% | 2.3% | 8.3% | 12.8% |

Table 2. U.S. and international allocations of Space Station capabilities.

Another space flight first occurred in April and May 2001. Over strong NASA objections and criticism from members of Congress, Russia, in return for $20 million, carried a tourist to the Space Station on a Soyuz crew change-out flight. Dennis Tito, a businessman and former JPL engineer, originally negotiated the deal with the cash-strapped Russians to fly to *Mir*. But when *Mir* was deorbited he went instead to the Space Station, staying for six days. Because of NASA's objections, he was forced to stay in *Zvezda* and was reported to have spent most of his time listening to opera on CDs and looking out the windows.

When installed in April, the Canadian RMS, now designated the Space Station remote manipulator system (SSRMS) and called the Canadarm II, ran into immediate problems. During checkout it did not respond correctly to backup commands. Two problems were found, the wrist joint malfunctioned when it was operated and communication dropouts occurred during control commands. Software patches were designed to fix the problems and JSC then rejuggled the assembly flights in order to install the patches.[21] Repairs were made in July during the STS-104 mission that also delivered the joint airlock module named *Quest*. All the EVAs conducted by the STS-104 crew during the mission were assisted by the Station Expedition 2 crew operating the Canadarm II from within *Destiny*. With this mission, Phase II of Space Station assembly was completed with only a few problems encountered, all of which were resolved. In August the Expedition 3 crew was ferried to the Space Station along with the Italian MPLM carrying supplies and several new life sciences investigations.

NASA was not alone in having problems containing costs. ESA announced in June that it was abandoning its fixed-price contract with EADS to develop the Automated Transfer Vehicle (ATV).

---

21. Persistent Station Arm Woes May Force Shuttle Mission Shuffle, Frank Morring Jr., *Aviation Week & Space Technology*, June 4, 2001.

Jorge E. Feustel-Buechl, ESA's space station director, said, "We are going to be substantially over the original price," attributing the problem to design hurdles and mismanagement at both the contractor and ESA.[22] Designing and building space hardware has never been and never will be a simple undertaking, as all who have taken up the task have discovered.

By the end of summer 2001 the estimate of the Space station cost overrun had ballooned to $4.8 billion above the $25 billion ceiling set by Congress. Aware that he would soon have to announce the cost overruns, Goldin established in July a management and cost evaluation (IMCE) task force chaired by Tom Young. Program and management changes that took place after the redesign in 1993 were analyzed. Problems reviewed included those that arose from the earlier belief that NASA could "ship and shoot" Space Station elements delivered to KSC. The result was an underestimation, by a factor of four, of the amount of software code needed to provide integration and verification. NASA Comptroller Malcolm L. Peterson told the task force that ISS managers took the approach that they would pursue a program to stay within the yearly appropriation rather than concentrate on total program cost. Uncertainties caused by the delay of the Russian SM were cited as a major reason that recent budgets had been difficult to project.

Young was quoted as remarking after the Peterson briefing, "What you are describing is a program that's managed technically only – not managed to schedule, not managed to cost. In this day and time . . . it comes across to me as the dark ages of program management."[23] Sean O'Keefe, OMB Deputy Director, and soon to be appointed as NASA's next administrator, told the task force, "Know that there is nothing else out there to pay for these [overrun] costs."[24] Faced with the requirement to quickly come up with new estimates after the redesign, ISS managers had been pushed into a cost estimation corner. OMB had been alerted in the fall of 2000 of the potential need for additional funding, and the NASA comptroller's office had begun a new, detailed, cost estimation process. Responding to the new costing process, JSC managers realized that in the rush to come up with a new estimate they had "left out" some items and resubmitted much higher numbers. With flat budgets mandated in future years, NASA had no choice except to manage the program within the yearly appropriations regardless of the impact on program decisions.[25] Recommendations from Young's task force were due after the November elections.

The terrorist attacks of September 11, 2001 changed many of the nation's funding priorities. However, Congress continued to push NASA to get the Space Station back on track. Of continuing concern were the decisions by the Administration to reduce crew size and the reduction of funding for research. Prior to September, both House and Senate appropriations committee bills added funds to the Administration's overall NASA FY 2002 request. They included an additional $35 million by the House and $50 million by the Senate earmarked for research.[26] After the two bills went to conference, summing all the Space Station accounts, an appropriation of $2.094 billion was passed, $7 million more than originally requested.

While NASA was wrestling with its funding problems, international partners at ESA and Japan were trying to understand how they would fit into the downsized program. The many changes were viewed by the partners as threatening the schedule and causing a protracted delay for their contributions, the *Columbus* module and JEM. In addition to changes such as the termination of the U.S. hab module and

---

22. Europe Abandons Fixed-Price Deal For ISS Cargo Ship, Peter B. de Selding, *Space News*, June 11, 2001.
23. The IMCE team report was derived from emails sent to the author at this time and is based in part on excerpts from *Aerospace Daily*, August 21, 2001.
24. OMB Official Tells NASA Station Bailout Unlikely, Brian Berger, *Space News*, August 27, 2001.
25. Young's criticism of NASA's management of the Space Station energized NASA to improve its ability to forecast the program's costs. Cost estimators were brought in from DOD and the Aerospace Corp. to beef-up NASA staff. In interviews with Mal Peterson in June 2005 he described how immature some of the program's assumptions were leading to low cost estimates. As described, when the new JSC estimate finally arrived and included the "missing items," the cost runout proved to be much higher than expected. The program was still trying to catch up with the fallout from the redesign and "the marching orders" given by the Clinton Administration to control costs. Once again, one of the frequent changes the program was forced to accommodate had tripped up NASA managers never permitting the program to settle down.
26. For a complete review of the language in the two appropriation bills see House Report 107-159 and Senate Report 107-43.

CRV, the number of shuttle flights to support the Space Station would be cut to four a year. Crew size, reduced to three for the immediate future, meant there would not be enough crew time available to conduct all of the investigations that the partners were planning. As a result, ESA decided to freeze non-essential spending and await the outcome of NASA's budget crisis.[27]

Goldin's tenure as NASA Administrator came to a conclusion on November 17, 2001. As neither the Clinton or the first Bush administrations had appointed a Deputy Administrator during Goldin's tenure (both Aaron Cohen and John R. Dailey only held acting positions and the position had been vacant for two years since Dailey's departure), Daniel R. Mulville became the Acting Administrator for one month. Sean O'Keefe was confirmed as the new administrator at the end of December. Coming from his previous job at OMB, O'Keefe's mandate was first to control costs, primarily on the Space Station, and then bring in managers who would help him get tough and solve other agency problems.

Goldin's departure coincided with the delivery of the recommendations from Tom Young's IMCE task force. Briefly, the IMCE report began with the obligatory praise finding that: "The ISS Program's technical achievements to date, as represented by on-orbit capability, are extraordinary." But from that point on it was all downhill. It stated: "The existing ISS Program Plan for executing the FY 02-06 budget is not credible." The Task Force then went on to recommend major changes in how the program was managed. It proposed continuing with a less capable Space Station for the next two years, until NASA showed that it could manage the program and control the costs. If NASA made progress on these concerns, the program could continue and the Space Station would be completed with most of the capabilities envisioned in 1993.[28] If the Task Force recommendations were followed it would result in an early Space Station configuration called "core complete" that would support a three person crew but omit the CRV, Node-3, and hab module – big, perhaps insurmountable ifs! The projected near-term savings were $1 billion.

Release of the IMCE recommendations brought immediate criticism from the ISS partners. At a meeting in early December with the NAC, each partner presented its concerns with the new direction that the Space Station was taking. The unilateral decision by the U.S. to reduce crew size and delay assembly were the chief sticking points. The Canadian representative reminded NASA that the program was based on a binding "Treaty Level" Intergovernmental Agreement. The ESA representative said the changes would penalize the international partners. Responding to these concerns Young told the gathering, "What NASA should do is put their noses to the grindstone and excel for the next two years, and earn back respect, credibility and confidence."[29] In spite of the displeasure expressed by the international partners, the IMCE recommendations were adopted. The international partners would have to wait in the wings and hope NASA would be able to recover from its management problems. If NASA was not successful, the partners would be left in a untenable position, with little recourse to salvage all their investments.

## 2002

At the beginning of 2002, the NASA Aerospace Safety Advisory Panel (ASAP) released its annual report for 2001. In Section II, Pivotal Issues, it addressed current NASA safety activities. Two concerns deserve mention. ASAP stated that since last year's report was prepared the long-term situation had deteriorated. Budget constraints had forced the space shuttle program to adopt an even shorter planning horizon in order to continue flying safely. As a result, items that should have been addressed were being deferred, adding to the backlog of required improvements. ASAP ended this section by saying it had "significant concerns" that the growing backlog "delayed or eliminated" identified safety improvements. In the section labeled Mishap Investigation it stated: "a mishap resulting in small economic loss but having *potential* [report emphasis] for significant loss of life or assets would not necessarily result in an investigation at the highest level. NASA managers do have

---

27. NASA's ISS Woes Stymie Partners, Frank Morring, Jr. *Aviation Week & Space Technology*, October 15, 2001, and ISS Budget Crisis Prompts European Spending Freeze, Peter B. de Selding, *Space News*, November 19, 2001.
28. Report by the International Space Station (ISS) Management and Cost Evaluation (IMCE) Task Force, to the NASA Advisory Council, November 1, 2001.
29. ISS Partners Criticize Task Force Report, Leonard David, *Space News*, December 10, 2001.

the prerogative to elevate an investigation to whatever level they deem appropriate, but this is seldom done, as they are not required to do so."[30] Did any of the "delayed or eliminated" improvements or slow pursuit of a small mishap contribute to the *Columbia* accident one year later? Certainly some of the warning signs were there.

Debate on the FY 2003 budget, the first opportunity for the Bush Administration to submit a budget fully reflecting its priorities, began in February 2002. Fallout from the September 11, 2001 attacks, as reflected in agency budgets, would result in some changes in the months ahead. For NASA, the $15 billion submission contained some new priorities, such as funding a major effort to develop nuclear power systems in conjunction with DOE. However, it also left open questions on funding shuttle safety upgrades and whether or not shuttle flights would be scheduled to service the Hubble Space Telescope. NASA, now with OMB's former deputy director in charge, could anticipate a number of new directions in the coming year.

As in past administrations and Congresses, the question was: "Would the Congress, controlled by the President's party, follow the President's leadership?" The Administration's request of $1.839 billion for the Space Station (that included $347 million for research) was some $200 million below the FY 2002 appropriation. The first hearings did not bode well for the program. Questions were raised on the perceived shortfall of $600 million needed to achieve the scaled-back Space Station "core complete" configuration and terminating $140 million of ISS research contracts. O'Keefe attempted to assure members of the House Science Committee that there would be sufficient funds for the ISS and, in any case, he had asked for a new cost assessment that was due in the summer. O'Keefe ran into even tougher questioning when he appeared before the House Appropriations Committee. Tom Delay (R-TX) was described as "incensed" that the X-38 program, which would have led to a seven passenger CRV, was terminated.[31]

The April mission of *Atlantis* succeeded in installing the center truss section (S-Zero), the element to which future trusses, cooling systems, and solar arrays would be attached. The other key element installed was the TRW mobile transporter that would eventually hold the Canadian RMS. The launch was not without incident; there was a problem and a rapid solution. At T-5 minutes and 30 seconds software glitches were noted. With the five minute launch window staring at them, United Space Alliance technicians were able to reload the affected software and the launch took place with 11 seconds to spare.[32]

A June mission by *Endeavor* relieved the Expedition 4 crew that had spent six months on the ISS. The Expedition 5 crew, two Russians and one American, was scheduled to stay for four and a half months. During their mission some new elements were installed, 5,600 pounds of cargo were transferred, and 3,000 pounds of trash and equipment to be refurbished were off-loaded for return to Earth. The most important piece of equipment mated to the ISS on this mission was the Canadian mobile base system and arm, with a wrist joint replacement for the original joint that had experienced the already described malfunctions. Debris shields were attached to the Russian SM. A new experiment facility, the ESA Microgravity Science Glovebox for materials, fluid, combustion, and biotechnology investigations, was transferred from the MPLM to the lab module.[33] The glove box would have an electrical malfunction five months later and the electronic panels believed to be the probable cause were removed and returned for examination in the hope that a repair could be made.[34]

On July 10th, the task force appointed by O'Keefe in March reported to the NAC and senior NASA officials. Its charter had been to examine the Space Station's ability to support research that would enable human exploration of space and have intrinsic scientific merit. Chaired by Dr. Rae Silver of

---

30. Aerospace Safety Advisory Panel Annual Report For 2001 (no release date), Code Q-1, NASA Headquarters, Washington, DC 20546.
31. Hearings background from: NASA's O'Keefe Tangles with Texans, Andrew Lawler, *Science*, Vol. 296, 26 April 2002; and Frustration Surfaces at NASA Budget Hearing, Brian Berger, *Space News*, March 4, 2002.
32. Fast Launch Teamwork Propels *Atlantis* to ISS, Craig Covault, *Aviation Week & Space Technology*, April 15, 2002.
33. Based on prelaunch article: International Operations Mark *Endeavor* Flight, Craig Covault, *Aviation Week & Space Technology*, May 27, 2002; and Space Shuttle Delivers New Crew, Supplies to Station, Brian Berger, *Space News*, June 17, 2002.
34. Engineers Rush To Diagnose Station Glovebox Malfunction, Ben Iannotta, *Space News*, December 16, 2002.

Columbia University, the task force recommended that 15 of the 32 research areas reviewed be labeled "first priority," including radiation health, cell and molecular biology, and fluid stability and dynamics. It urged NASA to provide the centrifuge facility as soon as possible so that high priority experiments could begin. But in a sharp rebuke, the task force also stated that NASA should stop claiming that the ISS was "a science-driven program" until crew size and facilities were expanded.[35] The report received decidedly mixed reviews. Some saw it as justifying expanding the Space Station's capabilities. Many feared the recommendations would be used to kill the program, because it would be some time before crew size could be increased and the needed facilities would be available for experiments.

Under tight security, beefed up after the 9/11 attacks, *Atlantis* was launched in October to add the next truss section to the ISS. It was followed at the end of November by the flight of *Endeavor* carrying the third truss segment. In addition to mating the trusses to the S-Zero truss, the crews on both missions displayed their abilities as plumbers and electricians, connecting the cooling system piping and cabling that would allow the Space Station to expand the cooling system and add 75 kW of solar arrays. With the two new truss sections the main truss now measured 135 feet and was scheduled to expand in the coming year to 356 feet with additional segments. *Endeavor* also carried the Expedition 6 crew replacing the Expedition 5 crew.

*Endeavor* departed the ISS leaving a small problem to be worked out in the weeks ahead. While attempting to reposition the mobile transporter (MT) during the third EVA, it was stopped when it ran into a folded UHF antenna. The antenna was scheduled to be deployed at a later date, but configuration control had missed the fact that while repositioning the MT the antenna was still in the way. *Endeavor*'s crew cleared the antenna, but the complete repositioning of the mobile transporter was deferred to a later date, until it was understood why the antenna obstruction was not taken into account before moving the MT.[36]

In between the two shuttle launches, Russia dispatched an improved Soyuz spacecraft to the ISS replacing the older version that served as the CRV. The modified Soyuz, funded with $40 million provided by NASA, would accommodate larger U.S. astronauts, who were, in general, larger than their Russian counterparts. Other changes included improving the environmental control, data, and television systems. The contract to build the new Soyuz also provided that Russia would furnish eleven Soyuz to be used as CRVs through 2006.[37]

NASA amended its FY 2003 budget request in early November to shift $2.325 billion within budget line items to pay for shuttle enhancements, ISS activities and preliminary studies for an Orbital Space Plane (OSP). By the end of 2002, the FY 2003 budget was still awaiting Congressional action. The two teams chartered by O'Keefe in the spring to audit the Space Station books issued their reports. One team, composed of NASA employees and contractors, concluded that the estimate of $4.8 billion to operate the program until 2006 was still some $500 million low. The second team, from DOD, decided that there was only a $117 million shortfall. Taking the high estimate and then adding to it, NASA planned to ask for $600 million more than originally projected for the Space Station program over the next three years. The request was not very likely to be approved by Congress in view of the reduced FY 2003 request and projected deficits. The international partners also continued to experience their own funding problems, attempting to maintain other space programs while still supporting the ISS.

In spite of these issues, NASA began discussions with the international partners at the end of the year to determine their interest in participating in developing the Orbital Space Plane (OSP). The overture was well received because the plan to restrict crew size to three persons had been a major concern since

---

35. Report by the NASA Biological and Physical Research, Research Maximization And Prioritization (ReMAP) Task Force to the NASA Advisory Council, August 2002. Reactions to the task force report were drawn from: Bigger Is Better for Science, Says Report, Andrew Lawler, *Science*, Vol. 297, 19 July 2002; and O'Keefe: Science Goals Setting ISS Capability, Frank Morring, Jr., *Aviation Week & Space Technology*, July 15, 2002.
36. The *Atlantis* / *Endeavor* missions were covered in three reports: Atlantis Crew Installs Truss, no author, *Aviation Week & Space Technology*, October 21, 2002; ISS Crews Exchanged; Outpost Nears 200 Tons, Craig Covault, *Aviation Week & Space Technology*, December 2, 2002; and Robotics Hang-up Provides ISS Learning Experience, Craig Covault, *Aviation Week & Space Technology*, December 9, 2002.
37. Improved Soyuz Launched to ISS, Craig Covault, *Aviation Week & Space Technology*, November 4, 2002.

it was first announced. If the OSP were used as a CRV it would allow the ISS to accommodate crews of seven or more, rather than the three that could be accommodated by the Soyuz CRV in an emergency return. Negotiations were expected to be completed by March 2003.[38] The major drawback to developing the OSP was that its earliest availability was projected to be 2010, a penalty that reduced its attractiveness to the partners, who needed more crew time as soon as possible. If launch schedules held, the ESA and Japanese laboratories would be in orbit but would only be fully utilized for short periods when the shuttle was docked and could carry larger crews.

## 2003

Space shuttle *Columbia*, STS-107, lifted off from KSC on the morning of January 16, 2003. During their seventeen days in orbit, the crew had trained to conduct the most extensive series of science experiments ever attempted on a shuttle mission. In the cargo bay was the Spacehab Research Double Module carrying many of the experiments. To complete all of the work, a 24-hour day, two-shift operational workload had been scheduled. Before the crew settled in for reentry on February 1st, their reports indicated that they had successfully accomplished their objectives. However, the results of all their work would never be completely known as *Columbia* disintegrated during reentry with the loss of the crew and experimental data. An accident board (*Columbia* Accident Investigation Board – CAIB) was immediately established to determine the reasons for the tragedy.[39]

With the loss of *Columbia*, the future of the Space Station became difficult to predict and planning efforts effecting future missions were placed on hold. Between 2003 and 2008, twenty-two more shuttle assembly flights had been scheduled. Critical elements that remained to be ferried to the ISS included solar arrays, four more truss sections, the JEM facilities, the ESA *Columbus* lab, the centrifuge module, another node, a cupola, and the special purpose dexterous manipulator. Being only about half way through the assembly and logistic flights before the *Columbia* accident, the Space Station was an imposing collection of hardware, but it was still far from being completed (Figure 32). Michael Kostelnik, Deputy Associate Administrator for Space Shuttle and Space Station, said, "If worse comes to worst, we could de-man the station."[40] How long the shuttle fleet would be grounded was anyone's guess, but using the *Challenger* experience as a guide, two years was not out of the question. It was soon decided that the launch of ESA's *Columbus* laboratory and the JEM would be delayed at least two years.

All of these uncertainties led NASA to develop new operating plans that relied heavily on the Russians to keep the ISS functional until the shuttle returned to flight. Russian Soyuz and Progress vehicles would continue to bring supplies, change-out crews and provide the propulsion for reboost. The one critical crew expendable that would not fit into this planning was water. Normally the water was resupplied during each shuttle mission, as it was available in large amounts from the operation of the shuttle's fuel cells. A three person crew used 4-6 liters a day and the Russian spacecraft were not configured to deliver 500 liters or more of water on their flights. One of the solutions selected to reduce the need for expendables was that when the three-person Expedition 6 crew was brought home in April it would be replaced by a crew of two.

NASA's FY 2003 budget of $15.389 billion was finally approved in February 2003. The overall budget was increased by $389 million above the administration's request, but the Space Station would receive only $1.845 billion, the lowest amount since FY 1991. With inflation factored in, it was the smallest budget since FY 1989. A positive development was language in the bill, patterned after the Hubble Space Telescope Science Institute, allowing NASA to establish a private institution to oversee ISS science. However, it would not have as much authority to select all the experiments to be conducted as the Space Telescope Institute. NASA would control the peer review process and the ISS

---

38. Europe Pressing NASA For Space Plane Role, Frank Morring, Jr., *Aviation Week & Space Technology*, December 23, 2002.
39. The report by the *Columbia* Accident Investigation Board (CAIB) is available on the NASA and other web sites. For a succinct synopsis of the CAIB findings, the author recommends the September 1, 2003 edition of *Aviation Week & Space Technology*.
40. NASA Has Few Options for ISS During Shuttle Grounding, Brian Berger, *Space News*, February 10, 2003.

Figure 32. International Space Station NASA photo showing configuration at the end of 2005. From bottom to top: *Destiny* lab module, *Unity*, *Zarya*, *Zvezda*, and Soyuz. STS114-304-015.

institute would manage the research. The late approval of the FY 2003 budget coincided with the start of the debates for FY 2004. President Bush's OMB continued to be generous with NASA, requesting $15.469 billion; however, the Space Station request was only $1.707 billion, reflecting the halt in assembly until the shuttle returned to flight.

If all the recommendations from the CAIB report released in August were accomplished (15 had been identified as prerequisites for returning to flight), it projected a long shuttle fleet grounding. The general condemnation of how mission managers reacted to concerns expressed by the lower level engineering staff showed a need to establish an independent engineering authority to review and act on such concerns. Of all the CAIB's many recommendations, how to inspect and repair damage in flight was the most daunting charge NASA would have to address.

Meanwhile, life onboard the Space Station continued, and hardware was being delivered to KSC for inspection and subsequent storage until flights began again. In October the first two-man crew was changed out and a new two-man crew took over. NASA decided to continue calling these crews "Expeditions," so it was the Expedition 8 crew relieving the Expedition 7 crew. The new crew was scheduled to remain on the ISS for six months. Some science was still being conducted on the Space Station and by the end of their stay the Expedition 7 crew was able to devote 15 hours a week to experiments. It was hoped that the new crew would do even better. The rest of their work days would be spent maintaining the ISS and unloading and loading Progress supply vehicles whenever they were docked. The Expedition 8 crew would also have to deal with a problem discovered before they were launched. The equipment monitoring air and water quality was not functioning correctly. In November NASA announced it was preparing *Atlantis* and *Discovery* to return to flight as early as September or October 2004.

## 2004

Before a packed auditorium at NASA Headquarters, President Bush, on January 14th, unveiled his New Vision for Space Exploration.[41] He listed a number of goals that would result in extending "a human presence across our solar system." The first goal was to return the space shuttle to flight as soon as possible, complete the International Space Station by 2010, and then retire the shuttle. His second goal was to develop and test, by 2008, a new spacecraft, the Crew Exploration Vehicle (CEV), a different concept and design from the recently canceled OSP. The first manned flight of the CEV was projected for 2014 and human missions might go to the Moon as early as 2015. The third goal was to use the Moon, by 2020, as a launching point for further missions. Returning to the Moon would be done in conjunction with robotic missions to Mars as "trailblazers" for human exploration. He stated, "We do not know where this journey will end, yet we know this: human beings are headed into the cosmos."

His vision was greeted with support in most quarters, especially at NASA; however there were also critics. The President invited other nations to participate. His invitation was endorsed by all the Space Station partners when they heard that his first goal was to complete the ISS. There were issues, such as the four year gap between the shuttle's retirement and the availability of the CEV that might be used to service the ISS, but it was hoped that other options could be developed by that time.[42] ESA was projecting that its automated transfer vehicle (ATV), launched on the Ariane-5, would be ready in late 2005 and be able to bring supplies to the ISS. If it was available, it could reduce the number of projected shuttle logistic flights.

An *Aviation Week & Space Technology* editorial titled: "Bush's Space Plan: Bold Vision or 'Moondoggle'?" made this critical observation: "Sadly, Bush's vision for space wasn't important enough for him to mention in his State of the Union message last week. Not a word about it. That instantly calls into question his administration's commitment to this endeavor. Failure to win wide national support for Moon-Mars will spell its doom."[43] In view of the history of previous administrations' grand announcements on where the nation's space program was headed, only to be quickly extinguished, the editorial was on the mark.

The first question that came to everyone's mind was how much would the new program proposed by the President cost? Without a precise time frame for many of the goals, this would be difficult to estimate. Despite this uncertainty, Bush asked aerospace executive and former U.S. Air Force Secretary Edward C. Aldridge, Jr. to chair a commission to recommend how to implement his vision. While Aldridge's commission was deliberating, NASA's FY 2005 budget was submitted. It requested $16.244 billion, an increase of $775 million over FY 2004. For the five year projection, annual growth of about 5% was factored in with the budget leveling off at approximately $18 billion by 2009.[44] By shifting funds within NASA's budget, deferring or terminating programs, the White House expected to be able to fund the cost of meeting the President's goals without greatly increasing NASA's annual budgets. For example, after ISS assembly was completed, the U.S. would reduce its involvement, leaving the majority of funding needed to continue operations to the international partners. In view of the partners' budget problems, that appeared to be a highly speculative premise on which to go forward. If the Space Station was not supported by the international partners in this way, its use would be reduced to whatever level all the partners agreed to.

One of the first casualties of this new approach was the proposed private sector institution to oversee ISS science that had been authorized in NASA's FY 2003 budget; it was canceled. The amount and type of science that NASA would support on the Space Station would be reduced. The new policy meant that the Space Station would no longer be used as a base for going on to the Moon and Mars as the original planners had contemplated. It was proposed that the Space Station would now have a very specific and limited role to play. In preparation for long-duration human space flights, U.S. space station researchers

---

41. President Bush Announces New Vision for Space Exploration Program, The White House, Office of the Press Secretary, January 14, 2004.
42. Bush Vision Greeted Overseas With Enthusiasm And Relief, Peter B. de Selding, *Space News*, January 19, 2004.
43. *Aviation Week & Space Technology* Editorial: "Bush's Space Plan: Bold Vision for 'Moondoggle'?" January 26, 2004.
44. 2005 NASA Budget Request Reflects New Priorities, Brian Berger, *Space News*, February 9, 2004.

would concentrate on studying human physiology, such as the effects of exposure to space radiation and loss of muscle tone and bone density resulting from extended stays in low "g" environments.

Claiming concern for astronaut safety and cost, O'Keefe announced that the shuttle flight to service the Hubble telescope would be canceled. This brought immediate criticism from many sides. The appearance that science would once again take a back seat to human space flight in the President's new program created considerable complaints in the space science community. The President's science advisor, John Marburger, urged patience stating, "This vision has greater scientific relevance than past missions and science will be more productive with it than in its absence."[45] Only the passage of time will confirm that his assurances will come true. Based on the announced major decreases in funding for science to be conducted on the Space Station, and the possible reduction or termination of Hubble operations before the James Webb Space Telescope would be in operation, the Administration would have to increase, significantly, science funding in other areas to make up for scaling back these programs.

Four months after announcing that the shuttle would return to flight by September or October 2004, at the end of February NASA announced that the first flight would most probably be postponed until March 2005. The delay was caused by continuing concerns on how to predict the behavior of the foam insulation on the external tank during launch. A piece of foam insulation that had fallen off the external tank during launch had been identified as the culprit that damaged *Columbia*'s wing allowing hot gases to penetrate a crack in the tiles during reentry and melt the structure. A new computer model would be developed and wind tunnel tests conducted to better understand the problem. Also, a CAIB recommendation required that the shuttle be launched during daylight in order to take high-resolution photographs. This requirement reduced launch window opportunities for rendezvousing with the ISS. March 2005 became the best date to schedule the next launch.[46]

The Expedition 9 crew arrived at the ISS in April. Shortly after they arrived a problem occurred with one of the four attitude control gyros. With one gyro already off line because of an earlier bearing failure, it was crucial to effect a repair of the latest problem to ensure that the ISS had a redundant gyro system in case of another failure. Gennady Padalka and Michael Fincke, after aborting two EVAs due to suit malfunctions, were able to conduct a successful EVA on June 30th and repaired the unit.[47]

Completing four months of hearings and deliberations, the Aldridge "Moon, Mars, and Beyond" Commission reported to the President.[48] Released with great fanfare on June 16th, it contained many familiar and a few controversial recommendations on how to proceed with space exploration. In the transmittal letter Chairman Aldridge wrote: "The Commission's web site received more than 6 million hits and over 6,000 written inputs. Public comments strongly supported the new space vision by a 7-to-1 ratio." He ended the letter: "This national effort calls for a transformation of NASA, building a robust international space industry, a discovery-based science agenda, and educational initiatives to support youth and teachers inspired by the vision." All worthwhile objectives. One that raised many questions and doubts was the call for the "transformation of NASA." After being "reinvented" by the Clinton administration in 1993, the questionably reinvented NASA was now to be "transformed."

Perhaps a more important question was, did the high support ratio reported on the web site really indicate the national consensus that will be needed to sustain the effort for decades? Addressing this issue in a press conference, Marburger explained that the program entails a long-term buildup of infrastructure that will lower the cost and risk for all space missions. He said: "I think it's not an easy vision to understand if you are coming out of the traditional one-shot-at-a-time NASA exploration history. People want to know how much it's going to cost. I believe that the vision is a sustainable vision and I don't care how long it takes. I think that Congress will eventually come around."[49] Progress in many areas would have to be demonstrated before his prediction could come true, most importantly changes in NASA's management culture.

---

45. How Much Space for Science, Andrew Lawler, *Science*, 30 January 2004, Vol. 303.
46. Next Shuttle Launch Postponed Until March 2005, Tariq Malik, *Space News*, February 23, 2004.
47. Station Gyration, Craig Covault, *Aviation Week & Space Technology*, July 5, 2004.
48. Report of the President's Commission on Implementation of United States Space Exploration Policy, A Journey to Inspire, Innovate, and Discover, June 2004.
49. Marburger Confident Vision Survives for Long Term, Brian Berger, *Space News*, May 24, 2004.

The "transformation of NASA" took a small step forward on August 1st. According to a NASA Press release the "transformation fundamentally restructures NASA's Strategic Enterprises into Mission Offices. Headquarters support functions also have been realigned to better clarify organizational roles and responsibilities." The press release went on to say: "The agency has redefined its relationships with the NASA field centers by developing clear and straightforward lines of responsibility and accountability. Specific Mission Associate Administrators are now assigned as Headquarters Center Executives."[50]

In mid-October the Expedition 10 crew arrived on a Soyuz spacecraft. In preparation for their return to Earth the Expedition 9 crew had been tidying up the ISS. Various kinds of "garbage" had accumulated since the shuttle missions had been canceled. Some of the items that needed to be discarded included broken exercise equipment, worn out EVA suits, and rendezvous and docking equipment that needed repair. Russian Soyuz and Progress vehicles, with a limited capacity to offload such items, had not been able to keep up with the amount of material over the last two years and the buildup and lack of storage space had begun to affect the living and working conditions on the ISS. When the shuttle returned to flight, the first mission would have a primary objective of returning worn-out equipment and other trash to free up space on board the Space Station.

At the end of October NASA announced that the shuttle's return to flight would slip at least another two to three months to May or June 2005.[51] The new delay was attributed to the need to repair hurricane damage at KSC. Further delays were possible if problems were encountered as NASA responded to the recommendations of the CAIB and its own internal technical and safety reviews that, in some cases, went beyond the fixes called for by the CAIB. Each delay added to the larger problem of completing assembly of the ISS. The latest estimates concluded that, to complete assembly, another twenty-eight shuttle flights would be needed rather than the twenty-two announced earlier. That would require a launch rate that probably could not be sustained until 2010 when the shuttle was scheduled for retirement. Dan Murphy, president and CEO of ATK, the solid rocket booster manufacturer, was quoted as saying, "All outside experts . . . believe the existing program has to fly [the shuttle] through 2014."[52] This problem would have to be addressed soon, as NASA's budget projections did not carry support for shuttle flights out that far. Making all the changes and improvements to the shuttle called for by the CAIB was already putting pressure on NASA's budget projections. Original estimates were that approximately $1.2 billion would be needed to return the shuttle to flight. In September O'Keefe told Congress that additional needed modifications to the shuttle fleet would raise the overall cost to $2.2 billion.

Congress rescued NASA from its immediate budget dilemma when it passed the FY 2005 appropriation on November 20th. It was one of the few government agencies to see a budget increase over the preceding fiscal year. Congress approved $16.196 billion, $818 million more than it had approved in FY 2004. Congress had given the President the first installment for his space vision.

Just as important as the overall number was language in the bill, which permitted NASA to shift money from one account to another as the need arose, as it undoubtedly would. The only bad news was that Congress had earmarked over $400 million for more than 160 pet projects that would consume almost half the increase. The Appropriations Bill required NASA to submit a number of reports within 60 to 180 days of the bill's enactment. First due was a report on the proposed capabilities of the Crew Exploration Vehicle (CEV), the new manned spacecraft that would replace the shuttle and perform wide-ranging missions, including returning astronauts to the Moon. Funds had been approved for two competitive contracts to begin the studies. Congress did not believe that NASA had clearly defined the criteria for its development. Other studies requested included detailing how the ISS would be completed.[53]

Having successfully shepherded the FY 2005 budget through the Congress, Sean O'Keefe, NASA's tenth administrator, announced his resignation on December 13th. In his resignation letter he told the President he would continue to serve until a successor was named.[54] No front runners emerged to replace O'Keefe; however, several former astronauts were reported to be interested in receiving the

---

50. NASA Press Release 04-253, NASA Transformation in Effect, August 2, 2004.
51. In Orbit – It's Official, edited by Frank Mooring, Jr., *Aviation Week & Space Technology*, November 8, 2004.
52. Debate About Shuttle's Future Heats Up, Brian Berger, *Space News*, November 1, 2004.
53. Congress Gives NASA Plenty of Post-Budget Homework, Brian Berger, *Space News*, December 6, 2004.
54. NASA Press Release 04-400, NASA Administrator Sean O'Keefe Resigns, December 13, 2004.

appointment. The President's choice would send a clear message about how serious he was that NASA, during the remainder of his term, pursue an aggressive space exploration policy. An administrator with strong management credentials and a broad background in space would be needed to keep the President's vision, on course and to steer the far-flung NASA empire in support of the effort. Someone having the qualifications called for by The Center for Strategic and International Studies in 1989 would be needed.

A pair of hungry Space Station inhabitants were the final story in 2004. Expedition 10 crew, Leroy Chiao and Salizhan Sharipov, had been placed on short rations since the beginning of December because of the problems of resupplying the ISS with food and water. The food shortage was caused in part when it was discovered that, before they were relieved, the Expedition 9 crew had eaten some of the food that was reserved for Chiao and Sharipov. A Progress resupply vehicle docked on Christmas day, carrying food, water, fuel, research equipment and presents. An emergency situation did not exist, the crew had enough food to last another week or two, but if the Progress vehicle had been delayed, plans were in place to return the crew on the docked Soyuz and leave the ISS unmanned until another crew arrived with their supplies.

At the end of 2004, with NASA now looking to a May or June 2005 return to flight, a number of open items remained before the shuttle would be cleared for launch. The independent Shuttle Return to Flight Task Group, chaired by former astronauts Thomas Stafford and Richard Covey, would have to review all of the CAIB recommendations, and the remediation tasks that NASA had performed, and then give its "blessing" to launch the shuttles again. The Task Group's report was required 30 days before launch.

After President Bush announced his "new vision" for space exploration in January 2004, Space Station planners were faced with the need to resolve many difficult issues. They included the apparent disconnect of how to complete Space Station assembly considering that the shuttle was scheduled to retire in 2010 and 28 more shuttle assembly and logistics flights were needed. Almost as important was what to do about crew size and the directly related problem of what science could be carried out on the Station if restricted to a three-person crew. How to accommodate the international partners in view of the narrowing of the Space Station's mission from the perspective of U.S. goals would be another delicate issue. All of these problems would have to be addressed in 2005 when the shuttle returned to flight and a new administrator was appointed. What the Space Station would look like by 2010 and what types of research it could support were questions that might not have immediate answers.

## 2005

Thus, NASA's plate was full as the year 2005 began. In addition to the high workload associated with returning the shuttle to flight and restructuring programs to support the President's space exploration vision, a reduction in the civil service work force was in the offing. How many would be affected was not known; however, rumors circulating placed the new ceiling at roughly 15,000, a reduction of some 2,600 full-time employees over the next eighteen months. Downsizing a federal agency is always a traumatic experience for those involved. The new administrator, when appointed, would have a difficult job keeping morale at a high level during the transition.

Preparing *Discovery* (STS-114) to return to flight after incorporating all the CAIB recommendations consumed the workforce at KSC and JSC in early 2005. In addition, parallel processing of *Atlantis* was also under way, since the new flight rules required that for every shuttle launch another shuttle must be almost ready to launch in the event of an emergency. If a rescue flight was needed, round-the-clock processing of *Atlantis* would then take place and a reduced crew would be launched on *Atlantis* as soon as possible to rendezvous with the stricken *Discovery*. Michael Leinbach, leader of the KSC shuttle launch team, was quoted at the beginning of February: "We can't launch *Discovery* without being prepared to take on that mission [*Atlantis*]. So we have to have both ready."[55] Depending on when an emergency might occur and the type of problem encountered during *Discovery*'s mission, one of the safeguards that could be used would be to have *Discovery* rendezvous with the ISS to await rescue. The

---

55. Countdown ticks for shuttle liftoff, Todd Halvorson and John Kelly, *Florida Today*, February 1, 2005.

other alternative was to have *Discovery*'s crew repair in orbit any damage to the external structure that might jeopardize the shuttle during reentry. This backup solution was under intensive study at JSC.

The Administration's FY 2006 budget request for NASA was $16.456 billion, an increase of $200 million above FY 2005. Included within that sum was $6.763 billion for Space Operations, the line item that covered shuttle and Space Station costs. In an attempt to address the inherent problem of considering NASA's annual budget in the context of the VA, HUD and Independent Agencies Subcommittee, Congressman Thomas Delay, (R-TX), proposed establishing a new appropriation subcommittee that would consolidate many of the agencies heavily involved in R&D programs. The new committee's oversight would include NASA, NSF, NIST, NOAA, and the Weather Service.

At the beginning of March, Michael D. Griffin was nominated to be the next NASA administrator. At the time of his nomination he held the position of Space Department Head, Johns Hopkins University Applied Physics Laboratory. He had a broad background in space, including a stint at NASA, and had been a senior executive in the private sector. His nomination was greeted with approval by members of Congress. Once endorsed by the Senate, he took office in the midst of the FY 2006 congressional budget debates, negotiations with OMB on the FY 2007 budget, and the start of implementing the President's space vision. In many respects, his first days in office mirrored the issues faced by Jim Fletcher almost twenty years earlier. In addition to the budget debates, he faced staff reductions, returning the shuttle to flight, and putting his stamp on the troubled Space Station program. It was a most difficult time to take control of the agency.

During his first term at NASA in 1989, Griffin was Associate Administrator for Exploration, at the time that the first President Bush announced his Space Exploration Initiative. That initiative, for which Griffin had a key implementation role, was abandoned. Now he would have the opportunity to try again to make a similar presidential "vision" successful. Does he possess all of the "attributes" needed by a NASA administrator as recommended by CSIS in 1989? In particular, will he "have access to the President" and "be adept in interagency politics, be willing to delegate, and have a good instinct for dealing with the press." Before his nomination he was on record as advocating retiring the shuttle sooner than 2010 and the need to develop a heavy-lift launch capability.[56] The two positions are not necessarily contradictory, but either one, if implemented, could have a major impact on the completion of Space Station assembly. The CEV and Crew Launch Vehicle (CLV) will not be available in time to support timely Space Station assembly and operations unless their development could be significantly accelerated. Retiring the shuttle early would require a major change in the Space Station flight manifest and probably affect its configuration and utilization. Griffin would undoubtedly be asked to clarify these positions during his confirmation hearings.

On April 14th, in record time, Griffin was confirmed and became the eleventh presidentially appointed NASA Administrator. His confirmation hearings and subsequent appearances in April and May before congressional committees could be described as a "love fest." In a gentle way, during an appearance on May 12 before the Senate Appropriations, Commerce, Justice and Science Subcommittee, he was reminded during his questioning of the festering problems he inherited. Without making too many promises, Griffin deflected most of the questions and focused many of his answers on the need to develop the CEV. Senator Mikulski, reflecting the general attitude of Subcommittee members, said, "We're off to a good start even if some of the things you are telling us are giving us heartburn."

On July 15th, the Return to Flight Task Group released its final report. It included a series of recommendations and observations. One of the lengthy observations, considered a "minority report" as it differed in tone from the full Task Group's report, was highly critical of the actions taken by NASA to return the shuttle to flight.[57] Included among the observations was the following statement: "NASA implementation of the CAIB return-to-flight recommendations may leave an impression of accomplishment that we believe does not represent a comprehensive picture of NASA's return-to-flight effort." Further clarifying their observations they stated: "This is not a set of conclusions, but is a detailed summary of persistent cultural symptoms we observed throughout the assessment process . . . NASA leadership . . . missed opportunities to address the enduring themes of dysfunctional organizational

---

56. Bush Taps Griffin To Lead NASA, Implement Exploration Vision, Brian Berger, *Space News*, March 14, 2005.

behavior that the CAIB and other external evaluators have repeatedly found." For those who have followed NASA's management problems for the past twenty years, these statements suggest that after two shuttle disasters NASA has not significantly improved its oversight of critical operations. *Discovery*'s upcoming flight would demonstrate the truth of these observations. Efforts to improve NASA's engineering culture still had a long way to go. The obvious question is: What are the implications for NASA as it prepares to move ahead and develop the next generation of manned spacecraft?

When the shuttle *Discovery* finally returned to flight on July 26, 2005 (STS-114), the problem that had doomed *Columbia*, external tank foam shedding, had not been fixed. Large pieces of foam were photographed flying off the ET before the shuttle reached orbit. Fortunately, some said luckily, there was no damage to *Discovery*. After rendezvousing and docking with the ISS, *Discovery* carried out a number of housekeeping activities. Repairs were made to the Space Station including activating the forth CMG, supplies and equipment transferred, and two and a half years of accumulated trash off-loaded. The Italian MPLM served as the "mini-van" for the latter two activities. The ISS tanks were filled with water generated by *Discovery*'s fuel cells, a fill-up that had not taken place for two and a half years. The Space Station was almost back to normal. The major deficiency was that it would still be home to only two crewmen, and future assembly and logistics flights were once again uncertain because of the continuing foam-shedding problem.

One of the CAIB's recommendations, to be able to repair damage to the shuttle while in orbit, was considered the most difficult to satisfy, depending on where the orbiter was damaged and what type damage it might sustain. During one of the EVAs on *Discovery*'s mission, astronauts Stephen K. Robinson and Soichi Noguchi of the Japanese Space Agency successfully used repair kits to simulate making a repair. In addition, on another EVA, a real repair was made by Robinson when he removed tile gap fillers that had come loose on *Discovery*'s heat shield during launch. STS-114 achieved all of its pre-mission objectives and landed without mishap at NASA Dryden Flight Research Center, California, after 14 days in orbit. All in all, a successful mission, but one that ended in controversy. Because of the foam shedding, the shuttle was once again grounded until a solution could be found, thus compounding the Space Station's assembly woes. The earliest that the next flight could be scheduled was tentatively the spring of 2006.

After his confirmation, Administrator Griffin began the task of sorting out the agency's priorities and restructuring NASA to concentrate on the President's space vision. Debate on NASA's FY 2006 budget continued in Congress through the summer. In a series of meetings and briefings with NASA staff and outside organizations, Griffin outlined his approach to achieving the agency's future goals. A most difficult task was explaining how NASA would downsize staff while gearing up for new starts to implement returning astronauts to the Moon and prepare to go on to Mars. All the centers and Headquarters would feel the ax of the reduction in force (RIF). Congress intruded into the RIF process and requested that Griffin submit a "comprehensive coordinated restructuring plan" before the RIF began. It noted that the centers contained "impressive core competencies" and that NASA must "maintain world class scientists and engineers at its field centers."[58] Griffin had stated that some of the expertise found at the centers would not be needed in future programs, setting up a potential confrontation with Congress if their constituents made the case that the RIF would reduce "core competencies." RIFS are never easy or without controversy.

Griffin also took the opportunity in several forums to discuss how NASA would return to the Moon, and described how the proposed Crew Exploration Vehicle (CEV), the multipurpose manned spacecraft, would be developed. The program to return to the Moon was characterized as "Apollo on steroids." It was similar in many respects, but the CEV would permit four astronauts to land on the lunar surface versus two during Apollo. A little larger but still Apollo-like, the CEV would be capable,

---

57. Final Report of the Return to Flight Task Group – Assessing the Implementation of the *Columbia* Accident Investigation Board Return-to-Flight Recommendations, July 2005. The quotes are from the A.2 Observations by Dr. Dan L. Crippen, Dr. Charles C. Daniel, Dr. Amy K. Donahue, Col. Susan J. Helms, Ms. Susan Morrisey Livingstone, Dr. Rosemary O'Leary, and Mr. William Wegner.
58. The quotes are from language contained in Conference Report in H.R. 2862, Science, State, Justice, Commerce and Related Agencies Appropriation Act 2006.

upon returning to Earth, of landing on the ground, rather than in the ocean, and be reusable. Details of what the astronauts would accomplish during their missions were scanty.

Preparing to send the FY 2007 budget request to OMB, Griffin provided guidance to William Gerstenmaier, Associate Administrator for Space Operations.[59] Options were outlined for Gerstenmaier to follow that would affect the Space Station. He was told to reserve one shuttle flight to service the Hubble Space Telescope and, in view of the limited number of flights available by the end of 2010 (the targeted date to stop flying the shuttle), "recommend an ISS assembly sequence in which the highest priority is placed on flights to meet International Partner commitments." The total number of shuttle flights that would be available to service the ISS was projected to be 18.

Congress finally passed the NASA 2006 budget in November. Considering the pressures on the nation's taxpayers to support the war on terrorism and assist in repairing damage done by hurricanes in many states, Congress had been generous. The final number was $16.457 billion, one million more than the administration requested. The total included $3.1 billion to support programs to return to the Moon. That was the good news. The bad news was that Congress in its wisdom had earmarked $275 million for pet projects that had little or nothing to do with NASA's mission and $260 million in targeted increases for some programs.[60]

Some things never change in Washington. At the top of the list is the ability of members of Congress to satisfy their desire to meddle in the minutiae of executive branch agency programs. It will probably never cease, to the detriment of good program management and the wise expenditure of taxpayer monies. Dr. Griffin may find that NASA's return to the Moon will be sidetracked to make a short swing by the newly named asteroid Seanokeefe, to satisfy the whim of a congressman or senator. Budgets projected for the next few years are already considered too low to accomplish all the programs that NASA had on its plate. Siphoning off large chunks for frivolous programs will only add to the shortages.

During the restructuring mentioned above, Griffin began to pare down or eliminate science programs that expected to utilize the Space Station. At the end of November the National Research Council released a report highly critical of the result of the reorientation of Space Station science programs.[61] The report's Executive Summary concluded: "The panel saw no evidence of an integrated resource utilization plan for use of the ISS in support of the Exploration Missions [to the Moon and Mars]." Specifically, in regard to Mars missions, the report stated: "The ISS represents a unique platform with which to conduct operational demonstrations in microgravity . . . the ISS may prove the only facility with which to conduct critical operations demonstrations needed to reduce risks and certify advanced system." To stay within projected budgets Griffin had been forced, as had NASA administrators before him, to compromise long-established research goals that only a robust Space Station could have achieved. With all of the changes the Space Station program was undergoing, twenty years of effort by thousands of workers, and more than $45 billion dollars of U.S. and international partner expenditures, has produced a laboratory with a severely limited future.

Closing out 2005, the NASA Authorization Act of 2005 was passed on December 22nd and sent on to the President for his signature. Language in the Act directs NASA: "Not later than one year after the date of enactment of this Act, the Administrator shall transmit . . . a plan describing how the national laboratory [the Space Station] will be operated." The Act contains other language concerning microgravity research and the need to install a centrifuge to validate in space work performed in ground-based facilities. Will the current direction taken by NASA be reversed to conform with congressional direction, or will the International Space Station witness a further reduction in its ability to advance space exploration? A decision may be known sometime in 2006.

---

59. NASA internal memo: to Willam Gerstenmaier, Associate Administrator for Space Operations, from Michael Grifin, Administrator, Subject: In-Guide for FY 2007 Budget, dated 4 October 2005.
60. The total of the earmarks varies depending on how one assesses the budget. In an April 30, 2006 editorial, "NASA Greedy Overseers," the *New York Times* stated that the total was $568.5 million.
61. National Research Council: Review of NASA Plans for the International Space Station, November 28, 2005.

## — CHAPTER 10 —
## WHO WAS IN CHARGE? – NO ONE!

The preceding pages have presented, in a very condensed form, the story of how the Space Station program evolved and describe the conflicting interests and personalities that shaped its current status. It reads somewhat like a mystery novel. As the program progressed you were never quite sure how it would ultimately emerge. Some authors of mystery stories will write at an appropriate point: "Now you have all the information you need to solve the crime; the characters that conspired to kill or cripple the victim and those who tried to keep it alive and whole." We have arrived at that point in the description of the Space Station program; a program that suffered through twenty years of constant change and as a result vast sums of taxpayer funds were spent without achieving the original goals. What was done right, what went wrong? Can we learn any lessons that would benefit future complex government programs? These questions are difficult to answer definitively as the program is still ongoing and based on past experience, further changes can be expected. However, that is the purpose of relating the history of the program – what can be learned? Are the conclusions that follow justified?

Soon after NASA was established in 1958, it earned a reputation of trust and respect among most members of Congress, the Eisenhower and Kennedy administrations, and the public. NASA moved rapidly and usually, considering its unique charter, performed with great success. Trust in NASA's ability to carry out difficult missions prevailed and NASA delivered. There were, of course, some lawmakers, members of the public, and especially the media who disagreed with how NASA pursued its programs.

When the last Apollo mission splashed down in the Pacific Ocean in 1972 the Viet Nam war was winding down but was still a very divisive issue. A new mood prevailed in Congress and the public. Trust in government programs was at an all-time low. Political differences became more contentious and personal and almost all budgets submitted by sitting administrations were declared "dead on arrival" by one faction or another. Comity disappeared from Capital Hill's lexicon. The "Watergate" scandal added to the general malaise. New space initiatives were difficult to start. In the intervening years, from the end of Apollo until the Reagan administration, NASA's leadership muddled along responding to administrations with no clear-cut space priorities. Convincing the Nixon Administration and Congress to develop a space shuttle, based on a degraded design, did not represent a new NASA mandate to accomplish great goals. On the contrary, having the shuttle left a void in what to do next with this new capability.

### MODERN TIMES IN WASHINGTON

What is required today in Washington to play in the game of promoting large government programs? Successful political leaders survive by testing the winds of public opinion, usually confirmed through extensive polling, and then put forward programs they expect will receive wide support. Such leaders seldom venture into uncharted waters. This approach is especially important when proposing new social programs that will effect large segments of the populace. To satisfy all of the competing interests, the resulting programs are usually quite complex and undergo continuous modification, reflecting the influence of one powerful lobby or another, changing priorities, or worse, improper design. Both sitting administrations and members of Congress play the game in this manner. However, to initiate a large government program outside the realm of common experience requires a different type of leadership, characterized by having great vision and the skill to verbalize the vision to garner public interest and support. And most importantly, it requires leaders with the courage to take on risk and the possibility of failure to achieve their goals. This latter attribute is usually in short supply in Washington.

Once committed to a difficult course of action these leaders must then find managers who, based on their experience and past records, have the ability to lower risk, avoid failure, and achieve success. Without the combination of both leadership and experienced management, a new initiative that would stretch the frontiers of science and engineering is almost guaranteed to fail. There are many examples that would illustrate this point: a success, the Manhattan Project; a failure, the Superconducting Super Collider. A large program in limbo, Prometheus Nuclear Systems and Technology, is currently suffering from leadership indecision.

Developed in great secrecy, the Manhattan Project brought together a small circle of brilliant physicists under a military manager and avoided close review by Congress. One might argue that this latter point was the prime reason for its success. Also, because of secrecy, it was never put in competition against other favorite programs for scarce funds. The Super Collider had a very different history. Almost from the day it was proposed it was criticized and nit-picked by the science community and members of Congress until it was canceled. The Prometheus nuclear power program, begun in 2003, is a sterling example of spending tax dollars, about $200 million to date, and then suddenly discovering there might be a $10 billion plus price tag to complete the program. As a result, it has been moved to the back burner. Space nuclear systems, beyond question, are required for continued robotic or human solar system exploration. Will the $200 million eventually be wasted? Let us hope this program will not meet the same fate as the SP-100 space nuclear program that spent almost $500 million during its eight year lifetime with little to show for the effort. All good programs cannot be funded. Nonetheless, Congress and the White House must realize that there is a long list of valuable programs that were started and stopped, wasting billions of dollars, and resolve not to repeat their past practice of indecisive leadership.

Experience tells us that we cannot depend on visionary leadership from a Congress dominated by lawyers with little or no understanding of complex technical programs and constantly faced with the need to appeal to the voters back home to be reelected.[1] A major new program such as the Space Station needed more than President Reagan's initial endorsement; it needed constant attention and support to overcome ceaseless Congressional opposition.

One year after receiving Reagan's approval, OMB tried to kill the program leading to Phil Culbertson's protest described in Chapter 1. Two years later, when NASA was signing contracts with the Work Package contractors, OMB struck again. Stofan had a letter signed by President Reagan "guaranteeing" a Space Station budget that included $767 million in FY 1988 and $1.8 billion in FY 1989 for the Work Package contractors. For FY 1988 OMB proposed less than $600 million. What would be allowed in FY 1989 was just a guess. Stofan was quoted as saying the letter should have been "as good as gold" and finally "that piece of paper and fifty cents will get me a cup of coffee."[2] Faced with continuous budget problems, Space Station managers added to the problem by failing to show that they were proceeding in a manner to meet cost and schedule. By not demonstrating these qualities, the sharp knives of special interests successfully carved the program up and left it gasping for breath on the banks of the Potomac.

Bounded by all the above observations, the problems confronted by the International Space Station program are crystalized. Congressional meddling, emboldened by the absence of consistent administration leadership, reflected in part by the constant battles with OMB, and lackluster program management, crippled the program, wasting valuable, national resources.

Congress, beyond question, controls the country's purse strings. Each administration proposes yearly budgets and the Congress, in its wisdom, appropriates the funds it deems necessary to carry out the government's functions. Most of the nation's annual budget is earmarked for entitlement programs, national defense, and to pay interest on the debt. Again, no question, because of their political sensitivity and size, the content and direction of these programs deserve and receive close Congressional oversight. The poor cousins to these mega-programs are those included in agencies that spend the ever decreasing percentage of the budget categorized as "discretionary." The budgets for

---

1. At the end of the 108th Congress, based on a search of members' biographies, only two held doctorates in science or engineering, Rep. Vern Ehlers (R-MI), and Rep. Rush Holt (D-NJ), both in physics. There are several senators and congressmen who are/were MDs. Members with undergraduate degrees in science and engineering were also in short supply.
2. NASA Selects Station Contractors Despite Funding Uncertainties, story by a team of writers from *Aviation Week & Space Technology*, December 7, 1987.

these agencies have become a yearly battleground. Most members of Congress are dependent on staff or outside "experts" to analyze and provide recommendations on whether or not to support programs that requires an understanding of their technical or scientific content. The same statement holds true for most of a president's immediate senior staff. Yet, many of these programs hold the promise of moving the nation forward in new and challenging directions with the potential to improve our standard of living in unimaginable ways. The Department of Energy, National Institutes of Health, National Science Foundation and, yes NASA, to name just a few, hold the keys to a better life or exciting future for our citizens. Almost all the programs of these agencies have strong technical foundations and success is dependent on their managers having the required engineering and scientific experience. By their very nature success is not assured for all these programs and that must be recognized and accepted by our lawmakers.

## Program Oversight

How can these conflicting interests and power centers master working together to assure success for large, complex programs? Before answering this question it is necessary to ask other questions. What role, if any, should Congress play in the day-to-day outcome of the myriad programs it funds? Congress may deny that it inserts itself into the daily management of programs but in fact it does. It occurs in many forms. Members and their staffs holding meetings with agency managers, GAO conducting probes reflecting the political bias or position of the member requesting a study and, of course, there are the never ending hearings. Quite often, hearings result in questions that might require an agency to undertake days or weeks of research and analysis to provide answers. Unfortunately, the only conclusion one can reach is that many of the hearings and GAO studies are for "show" and media coverage for voters back home. Serious debate and information exchange is not always an objective. Similarly, what should be the role of those working in the Executive Branch who are removed from daily accountability, but whose support is needed if a program is to stand a chance of survival? This is a conflict that is continuously fought with sister agencies, and especially OMB, in the back rooms of the White House complex. Answers to these questions are offered in the following pages.

Regardless of one's position on the need or appropriateness of congressional or White House oversight, it must be understood that responding to their demands for program detail, or other requests for information, means effort diverted from the job at hand by the agency managing the outcome of a program. At times, these requests for information can become overwhelming and completely sap the energy of a program office. For example, not too long ago, NASA experienced so many requests for GAO and IG audits that it was forced to put in place a special office just to track the status of required reports. The statement made in 1993 by Congressman George Brown (D-CA, now deceased) bears repeating, and is as true today as it was then: "it struck me as particularly disingenuous . . . to hear critics [in Congress] of the space station decry program stretch-outs, escalating budgets, and management disarray at NASA – delay, confusion, and costs for which they are at least partly responsible. No federal program can long endure the kind of annual sniping, redrafting and retrenchment to which they have subjected the space station."[3] All who have been associated with the Space Station program would say "Amen" to the congressman's observations.

What should oversight entail? Is there a way to arrive at a balance between the many competing interests? Agencies are reluctant to challenge members of Congress or their staffs who believe they are sufficiently conversant with the inner workings of a program to provide direction. If an agency wants its programs to succeed it knows it must at least give lip service to Congressional suggestions or orders. Within a given administration, arguments with White House staff and OMB are even harder to win. Agency managers are all working for the same boss, but those in the White House and OMB are first among equals. The result: budget and political demands often prevail over good program decisions, and programs flounder or fail.

Conflicts between agencies and the OMB are legendary but seldom publicly discussed in detail for obvious reasons: complainers almost always lose. Tommy Thompson, former Health and Human Services Secretary in the Bush administration, broke the rules recently and described his conflicts with

---

3. Commentary, George Brown Jr., *The Washington Times*, July 11, 1993.

OMB. In discussing his unsuccessful attempts to reform Medicare and Medicaid, he called OMB a "super God." Explaining this title he said, "They turn you down nine times out of ten just to show you they are the boss." And as for winning a debate with White House advisors, he explained that they, "do not believe that anything smart or original can come from a secretary or a department."[4] At the start of the Space Station program, NASA won one of these arguments, but only succeeded in alienating powerful forces with long memories that carried over to the next administrations. Senior OMB appointees change with each new administration, but the working staff remains.

If these conflicts are to be avoided, Congress and each administration must learn to depend on the track record of the agencies and their managers in order to be comfortable providing the needed resources, especially in times when resources are tight. There is a flip side of course to this issue; agencies must earn this trust. For cabinet level or sub-cabinet level management positions that turn over with each new administration, often several times, familiarity with the job, experience as a manager of a large organization, or technical expertise, are not a prerequisite in order to be appointed. Regardless of their backgrounds, these senior appointees are dependent on the career civil service workforce that remains in place to carry out old and new programs. Fresh directions and priorities are expected with each new administration. But programs already on the books will almost always dominate the daily workload and provide the foundation upon which each administration gains confidence that its programs will be well managed. For NASA, it must strive and succeed in being viewed by Congress and administrations as it was forty years ago.

### NASA Top Management Turnover

In NASA's first twenty-three years, five administrators held the top position. When Beggs left NASA in 1985 it started an extensive management turnover. During the next seven years, NASA would be led by four administrators, performing in either an acting capacity or as full-fledged presidential appointees (in sequence, William R. Graham, James C. Fletcher, Richard H. Truly and Daniel S. Goldin). Could any Fortune 500 company, NASA's equivalent in size and budget oversight, survive such rapid turnover at the top? These frequent changes had a lasting, debilitating, effect on the Space Station program.

With each new NASA administrator came reorganizations, major or minor, and turnover in the next level of management below the administrator; but regardless of the magnitude, a reorganization. Some corporate memory was always lost when senior managers left. Frequently some of these individuals would be picked up by a major aerospace contractor thus, from a larger perspective, it wasn't a total loss. Some key Headquarters positions remained vacant for long periods of time, or were staffed by acting managers often holding more than one job. Center directors were also riders on this merry-go-round and not immune to turnover of these critical positions. For one reason or another, center directors were replaced frequently during Goldin's tenure. During the Apollo era, center directors were the most stable of any of NASA's senior managers, and for good reason: that was where the hardware met the launch pad. The slightest mistake could result in failure. Continuity and experience at the centers was essential to success.

Fortunately, the tenure of the first administrator (acting) who succeeded Beggs, William Graham, was brief. But during the six months before his successor was appointed, Graham managed to perform some mischief. He removed Phil Culbertson from his position as the senior Headquarters Space Station manager and alienated other senior managers. After the *Challenger* accident, and because he was not well received during his few appearances before Congressional committees, the Reagan administration decided that Graham would have to be replaced. They turned to Jim Fletcher, a respected former administrator and gentleman from the old school, who could be counted on to be a team player. NASA needed a firm hand on the wheel if the agency was to recover from its most publicized tragedy, the *Challenger* accident.

Fletcher was appointed in May 1986 and Graham returned to his position as Deputy Administrator, where he remained until October when he was replaced by Dale Myers. Myers was also coming back

---

4. Thompson candid about Medicaid, Tony Pugh, Knight-Tribune News Service as printed in *The Florida Times-Union*, March 25, 2005.

to NASA for a second time, having served as Associate Administrator for Manned Space Flight from 1970 to 1974. Together, Fletcher and Myers moved quickly to try and restore NASA's tattered morale, brought on by the *Challenger* accident and the uncertainties and changes created by the rapid turnover in top management. Culbertson returned for a short period as general manager and then moved to another management position. He continued to provide some Space Station oversight but was not in the direct chain of command. Andrew J. Stofan transferred from the Lewis Research Center to Washington to be Associate Administrator for Space Station, and former astronaut Richard H. Truly was transferred from Houston to Washington to be Associate Administrator for Space Flight. Management stability appeared to be restored. But this would be a short respite from the many management and program changes that soon followed; change became the norm rather than stability.

## Early Space Station Management Problems

When Beggs established the Space Station Task Force in 1982, he was able to assign experienced NASA managers such as Phil Culbertson, John Hodge, and Bob Freitag to key positions. They had been with NASA in senior management positions since the early Apollo days. In addition, other experienced mid-level managers such as Daniel Herman, Richard Carlisle, and Terry Finn were transferred and joined the Task Force. Managers with good pedigrees, such as Luther Powell, were brought in from the NASA centers to round out the team. Their broad experience in dealing with the Washington bureaucracy and congress, as well as the political intrigues of the NASA Centers, assured that the early efforts to gain support for such a huge undertaking would be successful. Planning proceeded in good fashion; the major deficiency as the program progressed was the inability to secure all the early funding needed to move the program ahead efficiently. The lack of funding would constrict all subsequent activities.

Soon after the Phase B studies began and a Space Station management structure was put in place, the need to fill additional senior- and mid-level management positions could not be satisfied. At JSC, where day-to-day oversight and contract management would be exercised, the pool from which to select managers for Level B was limited. Some members of the astronaut corps were transferred to fill the void. Depending on astronauts and payload specialists, beyond question smart, hard working and dedicated individuals, to manage programs at middle management or higher levels was not a good move for Headquarters or JSC's center directors. Astronauts join NASA to fly in space and their backgrounds confirm that choice; experience in program management is not a requirement in order to be selected. A program manager has to start with some program, but not at a critical position in a high visibility program. The situation at MSFC, LeRC, and GSFC was somewhat better. Nevertheless, the good managers were already working on important projects; raiding one program to support another was a difficult choice to make for center directors.

At NASA Headquarters, finding senior managers for high visibility programs is even more difficult. Adding to the need to deal with Congress and the White House are the ever present media, lobbyists and professional societies looking over a program's shoulder, each with a particular ax to grind. If a problem arises it is quickly discovered, usually magnified in importance, and then dissected in detail by the media over the next days, weeks, and months and blame assigned – rightly or wrongly. The further one is away from Washington, as at the NASA centers, the less intense is such scrutiny. In Washington, all programs are in competition with all other programs. Not necessarily bad, it just takes a well prepared manager to be able to compete and survive in that environment.

## NASA Attempts to Improve Program Management

The ability to be a good program manager is not an inherited virtue. It is achieved by dint of training, hard work, careful observation, and the ability to learn from occasional setbacks. Space Station managers were being asked to oversee the most complicated engineering and scientific program ever attempted, while enduring intense political scrutiny. And with every passing year it became more complex with the award of large contracts to aerospace companies and the addition of international partners, each with a unique way of doing business. Every NASA commission chartered to study the program or to recommend how NASA should proceed in the future commented on the complexity of the Space Station program and the management challenges it presented. NASA recognized this problem at the outset of the program but did not respond adequately to put in place and maintain a strong team to meet the challenge as the program progressed. Staffing problems plagued the program almost from the start as has been discussed. However, NASA cannot be assigned all the blame for this shortcoming. It was denied the ability to add experienced staff due to restrictions contrived by Congress and, at times, OMB.

After the *Challenger* accident in 1986, General Phillips led a team to review NASA's management practices. His team recommended that NASA "institute formal training and development program(s) for program / project managers." This recommendation, coming on the heels of one of NASA's most devastating management failures, was similar to recommendations developed by NASA in-house management studies as far back as 1975. The recommendations from these studies sounded good, and looked good on paper, but how were they followed up? A Program and Project Management colloquium held at Wallops Flight Center in 1980 (at which the author lectured) attempted to develop requirements for training program / project managers. This colloquium resulted in establishing the Project Management Shared Experiences Program and was one of the background studies briefed to the Phillips team.

Phillips' recommendations resulted in an October 1987 NASA study that concluded "the management of NASA programs and projects is becoming increasingly complex, and the demand for trained and experienced personnel is increasing as the available pool is being depleted."[5] This study led, in turn, to a NASA Model For Development And Training Of Project Management Personnel designed to assist program / project managers throughout their careers as they progressed to manage ever more complex programs. The problem had been identified and remedial actions developed. There were lots of studies, NASA's usual response to a problem, but how many Space Station managers had the opportunity to take the courses?

Apollo succeeded because NASA brought in senior managers who had earned their spurs on DOD and other agency's major programs. After the redesign in 1993, the Space Station program was not as fortunate; the management pool was "depleted." The new management team lacked experience. In a 1995 Space Station article in *Aerospace America*, Theresa M. Foley, a journalist who had been closely following the program, wrote: "The space station has a history of eating managers alive."[6] She did not go into detail of how she arrived at that conclusion; apparently she felt that listing the many managers who had come and gone was sufficient to prove her point. Two reasons for her observation are possible. Either the managers were not up to the job, or the job, as structured, was unmanageable. In reality, it was a combination of both. Finding experienced managers for NASA programs, not just for the Space Station, continues to be a challenge.

## CONGRESSIONAL AND WHITE HOUSE CONTROL

Oversight interactions with Congress for Space Station were quite different. Committee and Subcommittee chairpersons and membership remained fairly constant until 1994. Members preserved memories of past battles, slights (real or imagined), and missed opportunities. Seniority gives members their positions and, unless the electorate turns them out, their positions are secure as long as desired. Only the leadership of House committees is subject to change by a rule that limits the number of terms of a chairperson. Thus a Congress controlled by Democrats through the 80s and early 90s, and a White House controlled by Republicans, was a formula that often equated to impasse. The formula had new factors after the 1994 elections, but with similar results which continued through the launch of the first Space Station elements.

As was the case at NASA, senior White House advisors and officials turned over with great frequency. Lacking program background, decisions that they enforced compounded Space Station difficulties. A decision made by the Reagan White House would come back to hamstring the program. Discussed in

---

5. Quotes are from NASA SP-6101, Issues In NASA Program And Project Management, edited by Frank T. Hoban, Program Manager NASA Program and Project Management Training and Development Initiative, 1989.
6. Space Station: The Next Iteration, Theresa M. Foley, *Aerospace America*, January 1995.

Chapter 3, NSDD-42 restricted how future heavy-lift launch vehicles would be developed, leaving the Space Station with no alternatives after the *Challenger* and *Columbia* accidents.

The NSDD[7] was a classic example of how powerful conflicting interests influence decisions in Washington. The NSDD prohibited NASA from "maintaining an expendable launch vehicle adjunct to the Shuttle." Along with DOD, the NSDD added the Department of Transportation to the mix of agencies with responsibilities to procure expendable launch or unmanned launch vehicles. Although the NSDD called for NASA and DOD to cooperate in the development of "space transportation systems," in reality DOD was in the driver's seat and called the shots over the next years. What followed was years of debate and conflict between Congress and administrations with huge sums spent on programs such as the Evolved Expendable Launch Vehicle that kept two different launch vehicles in production, the Atlas 5 and Delta 4. Neither of these vehicles could be used to service the Space Station nor provide the heavy-lift capability needed today for the proposed, expanded solar system exploration. The problem that NASA faced following the *Challenger* accident in 1986 still exists. There is no alternative to the shuttle for carrying large Space Station elements and fully serving the logistics and crew requirements.

Underfunding the Space Station began with the FY 1985 budget and continued until FY 1991 (Table 3). In the first five years, FY 1985 through FY 1989, administration budgets submitted to Congress reduced the amount NASA requested by almost 40% ($2.5 billion versus $4 billion). Congress then further reduced the president's requests by over $400 million. During the next ten fiscal years the differences between what was requested directly for the program (not in other line items) and appropriated were not as large. The Congress appropriated $22.9 billion versus the Bush and Clinton administration's requests of $23.6 billion. However, administration budgets were some $500 million below NASA's requests to OMB. For the last six fiscal years (FY 2001-FY 2006) NASA's requests and the president's budgets have generally tracked each other. Congressional appropriations, however, have been $900 million lower than the president's budget requests. Opponents could make the argument that the program was never underfunded and, based on performance, received more funds than it deserved. Senator Bumpers and other critics had many grounds on which to justify their attempts to cancel the program.

Funding shortfalls were not just a failing of Congress. Every administration contributed to the program's problems. Although each administration claimed at one time or another that the Space Station was a priority, in actuality it was always low on their lists for budget emphasis. Disagreements with Reagan's OMB and cabinet-level departments, the Clinton administration's disposition to modify the program's content to achieve a political goal, and modifications initiated by the two Bush administrations, obstructed the Space Station's progress.

In the run-up to the presidential elections in 1988, Vice President George H.W. Bush's campaign headquarters established a number of "Issue Groups." Two separate groups, but with some common membership, dealt with military and civil space issues. Many recommendations were sent forward on positions the Vice President should take on civil space. Almost all were accepted. In a speech at the Marshall Space Flight Center in October 1987,[8] Bush unveiled a comprehensive list of space priorities to which he would be committed. He stated that he would support the construction of a replacement shuttle, a somewhat contentious issue at the time. He endorsed the Mission to Planet Earth, a program that had been proposed by the working group chaired by astronaut Sally Ride. He also endorsed the development of a transatmospheric vehicle and construction of the Space Station. And, in his most ambitious statement, declared that "we should make a long-term commitment to manned and unmanned exploration of the solar system."

---

7. Fact Sheet, The President's Space Policy And Commercial Space Initiative To Begin The Next Century, The White House, Office of the Press Secretary, For Immediate Release, February 11, 1988; and Fact Sheet, Presidential Directive on National Space Policy, The White House Office of the Press Secretary, February 11, 1988.
8. Speech delivered by Vice President George Bush at the Marshall Space Flight Center, Thursday, October 29, 1987.

## Table 3 - Total NASA & Space Station Budgets
(Totals in billions of dollars, may be rounded up or down to nearest million.)

| Fiscal Year | NASA Budget Recommended by NASA | President's Request | Appropriation | Notes | Space Station Budget Recommended by NASA | President's Request | Operating Plan -all accounts | Notes* |
|---|---|---|---|---|---|---|---|---|
| 1985 | 8.119 | 7.491 | 7.552 |   | 0.235 | 0.150 | 0.156 | Definition |
| 1986 | 8.152 | 7.886 | 7.764 |   | 0.280 | 0.230 | 0.200 | Def. plus FTS |
| 1987 | 8.139 | 7.694 | 10.775 | a | 0.600 | 0.410 | 0.433 | Def. plus b |
| 1988 | 10.569 | 9.481 | 9.002 |   | 1.055 | 0.767 | 0.393 | Plus FTS |
| 1989 | 12.843 | 11.488 | 10.801 |   | 1.872 | 0.967 | 0.929 | b |
| 1990 | 14.298 | 13.274 | 12.295 |   | 2.130 | 2.050 | 1.799 | b |
| 1991 | 15.864 | 15.125 | 13.868 |   | 2.693 | 2.451 | 1.913 | b |
| 1992 | 18.027 | 15.754 | 14.334 |   | 2.907 | 2.029 | 2.135 | c |
| 1993 | 15.992 | 14.993 | 14.323 |   | 2.265 | 2.250 | 2.255 | c |
| 1994 | 15.657 | 15.265 | 14.549 |   | 2.250 | 1.946 | 2.206 | c |
| 1995 | 14.319 | 14.300 | 13.996 |   | 1.916 | 2.121 | 2.233 | c |
| 1996 | 14.335 | 14.260 | 13.821 |   | 1.834 | 2.115 | 2.244 | c |
| 1997 | 13.896 | 13.804 | 13.709 |   | 1.782 | 1.802 | 2.449 | d |
| 1998 | 13.653 | 13.500 | 13.648 |   | 2.121 | 2.121 | 2.441 | d |
| 1999 | 13.247 | 13.465 | 13.653 |   | 2.138 | 2.270 | 2.300 | d |
| 2000 | 13.974 | 13.578 | 13.602 |   | 2.098 | 2.483 | 2.323 | d |
| 2001 | 14.101 | 14.035 | 14.253 |   | 2.386 | 2.115 | 2.128 | d |
| 2002 | 14.780 | 14.511 | 14.892 |   | 2.032 | 2.087 | 2.094 | e |
| 2003 | 14.985 | 15.000 | 15.389 |   | 1.492 | 1.492 | 1.845 | e |
| 2004 | 14.958 | 15.469 | 15.378 |   | 1.629 | 1.707 | 1.364 | f |
| 2005 | 16.043 | 16.244 | 16.196 | g | 1.566 | 1.863 | 1.591 | f |
| 2006 | 17.002 | 16.456 | 16.457 | g | 1.822 | 1.857 |   |   |

* For brevity, Notes b-f for "Operating Plan-all accounts" do not always name all line items in addition to development and operations.
a - Reflects Congressional action to replace Challenger ($2.1 billion). b - Includes funds in Operating Plan for CoF and FTS.
c - Includes CoF, research, and U.S./Russian cooperation.
d - Includes U.S./Russian cooperation, Russian Program Assurance, CRV, and research.
e - Operating plan includes research program.
f - Totals include "full cost" with civil service salaries and benefits, and reflect stand-down for Columbia accident.
g - FY 2005 included $126 million for hurricane repairs, FY 2006 included $3.1 billion for return to Moon.

Table 3. NASA and Space Station budgets from FY 1985 to FY 2006.

From the Issue Group's perspective, one of the most important statements in his speech was the acceptance of the recommendation that he reestablish a National Space Council. This was not a new idea, President Eisenhower was the first to have a National Aeronautics and Space Council, created by the 1958 NASA Act. When first formed it was chaired by the president, but in 1961 the Act was amended, and in the Kennedy administration Vice President Lyndon Johnson became chairman. However, the Council fell out of favor during the Nixon administration and was abolished in 1973. The Issue Group felt that because Nixon and the presidents who followed did not have trusted advisors on civil space matters, a steady erosion of support for NASA programs occurred. If civil space programs were to compete for administration support and necessary resources, the president needed an informed advocacy council in his inner circle.

The Issue Group believed that the Senior Interagency Group for Space (SIG Space), established during the Reagan administration, was not a substitute for a National Space Council. SIG Space's priorities dealt mostly with recommendations for national security. It could be influenced by one or more of the multiple security agencies that dominated its membership and, in particular, by DOD. Turf battles could prevail over good policy. Former NASA administrator Paine's warning in 1987 directly addressed this issue: "NASA's problem is confined to a tiny triangle bounded by the White House, Capital Hill, and the Pentagon." During the early debates within the administration on whether or not to support the Space Station, SIG Space, chaired by President Reagan's science advisor George Keyworth, reflected such turf battles. Many members of SIG Space did not support the Space Station and their positions never changed after Reagan gave the go-ahead. The only disappointment in how Bush proceeded was that the Issue Group had recommended that, when he was elected, he chair the Council. Bush decided to have it chaired by the Vice President. Bush's support for space programs was well known. The Issue Group reasoned that his direct involvement would assure a continued high priority. Senator Dan Quayle's (Vice President presumptive) positions on civil space matters were less clear, and how influential he would be when elected was another unknown.[9]

After his election, Bush moved quickly to implement some of his space promises. Fletcher, desiring to return to academia, tendered his resignation and it was accepted. In his place, Bush elevated a close acquaintance, Richard Truly, from the position of Associate Administrator to Administrator. The National Space Council (NSC) was established under the direction of Vice President Quayle and a small staff was assembled. As staff director Quayle brought in his friend Mark Albrecht, a Senate staffer. From NASA's perspective, the future seemed brighter. In theory having a space council close to the president was good; it had functioned with some success during the Kennedy Administration.

In practice, during the Bush administration, it was a calamity. Perhaps if it had been chaired by the President, as his Issue Group had recommended, it would have provided him with useful council. Instead, the inexperienced Vice President and the staff he assembled set the President up for a painful fall. The program announced by Bush on July 20, 1989 to return to the Moon and then go on to Mars was ill conceived and poorly timed. With large budget deficits and NASA's major program, the Space Station, in disarray, it was easy for a Congress controlled by the Democrats to deny the funding for the President's showcase program. All that happened was that a lot of time was wasted by many people attempting to convince Congress that it was a great idea. A sitting administration butted heads with recalcitrant House and Senate space committees, and lost. The obvious conclusion, a National Space Council is only as good as its leadership and staff, and its existence does not guarantee that a president will receive good advice.

## NASA ADMINISTRATORS' SPACE STATION LEGACIES

As you enter the NASA Administrator's office suite in our nation's capitol at 300 "E" Street, S.W., you walk down a corridor and on the walls hang paintings of seven past NASA Administrators. Since 1958, NASA has been led by thirteen men, but only eleven have been presidential appointees.[10] The faces of all but the last three appointees (current administrator Michael Griffin and his predecessors Sean O'Keefe and Daniel S. Goldin) stare out from the frames, some rather dourly, two with slight smiles. One can imagine that perhaps, while sitting for their portraits, they were contemplating the vast expanses of space through which they steered the NASA ship of state. Each brought to the job unique leadership and management skills and also, at times, some leadership deficiencies. All were responsible for leading an agency with an unprecedented mission in the vast government bureaucracy of large cabinet level departments and smaller independent agencies. NASA prospered and faltered at times under their direction.

In the 1970s, with the country absorbed in its number one problem, energy shortages, NASA faced an uncertain future, with no major programs on the immediate horizon. Annual budgets had been steadily declining and the Carter administration did nothing to reverse the trend. Just in time, President Reagan's election brought a ray of hope to NASA's beleaguered staff. Receiving some encouragement from members of the new administration that there would be support for a larger civil space program, James Beggs, during his confirmation hearings, indicated that a space station should be the "next step" for NASA.[11] Eight months later he established the Space Station Task Force and after two years of studies and lobbying his efforts were rewarded. The Space Station received its official blessing from the administration, and a somewhat reluctant Congress agreed to provide the first funds in FY 1985.

Beggs' tenure did not have any lasting impact on NASA's culture. He was viewed as a rather peripatetic manager who left most of the day-to-day issues to be resolved by his deputy, Hans Mark. Mark left a year before Beggs resigned and the position would remain vacant. Nonetheless, Beggs was

---

9. Statements attributed to the Issue Group are based on recollections and notes made by the author who served on both Vice President Bush's Military and Civil Space Issue Groups from October 1987 to October 1988.
10. The two Acting Administrators were William R. Graham and Daniel R. Mulville. Jim Fletcher served two separate terms, but is represented by just one portrait painted after his first term.
11. U.S. Senate, Committee on Commerce, Science, and Transportation, Nominations – NASA, 97th Cong., 1st session, June 17, 1981, p. 22.

instrumental in starting the Space Station program and left NASA with a strong team in place. Unfortunately, while attempting to sell the program, supporters tended to overstate the potential advantages of living and working in Earth orbit. From the beginning it was claimed that research on the Space Station would result in enormous benefits, from curing cancer, to extending life expectancy, to manufacturing materials with amazing properties not possible to make on Earth. Who could be against a program with such promise? As the program progressed many reputable researchers began to challenge the assumptions that led to the promised benefits of conducting research in microgravity. They were, after all, largely assumptions.

After Graham's short-lived career, his successor, Jim Fletcher, attempted to strengthen management in the agency and solve the many programmatic problems that he inherited. But both he and his deputy, Dale Myers, tried to manage NASA as they remembered it when Apollo was the centerpiece. It was a vastly different agency in 1986 and the Washington political world had also changed dramatically. They were not tough enough within the agency, especially with the centers, to affect needed changes. They also failed to muster sufficient support in Congress to put the Space Station program on a solid footing. Except for the FY 1987 budget, a relatively modest $420 million, the budgets approved in the next four years were well below administration requests, which in turn were less than NASA's requests to OMB.

When Fletcher decided to address the $8 billion cost estimate issue and provide OMB a new runout estimate of $14.5 billion with the FY 1988 budget request, that was the critical moment for the program. If the goals and design, still on the drawing boards, were too rich for the nation's pocketbook, the Reagan Administration and Congress had an obligation to call a halt, reassess the program in a businesslike way, and decide how to proceed – either stay with the program as proposed and accept the cost, or agree to a major revision. Although ordered by the Congress to rescope, firm direction was not given. By proposing that the Space Station, then in design, could be built while receiving flat out-year budgets, a proposal dictated by political pressure not good program management, Fletcher placed an unrealistic constraint on his managers. At this point, a well-managed program required rising budgets until most of the hardware was built.

The funding profile required to assure success for a major, high-risk program was never achieved. Both the Reagan Administration, led by OMB, and Congress rejected the new, more realistic, cost estimate. NASA, supported by some in Congress, tried to finesse the direction it was receiving to phase and rescope the program to stay within the funding restrictions but still maintain all the original capabilities. By dividing it into two phases, Phase I of $12.2 billion and Phase II of $2.3 billion or more, NASA thought it would be more palatable for the critics. In other moves, an expensive piece, the Polar Orbiting Platform, was transferred to an OSSA account to reduce the Space Station runout cost line and, following past practice, operational costs were not included. It was a losing cause; these bookkeeping ploys were not accepted. By the time Fletcher resigned at the start of the first Bush Administration, the Space Station was in trouble. In spite of all the slight of hand, costs continued to escalate. To deflect the critics, NASA continued to provide optimistic low estimates of runout costs ignoring, among other things, the warnings it was receiving from the prime contractors. NASA's "can do" culture deceived the program's managers into believing they would find a way around their cost dilemma.

Truly's conflicts with Congress and the White House staff, after replacing Fletcher, have been described. Perhaps Truly's biggest mistake was when he elected to try and find a way to accommodate the Congressional direction that accompanied the FY 1991 appropriation. This was an opportunity to warn the new Bush Administration and Congress that the program could not continue under the guidelines he received. (An easy criticism to make, in hindsight.) Such a declaration would have taken extraordinary courage but would have served as a wake-up call to stop all the micromanagement. Depending on the response to such a warning, the program might have continued at a reasonable pace, or he would have been taken at his word and it would have been terminated. He was already having problems at the White House so he had little to lose. At that point there was essentially only paper to show for the funds spent (less than $4 billion); the big commitments were still in the future.

Truly was never able to build a strong relationship with his far flung staff, and the manner of his dismissal, abruptly fired by the President, did not elicit great sympathy. Although firmly supported by the President early in his term, and an engaging personality, Truly alienated powerful Congressional committee members and White House staff, particularly those on the NSC. His background and interests were clearly skewed toward manned space flight programs. He was seen as being inflexible and unwilling to make the changes needed to appease a critical Congress whose support was required if the Space Station and President Bush's new Space Exploration Initiative were to succeed.

Just before Truly was nominated, The Center for Strategic & International Studies released their study of civil space policy.[12] It included a list of qualifications that a NASA administrator should possess: "The individual should: be on friendly speaking terms with the president and have access to the president; know 'Washington,' particularly the policy and budget processes; have experience and work well with Congress; work well with DOD; be adept in interagency politics; have a technical background (a desirable but not mandatory qualification); have experience in industry; understand the importance of the commercial use of space; be willing to delegate; and have a good instinct for dealing with the press." Truly's background did not include some of these qualifications. He left the agency with the problems that he inherited unresolved.

Truly's successor, Dan Goldin, was the most controversial of all of the NASA administrators who held the top job during the development of the Space Station. He has been both praised and maligned. Once in office, he immediately tried to put his stamp on the agency and in so doing created many enemies inside and outside the agency. Serving under three different administrations (for both political parties), he accomplished two diametrically opposite results. He kept the Space Station program alive during difficult times, but his management peculiarities and the problems they provoked complicated the program's existence and threatened its cancelation.

An notable example of his management style that created internal NASA criticism was how the monthly General Management Status Reviews (GMSRs) functioned after he was appointed. As described in Chapter 3, GMSRs were instituted during Administrator Webb's tenure and were the primary means by which senior management stayed abreast of all the programs. Each associate administrator and lesser department head was required to brief the administrator every month on the status of their programs. These reviews included activities at the centers as every center reported to an associate administrator. Progress, problems, program funding profiles, personnel activities, upcoming hearings and major meetings, and more were discussed. With tightly controlled attendance, the GMSRs provided a forum for the administrator to understand how the agency was performing. Although not an originally planned function of the reviews, he could, when necessary, solicit opinions from his most senior managers on how to solve a problem.

A former senior Headquarters' manager described to the author how the GMSR deteriorated under Goldin. In the beginning, programs were reviewed as they had been for thirty years. But very soon the information exchanged at the GMSRs shifted. The Associate Administrators realized that if they discussed any problems Goldin would get upset and give off-the-cuff directions on how to solve them. Not only that, but Associate Administrators found themselves in Goldin's "doghouse" if they brought problems to the GMSR. Eventually the only material presented dealt with completely bland items. Goldin considered GMSRs a waste of time, which they had become, and they were discontinued.

To replace the Headquarters' GMSRs, Goldin utilized the Program Management Council and when this Council held program reviews Goldin's deputy presided. Some programs were assigned directly to the Centers, bypassing the traditional Associate Administrator control, and reviews for these programs were conducted at the Centers. Another long-standing management committee, the Senior Management Council, consisting of Center directors and Headquarters managers, met on an infrequent schedule. It was primarily a policy making body and did not provide close program oversight.

As a result, Headquarters' senior managers lost touch with what was happening with agency programs. Yet they had the responsibility to represent the programs before Congress and defend them at budget time at OMB. Is it any wonder that Congress often heard divergent stories during testimony by senior

---

12. A report released by The Center for Strategic & International Studies, A More Effective Civil Space Program, May 1989, Appendix C discusses Attributes of a NASA Administrator, and lists eleven Proposed Qualifications for the Administrator. Of all the administrators before and after Truly only one, in the author's opinion, came closest to filling all these qualifications, James Webb.

managers on the same subject, or that stories changed in a very short time? Or that OMB was able to deny funding for important programs that NASA management had a difficult time defending? Since the day-to-day management of most NASA programs occurs at the centers, canceling the GMSRs consolidated center control of the programs. Goldin depended on personal briefings from the centers, and without any intervening oversight, center managers told him what they wanted him to know and nothing more.

For the Space Station, Goldin's management approach guaranteed problems. When he decided in June 1993, after the Space Station redesign, to move essentially all management functions to JSC, he lost control of the program. Instead of it being a NASA program it became a JSC program. Headquarters and the other centers danced to JSC's tune and problems, such as contract overruns, were easily hidden from view, only to surface at an embarrassing point. The statement made by Bryan O'Connor in 1996, is relevant: "Now the boss says we have a new NASA. We're not going to have disagreements between the centers as in the past. I disagree. In time the center-to-center rivalries will show up again."[13] O'Connor, a former Space Station program manager with arguably the strongest program management background, was proven right; analysis of the *Columbia* accident bore out the warning he had so succinctly described seven years earlier.

Before Space Station management moved to JSC, Dick Kohrs was beginning to get his arms around the problems he had inherited and was adding new management talent to his Washington team. It was a slow slog because of internal hiring restrictions and difficulties working with center management, but those who served on the SSAC could see that the program was turning the corner. Kohrs was never given the opportunity to prove that the program under his management was moving in the right direction. Goldin impulsively "fired" all the Space Station Freedom managers and replaced them with managers of lesser experience. Adding to the Space Station's management problems at that critical time was the decision by the Clinton Administration to bring Russia into the program. Perhaps Kohrs would not have been able to solve all the problems that resulted. For his successors the added complication of integrating the Russians into the program was daunting, and it was dropped in their laps before they had a chance to get on top of their new assignments.

What is the legacy of Goldin's long tenure? An article in the December 1997 issue of *Aviation Week & Space Technology*, half way through his term of office, summed up the pluses and minuses to that date. He was credited with saving the Space Station, pushing a "faster, better, cheaper" approach to developing new programs, and encouraging cutting-edge technologies that had languished during previous administrations. He was called "charming, politically astute and passionately dedicated, accessible, and polite to reporters." But then the article discussed "a darker side . . . his critics in NASA complain that it often doesn't make print. Stories of tirades, profanities, berating of subordinates and erratic behavior have come out of the space agency almost since he arrived."[14]

Goldin survived the bad press and internal dissension because of a political fluke. In the early days of the Clinton Administration he was retained because there was no alternative to head NASA. Aaron Cohen, the acting deputy administrator, had resigned seven months after becoming Goldin's acting deputy. Since it normally takes a new administration's nominees to high-level positions several months or longer before receiving Senate consent, Clinton chose to retain Goldin while addressing the selection of appointees to more critical positions in his cabinet.

Goldin then cozied up to his new boss and endorsed the controversial initiative to add Russia to the Space Station program. That is the kind of loyalty that is rewarded in Washington. By the end of 1997, when the *Aviation Week* article was published, he was comfortably settled and would be administrator for another four years. Many of the pluses cited in the 1997 article, such as privatizing the shuttle, never happened. Failures blamed on deficiencies in the "faster, better, cheaper" programs he endorsed had not yet occurred. In 2001, the NASA IG issued a report severely criticizing the faster,

---

13. See Chapter 7. Although O'Connor had broad management experience, he lasted only nine months as a senior Space Station manager, the last three as Acting Space Station Program Director. He was then appointed in April 1994 as Deputy Associate Administrator, Space Shuttle Program and held that position until his disagreement with Goldin in 1996 over shifting program responsibilities to JSC, at which point he resigned.
14. NASA's Paradoxical Daniel S. Goldin, Joseph Anselmo, *Aviation Week & Space Technology*, December 22/29, 1997.

better, cheaper approach. Goldin's legacy always will be a matter of opinion, but in the end his portrait will hang in the administrator's suite with all the other NASA administrators, each responsible for guiding the agency through demanding times.

Goldin's successor, Sean O'Keefe, served for three years. When he announced his intended departure in December 2004, a *New York Times* editorial opined: "Through no great fault of his own, Mr. O'Keefe is leaving the space program in worse shape than he found it. The remaining shuttles are still grounded for safety repairs, the space station they service is a shrunken shell, and the agency has been given a challenging long-term mandate for space exploration with little new money to carry it out."[15] Perhaps an overly pessimistic view of the overall condition of the agency but certainly true in all the specific observations. It would be hard to argue that O'Keefe succeeded in making any substantial changes in the way NASA manages its programs. He made a valiant effort to restructure NASA's imperfect bookkeeping, the root cause of many of the agency's problems, but was only partially successful. In 2003 he established the Integrated Management System to consolidate the agency's accounting practices into one format. Until that time NASA's major centers were using systems inherited from their past, each different from the other. It has been a difficult transition and as late as the spring of 2004 the independent auditors from PriceWaterhouseCoopers LLP and Ernst & Young reported that they were unable to sign off on NASA's books.[16] And it was on his watch that we witnessed the *Columbia* tragedy, now acknowledged to have been caused, to a large degree, by management failures.

In an interview in September 2002, when asked what his plans were for NASA after almost one year in office he responded, "We are in the Age of Sail right now in space exploration and we are aspiring to the Age of Steam. Everything I am pressing in terms of the agency's agenda right now is to overcome that and arrive at that Age of Steam."[17] A rather strange analogy for a NASA administrator, from biplanes to jets or some other aerospace comparison would have been more appropriate. However, this undoubtedly reflected his partial understanding of NASA's deep-rooted problems. Any progress he made to move the agency to the "Age of Steam" was obscured and blown away by the *Columbia* tragedy four months after the above statement.

Now there is a new hand on the throttle. Michael D. Griffin, with previous experience at NASA in senior management positions as well as in industry, is the new Administrator. He takes on his job with as many problems and unknowns on the table as any previous administrator had to face. And, as occurred with his predecessors, he has begun to make sweeping changes. He has been welcomed during his first months in office by all the warring factions. He will have to perform a delicate balancing act to keep the welcome mat in place as he decides how the Space Station and other programs will proceed as part of President Bush's space exploration vision.

## What Are the Lessons to be Learned?

With few exceptions, senior NASA Headquarters' and center managers had a difficult time dealing with the complexities of the Space Station program. After the 1993 redesign, managers departed after short careers. At this point, it was important that the program have management continuity until the first elements were launched or were close to launch. As a result, Space Station managers never had a chance to really get on top of the program. Without this continuity Congress and the White House, with the exception of Administrator Goldin, were debating the issues during critical times with new faces, almost on a yearly basis. That is not how you breed trust with policy makers, especially during the 1990s when the program was experiencing many serious problems. Rapid turnover of senior

---

15. NASA's Chief Bails Out, *The New York Times*, December 27, 2004.
16. Recent reports of NASA's financial management problems are extensive. See for example: House Science Committee Hearing Charter: Financial Management at NASA: Challenges and Next Steps, Subcommittee on Space and Aeronautics, October 26, 2005; Statement of: The Honorable Robert W. Cobb, NASA Inspector General, Financial Management at NASA: Challenges and Next Steps, October 27, 2005; Testimony before Congressional Requestors of GAO's Gregory D. Kutz and Allen Li, NASA Long-standing Financial Management Challenges Threaten the Agency's Ability to Manage Its Programs, October 27, 2005.
17. New Chief Hopes To Guide NASA to Its "Age of Steam" in Profile, an interview with Brian Berger, *Space News*, September 9, 2002.

Executive Branch appointees and managers was called out by the Grace Commission, which reviewed government operations during the Reagan administration, as a failing that adversely impacted the successful management of programs.[18] That assertion still held true ten years later and the Space Station is a perfect example.

Through the years, the one recurring problem that stands out was the inability of Space Station managers to provide accurate, out-year, cost estimates. There were many reasons for this deficiency, not the least being political pressure creating an environment that forced NASA to forecast optimistic, low numbers. Space Station managers forgot or never knew of the warnings contained in the 1981 Hearth Report (discussed in the Introduction) that: "NASA, OMB, and the Congress should expect up to a 30 percent cost growth even if the project is well managed and there are no major technical surprises." Tom Young's IMCE investigation highlighted this problem but NASA never mastered a solution. In the years ahead, NASA must conquer the enormous challenge of providing program cost estimates that will survive a program's lifetime! Full cost accounting must become the norm to avoid criticism that costs are being hidden.

Perhaps the Space Station was so complex that it defied any conventional management approach. Nevertheless, not enough thought was given to the end-to-end complexities of the program and the importance of continuity at all management levels. The management structure put in place and frequently modified, did not measure up to the job. Senior management responsibilities were added or subtracted at Headquarters and the centers during the contentious 1990s. At times, management integration really only occurred at the administrator level. General Phillips warned that this would be a problem and, even when finally recognized, all the elements were not addressed and consolidated at the proper level. A NASA administrator cannot be an effective day-to-day program manager with all the other responsibilities of the position. Another shortcoming: the management structure did not fully reflect the many interfaces that were needed to deal carefully and efficiently with all of the international partners, especially with the Russians, whose participation added a significant management complication. And the program never came to grips with how to service the user community, continuously shortchanged by budget problems.

Times change but certain principles endure, especially when it comes to managing large programs. As mentioned, one of the most important principles is management continuity. For the Space Station, NASA did not follow the successful model used to manage the Apollo program. After the Apollo program began, there was only one administrator change before the successful *Apollo 11* mission. However, almost all of the other senior Apollo managers at Headquarters and the centers stayed the course until the first landing was accomplished. For other successful programs during this era, such as Viking, Voyager, Surveyor, and Lunar Orbiter, the original management teams remained in place until the programs' objectives were achieved.

Where will NASA find the experienced managers needed for the next generation of major programs? Grooming them from within will be difficult. The luxury of having time for managers to learn on the job may not be possible or desirable. Returning to the Moon or expanding the exploration of Mars with combined robotic and human systems means that from the start the programs will be large and complex. Today there is only one potential pool of managers that have the broad experience required, DOD and DOD contractors who have demonstrated the ability to successfully manage large, complex programs. Those were the sources for Apollo managers. Granted, it may not be as easy or seamless to bring in senior managers from outside organizations as it was at the birth of NASA, but it must be done.

Prodded by the Clinton and Bush administrations to first "reinvent" and then "transform" NASA, some management changes have taken place. The "reinvention" in 1993 did not result in any major improvements in how NASA went about conducting its business. In particular, the problem of estimating program costs and overall bookkeeping issues persisted through the Goldin and O'Keefe eras damaging NASA's management reputation on Capital Hill and in the White House. The

---

18. The President's Private Sector Survey On Cost Control, January 12, 1984 (Also known as the Grace Commission). Although focused primarily on how to "suggest remedies for waste and abuse in the Federal Government," the Survey's 36 Task Force reports and 11 studies on special subjects, made 2,478 recommendations. The author participated in the Commission's work.

"transformation" of NASA, called for by the Aldridge commission that added detail to President Bush's vision, has begun. There is no question that the Headquarters organization chart resulting from the first steps of the transformation reduced the large number of offices that reported either directly or indirectly to the administrator during Goldin's and O'Keefe's tenures. But a simplified organization chart showing new titles does not guarantee efficient management. How effectively all the boxes will work together depends on the skills of NASA's two top appointees, Administrator Griffin and his deputy Shana Dale, and a newly created position of Associate Administrator.

The clear reporting lines of the field centers to Mission Associate Administrators were extolled in a NASA press release as one of the advantages of the "transformation" and new organization. For old NASA hands this change appeared to largely restore the management structure that had been so successful forty years earlier, with just the substitution of new titles. Who the centers reported to had been clearly defined during the Apollo era. But control had deteriorated over the years as administrators and associate administrators, little by little, reduced their center oversight. However, this initial "transformation" was short lived. Administrator Mike Griffin, as did all administrators before him, has begun the process of rearranging the chairs and appointing new managers. It appears these changes will be more extensive than those made by any previous administrator. Griffin quickly decided to remove center oversight from Mission Associate Administrator responsibilities. Center directors now will report directly to the new Associate Administrator position, the third highest ranking Headquarters' manager. This will be a tough assignment for any one individual to master.

Eventually, the managers selected to fill the new organization will be judged by how well they develop the chemistry to work as a team. In the waning days of a lame-duck administration, Griffin will have to convince managers with proven track records to come to NASA and reenergize the offices that will be responsible to carry out the new programs. Complicating his task, he will be faced with another problem: maintaining high morale throughout the agency during the transition as the overall size of the NASA civil service workforce is reduced. And unfortunately, using history as a guide, there is the probability that in 2008 a new management team will change the management dynamic again.

In terms of funds thus far expended on the Space Station, a simple comparison is useful. Over the decade of the 90s, every two years American taxpayers spent on the Space Station the equivalent of purchasing one Nimitz-class, nuclear powered aircraft carrier. Each carrier has a projected lifetime of 50 years and GAO estimates total construction, modernization, and maintenance costs over its lifetime will be some $22 billion. Between 1984 when the program began, and the end of 1998 when the first elements were placed in orbit, funds directly appropriated to the Space Station totaled over $21 billion. In addition, an average of $2 billion was spent every year from 1998 to 2005. These numbers do not include funds in other NASA line items for required Station activities, such as experiment development, shuttle upgrades and launches in support of the program, and, until 2004, civil service salaries. It would be difficult to completely account for all those funds. What is certain is the program will fail any cost-benefit analysis if the nation essentially walks away, as proposed, in 2015.

To hold up their part of the bargain the fifteen international partners have spent the equivalent of over $10 billion dollars. This sum does not include money spent by Russia. To date, the bargain can not look very good as a return on investment. Other than the high-tech jobs produced and pieces in storage, the return has been vanishingly small. Some Russian, Italian, and Canadian elements have reached orbit, but after twenty years the ESA and Japanese modules remain Earthbound and with an uncertain future. On March 2, 2006, NASA and its international partners announced a new agreement on how the ISS program would proceed. Instead of an earlier plan to provide 18 shuttle flights to complete ISS construction (reduced from 28 after the *Columbia* accident), only 16 assembly flights would be scheduled. Power available for experiments would be reduced and some elements would be eliminated from the manifest. The fate of the Japanese-funded centrifuge, originally the most important research facility on the Space Station, is particularly troubling. It was one of the elements that did not make the cut.

Depending on what will be the outcome of shuttle availability and launch rates, it is possible that some of the partners' elements will still be on the ground when the shuttle is retired, especially if the shuttle should run into any further problems. Crew time available for work on experiments in the partners' facilities, or any ISS facility for that matter, may increase but will be greatly reduced compared to original planning, because of the missing CERV. Perhaps the most succinct statement made during the March press conference announcing the changes was that by Anatoly Perminov of the Russian Federal Space Agency. In response to a question about the future of the program he said, "It is hard to say because each year brings corrections to the program. So right now, we are all looking forward to the second flight of the shuttle after the resumption of flights, and we will see after that how the program progresses in the future."

If the U.S. reduces its participation after 2015, the international partners will face hard choices on how to proceed, none of which will be very attractive. The concern expressed by Culbertson in 1984 when the first MOUs were in the process of negotiation, whether or not the international partners would sign on for the long-term use of the Space Station, has come full circle. Now the partners are concerned that the U.S. may not be around to fulfill its treaty obligations.

Through the years, a large amount of national and international resources have been wasted or poorly applied on the Space Station program. The reason: vacillating administrations, NASA management mistakes, and congressional intervention. All must take responsibility for a program that almost certainly will not meet expectations or justify the expenditures made. The good news is that the highly criticized decision to add the Russians has paid off in unexpected ways. In spite of all the recriminations and problems, Russia's participation undoubtedly saved the program after the *Columbia* accident. Without Russian logistic flights and using Soyuz to carry crews, the elements placed in orbit prior to the *Columbia* accident would have become a lifeless hulk. Orbital decay would have eventually ended its short life. But the small, two-man crews rotated on the Russian Soyuz kept it functioning until the shuttle returned to flight and they will continue to maintain the ISS until the shuttle once again overcomes its problems.

Former Secretary of the Navy and member of the 9/11 Commission, John F. Lehman, while discussing what the 9/11 hearings revealed, stated that he was "shocked" to find "a culture that had evolved in our government of total non-accountability. Nobody's responsible."[19] Is this what happened for the Space Station program? The *Challenger* and *Columbia* accidents seem to point that way. Urgent warnings on shuttle safety prior to both tragedies were on record from several sources. Space flight is dangerous. Those who take the greatest risks, the astronauts and payload specialists, come from a culture or join a fraternity that accepts the inherent dangers. Future administrations and NASA must meet the challenges of space flight and assure that the managers who develop the next generation of spacecraft have the experience to minimize those risks. The responsibility to find such managers starts at the top of the management chain in the White House and continues within NASA down to the lowest management level.

## NASA's Future

For many years, critics complained that NASA lacked a vision to lead the country in space exploration. From the start of the Space Station program, multiple studies and commissions were chartered to recommend how NASA should proceed; some of the reports contained great detail. If you have not been counting, since 1985 eight different long-range programs have been proposed. Almost all, if they had been adopted, would have authorized NASA to send astronauts back to the Moon and on to Mars.[20] In addition to manned missions, all included robotic exploration of the solar system and research to improve our understanding of the most important planet, Earth. The differences between

---

19. Remarks by the Honorable John F. Lehman at the FPRI Annual Dinner, November 9, 2004.
20. In order of appearance: NASA 1985 Long-Range Plan; The Paine Commission, 1986; The Ride Report, 1987; The NASA SEI Plan (chaired by Frank Martin) to implement President Bush's Space Exploration Initiative, 1989; The Augustine Commission report, 1990; The Stafford Synthesis Group, 1991; NASA Strategic Plan of 1992; and the Aldridge Commission's report released in June 2004 for President Bush's new space exploration vision. I may have missed one or two as NASA published other plans in 2002 and 2004 to return to the Moon, but note the twelve year hiatus from 1992-2004. It is the longest gap in commissions or studies projecting the nation's space future and included all the years of the Clinton administrations. In September 2005, Administrator Griffin unveiled the exploration plan and architecture that would implement President Bush's Space Exploration Vision with an estimated cost of $104 billion.

the reports are minor, usually just in what was emphasized. Placed end-to-end they would fill a very long bookshelf. What happened? None of the plans were embraced by all the required decision makers. What will be the outcome of the latest plan, the eighth, proposed by President George W. Bush, remains to be seen.

Over and over again the blue ribbon panels would come back to the same complaint. The nation lacked a consensus on how to utilize this newfound ability to send astronauts and robots to the far reaches of the solar system and beyond. Frequently during Congressional hearings a member of the House or Senate will declare, "I haven't heard any public support for this NASA program. Without such support I can't vote for program X,Y, or Z." If you have read the trade journals or followed press coverage of space programs over the past thirty years, NASA was continuously described as being in a state of crisis, or lacking direction and focus.

Why is that a common perception of an agency that accomplished so much in such a short time? The truth is that NASA's future is embedded within the 1958 Act that established NASA. It is not up to NASA to decide what the next program(s) should be, it is just one participant in the planning. If you tell NASA that you want to land astronauts on Mars or a robotic rover on Pluto, NASA will describe six different ways to do it. If you say you want to do it within a specific timeframe, some of the alternatives may drop out. If you should require a tight estimate of what a specific program will cost, then you will be confronted with an uncertainty that might give you pause. The direction the civil space program will take hinges on the unlikely possibility that future congresses and administrations, even if responding to a national consensus, will come together, agree on how the nation should proceed to explore and utilize space, and stick to it!

If a long-term consensus is ever reached, then those same leaders must agree that, once a direction is taken, they will find competent managers and then step back and allow them to get on with the task. In today's environment can you imagine what would have been the result if the Lewis and Clark expedition had been chartered twenty years ago instead of by a bold president in the nineteenth century? It would still be in St. Louis waiting for its first boats because senators from Illinois and congressmen from Missouri would be holding the program hostage, arguing over which state's boatyard was the lowest bidder for the boats that the expedition needed.

That is not to suggest that Congress should end oversight of NASA's programs. Nevertheless, once a program is approved that will challenge the technological state of the art in order to succeed, Congress' role should be circumscribed by clearly understood boundaries. If waste or failure to meet goals occurs and the administration in power is incapable of correcting the problem, then the Congress should step in. If national priorities should change, then Congress has an important role to play. However, oversight does not translate into approving such things as RFPs or program staffing levels. What must stop is second guessing details. If a program is worth pursuing then funds should be appropriated to assure the most cost effective outcome, Congress' primary responsibility.

NASA made an early, serious mistake while promoting the Space Station. It insisted on defending a low-cost ($8 billion) estimate for the program. Even taking into consideration costs normally not included in NASA programs, such as civil service salaries and operations, those missing items only partially explained why the estimate was so low. All knowledgeable reviewers knew it was unrealistic for the design that was emerging. As a result, the Space Station was easily susceptible to battering by opponents, leaving a contentious legacy that carried over from administration to administration, and congress to congress. When Fletcher finally provided a more realistic estimate it was too late in the game to overcome past miscalculations, and soon the new estimate also was proven to be too low.

President George W. Bush proposed in January 2004 a new Vision for Space Exploration.[21] It includes completing the Space Station by 2010, in some unspecified configuration, continuing robotic and human exploration, and returning astronauts to the Moon, and then sending them on to Mars. It sounds exciting, but if his vision is pursued it will require spending a great amount of national resources. Is it the right vision? Perhaps; however, at this time there is no assurance it is a vision the nation will

---

21. President Bush Announces New Vision for Space Exploration Program, Remarks by the President on U.S. Space Policy, For Immediate Release, Office of the Press Secretary, January 14, 2004.

enthusiastically embrace for decades. Such a huge undertaking cannot be a unilateral proposal made by an administration that will not be in power when the first missions lift off. Final, continuing, responsibility to preserve such a vision resides in Congress representing the nation's will. Congress does not have a good track record of supporting expensive programs that extend over long time periods, not even for such a basic need as national defense.

The commission President Bush appointed to provide detail for his vision mostly replowed old ground, similar in many respects to the Synthesis Group report in 1991, and is just one more study to add to the bookshelf.[22] The web site set up by the Aldridge Commission seems to indicate that there is support for the President's program, but it was only a small step toward obtaining a national consensus. If you think this is not true, ask your neighbors what they know about President Bush's Vision and see how many have a clue as to what is being planned. Further efforts are needed to explain the breadth and scope of the program that the nation is being asked to support for the next twenty or more years. What is still missing is an unambiguous national consensus and that will not be achieved without a much broader dialogue that includes Congress, private sector leaders and ordinary citizens.

The Aldridge Commission stated that "the space exploration vision must be managed as a significant national priority, a shared commitment of the President, Congress, and the American people."[23] One president or one Congress does not assure a "national priority, a shared commitment." Many future presidents and congresses will have to embrace President Bush's vision, and they can't be polled today. To obtain consensus for such an expensive program will require that there is the potential for economic rewards beyond just doing it and the thrill of exploration. It will be difficult to convince the private sector to invest in such a vision without a payoff, despite what some space enthusiasts will claim. The private sector and aerospace industry will gladly accept taxpayer money to build spacecraft and equipment, but it won't invest much of its own, betting on some undefined economic gain.

For NASA's next big undertaking, funding levels and sufficient reserves should be agreed to at the outset that will allow a program to develop efficiently and be able to solve the unexpected – because the unexpected always occurs. Multiyear funding must be appropriated so that managers can execute contracts with the best value for the government and taxpayers. This is not the standard ground rule for most programs, and is perhaps an audacious proposal, considering the unrestricted power that Congress enjoys and probably will never relinquish. But that is what leaders do. They set the rules, make sure they are followed, and then step aside unless there is clear evidence of malfeasance.

The key question that must be answered is: "can program goals and standards be defined that all stake holders will agree are reasonable?" We are not there yet for President Bush's vision. This Administration, Congress, and other key participants, must step back and lay out a detailed plan that all can agree is doable. There must be a reasonable time frame to work toward, as well as clear and measurable intermediate goals with cost estimates that recognize that there will be unforeseen problems to overcome before all the goals will be realized. It must also be recognized that the further removed in time you are from a goal, the less accurate will be the cost estimate. That is the function of a program reserve, and a large program will not survive without a sensible reserve.

Can NASA provide decision makers with cost estimates for a complex program that will hold up over the years? If the answer is yes, then NASA must be in charge to make the day-to-day decisions and not have them overturned based on external whims. The Space Station program is just another example of the pitfalls of trying to manage a complex program by responding to multiple masters, each with its own agenda and oblivious of good management practices.

At present it appears that the Congress will go along with the President's vision, and the first two funding installments were included in NASA's 2005 and 2006 budgets. Will that support continue in future congresses and administrations? The first installments are not budget busters. Is there a real consensus that will sustain the effort or will this new space initiative suddenly be terminated or allowed to die a slow death after wasting untold amounts of resources?

---

22. Report of the President's Commission on Implementation of United States Space Exploration Policy – A Journey to Inspire, Innovate, and Discover. June 2004.
23. Ibid. Finding 2.

If what President Bush proposed is not the right vision, where does the Nation's space future lie? What are the alternatives? Strong arguments have been made that future space exploration should emphasize robotic missions. Robots are becoming more capable with every passing year. This debate has gone on since the dawn of the space age and was an early point of contention for the Space Station. It is time to put this difference of opinion to rest once and for all and not waste time and money restudying the issue. Obviously it should be a combination of robotic and human exploration. The difficult question to answer is: what is the proper combination? Regardless of this debate, there is a larger cloud hanging over NASA's future. Administrator Griffin is restructuring and skewing the agency to support President Bush's vision. As a result, some of NASA's traditional strengths will be cast aside or severely reduced. The question must be asked: how will the agency function if the current emphasis is changed for one reason or another? In that eventuality, the nation will be in the unenviable position of needing to reinvent NASA.

The Bush Administration must develop a better endgame than that presently proposed for the Space Station. An agenda that indicates that the U.S. will essentially walk away from the program by 2015 would be an unconscionable waste of taxpayer dollars. Assuming that assembly will be completed by 2010, a questionable assumption, or that it will achieve a configuration that permits useful experimentation, another questionable assumption, that would mean that the U.S. would only share the facilities to conduct extensive research for five years. Under either scenario, how much useful knowledge would be gained during the five years is another big assumption, because the size of the crews onboard will be restricted. Many of the original high-priority experimental areas will be severely reduced or eliminated. The 2010 completion is driven by the decision to retire the shuttle by that date. Will the new Crew Exploration Vehicle (CEV) be available to replace the shuttle when it is retired, or will there be a hiatus in our ability to have more than two or three crew on board? Plans to accelerate the development of the CEV raise many questions. It would have to be designed, built, and have several flights under its belt within four years before it could take on the job of servicing the Space Station – an exceptionally short time for such a complex spacecraft, even if it incorporates some shuttle systems. Let us hope that this effort does not result in another vehicle, like the shuttle, that never attained its original design objectives, because Congress limited funding or compromises were dictated in order to achieve an arbitrary readiness date.

Perhaps the emergence of China as a major space power will provide the same type of impetus that competition with the Soviet Union did at the beginning of the Mercury program. Following the successful launch of its first astronaut in 2003, China launched and recovered a two-man spacecraft in 2005. This crew was equipped with Chinese manufactured space suits rather than using a Russian derivative, as was the case for their first launch – a clear sign that they are making progress with homegrown technology. China's plan sounds somewhat similar to the first steps taken by another space program that launched its first manned flight in the early 1960s. With its reliable Long March boosters, and the new heavyweight 5-500, a rocket equivalent to the Ariane-5 and Titan-4, China is capable of conducting interplanetary exploration, and have announced their intention to send taikonauts (astronauts) to the Moon. Will they go it alone? They could, or they could be part of a larger consortium, and initial discussions soliciting China's interest have taken place.

Is the only way to implement President Bush's vision a multinational coalition? Probably; however such coalitions are difficult to form and even more difficult to sustain. Other nations have been invited to join in his vision and there appears to be initial interest, but this interest should be viewed with caution. If countries use the Space Station program as a guide, there may not be many willing to finally sit down at the table and sign up.

The many hard lessons learned during the design and development of the Space Station are mirrored in the thirty years of vacillating cooperation in magnetic confinement fusion research. It is another example of how difficult it is to make multinational coalitions work for very large programs. Development and construction of the International Thermonuclear Experiment Reactor, first proposed in 1985, has been slowed down for years, confronting different problems and arguments, including the most recent: where it should be sited. In the meantime, the cost of the program has grown from $5 billion to $12 billion (and counting) and its success is dependent on continuing contributions from many countries, including the U.S. Large, multinational programs are just very hard to manage! Returning to the Moon and exploring Mars, if that is the course agreed to, can only succeed as a multinational effort if one country agrees to take a dominant position and the other partners agree to supporting roles.

Wasn't that the path followed by the Space Station? Yes and no. Adding Russia diluted the U.S. lead role, overburdening management and funding. To a large degree, Russia and the U.S. were equals, with Russia placed on the critical path for program success. There cannot be two captains on a program's bridge. If there are, it guarantees the program will run aground in hard times. Also, there are other downsides to a multinational project that will require the use of advanced technology. Not the least is the question of technology transfer, which complicated relationships between the U.S. and its international partners early in the Space Station program. Legislation that governs the export of sensitive weapons technology, such as the U.S. International Traffic in Arms Regulations (ITAR), can severely impede cooperative, international enterprises.

Why should a multinational coalition work any better for another, more complex, space venture? It won't, unless there is a clear recognition of the difficulties the program will face and a firm commitment is made by the dominant partner to see it through, and supply whatever resources are needed, regardless of the contributions of the other partners. If other nations choose to join, they should be encouraged to add their unique contributions within a well-defined framework, including being invited to the table when difficult decisions are made. Even after multiple Space Station Intergovernmental Agreements and MOUs were signed, there was a continuous undercurrent of discontent or need to change the terms. As problems of one type or another surfaced, they usually resulted in modifications to the agreements or unilateral responses. A program that continues for decades will encounter similar problems. All the partners must agreed to a process, up front, that will lead to a fair resolution of inevitable issues.

Determining the precise size of the Moon's core or the detailed geology and composition of its crust may be exciting to planetary geologists and geophysicists, but as ends in themselves will not justify spending tens of billions of dollars. We already have excellent first-order answers to those questions and additional detail will not add significantly to our understanding of the early history of the Earth-Moon system. All the primary scientific questions that were asked before the first Apollo landing have been answered to the satisfaction of most planetary scientists. Some may dispute this statement, but in the galaxy of major mysteries that remain to be solved that will shed light on our place in the universe, the additional insight that might be gained by returning to the Moon for answers will be minimal. In 2005, the American Association for the Advancement of Science listed 125 questions "that are driving basic scientific research."[24] None of the questions will be answered by returning to the Moon.

Using the Moon as a base for some types of experimentation would have unique advantages, but can the expense of building a lunar base be justified if it is primarily a test bed before we send humans on to Mars? We knew how to build Moon bases, relatively cheaply, forty years ago, but the nation turned down the opportunity. All the infrastructure was in place. Constructing Moon bases today will require starting over again in almost all areas. Detailed plans also existed at that time to send astronauts to Mars using modified Apollo hardware. Establishing a Moon base is not a prerequisite before sending astronauts to Mars or beyond, and cannot be justified from a cost / benefit standpoint compared to starting such missions from Earth and Earth orbit.

What has changed? Most of the arguments heard today to rationalize returning to the Moon are the same as those made in the 1960s and early 1970s to warrant continuing lunar exploration and building Moon bases. Those early arguments were based on a minimum of information, before all the Apollo data was digested and analyzed. Analogies were and are trotted out comparing 15th and 16th century explorers, such as Columbus and Magellan, to rationalize why we must continue sending humans to explore the new frontier of space. But the point always downplayed is that those early explorers were driven by the potential for economic gain. Human missions of exploration of the solar system, for as far into the future as is reasonable to extrapolate, will be confined to the inner planets and near-Earth

---

24. Science, Vol. 309, 1 July 2005.

objects. What will be the motivation to continue sending astronauts on long, costly, and perilous journeys – economic gain or knowledge? Obtaining new knowledge cannot be argued against as an objective. However, attempting to justify interplanetary missions on the basis of potential economic return is very difficult to imagine.

Space exploration must continue, but what should be the focus? The unanswerable question always asked is: would the money spent pursuing space programs yield a better return if spent in some other way, such as for medical research or social programs? Technological advances required to conduct space programs eventually find their way into the marketplace. Called "spin-off" by NASA, the benefits have been downplayed by critics of the space program as being exaggerated. Regardless of the critics, it cannot be denied that there is a long list of technology hand-offs that either originated with NASA research or matured in support of NASA programs. Programs that advance the frontiers of knowledge, including space exploration, will always inspire and challenge the next generations of scientists and engineers, thus, for this reason alone, they have great value. Because these programs explore the unknown, it is impossible to predict their outcome; however, the merits of many past government programs that increased our understanding of the universe around us are well documented.

Only one question may justify future, very large space expenditures to find its answer: did life evolve in some form elsewhere in our solar system? It is such a fundamental question, and one of the 125 basic scientific questions mentioned earlier, that perhaps it can be framed in such a way that an unassailable consensus can be achieved to pursue a long-term program to find the answer – unassailable in the sense that all necessary decision makers, including scientists and political leaders, have carefully defined the program, estimated its cost and schedule, and then presented it to the public at large for all to understand the magnitude of the quest. Sending astronauts to follow up on robotic investigations may be necessary to finally resolve the question and should be an element of the program. Nevertheless, human missions to Mars should await the return of data from comprehensive robotic exploration that would identify why, where, and how we should undertake such hazardous missions. We are just beginning to start extensive unmanned exploration of Mars, and a complete program remains to be defined. Wide-ranging robotic sample return should be a priority to complement remote testing and sampling. Without first conducting extended robotic exploration, the justification for sending astronauts to Mars will be difficult to defend.

While awaiting the results, we should continue research into how to safely send astronauts to Mars and other destinations. That was an original function for the space station, and should be continued. It will also be a way to keep the large astronaut contingent closely involved until the next human missions are launched.

Whether or not life forms are found on Mars, the implications will have an immense impact on future space exploration. It remains to be seen what should be the pace of a search for extraterrestrial life and if it should take priority over other national needs. Addressing major philosophical objectives, such as establishing colonies elsewhere in the solar system in case a catastrophe should befall life on Earth, will require a long debate and advances in technology far beyond today's capabilities. Nonetheless, having a consensus on a long-term focus is a must for NASA. Those who labored for many years to develop a post-Apollo program of lunar exploration and Moon bases know that programs and plans can disappear in the twinkling of an approved appropriation bill.

Although the search for life forms in our solar system may serve as the focus for future space exploration, it does not represent a quest that is time dependent. There is no criteria that dictates when or how fast the search must be conducted. Regardless of whether or not an answer is found in the next twenty years, or longer, life on Earth will continue. However, there is one time-critical space program that is not receiving its proper share of attention and resources. That effort is identifying and tracking near-Earth objects (NEOs) and other bodies in our solar system that have Earth-crossing orbits, and then devising a method to deflect the objects before they collide with Earth. Because of the need to identify these objects as quickly as possible, and then have a means to divert them if their orbital paths indicate a possible Earth impact, a national or international study is needed to recommend a course of action. This should not be considered a science fiction scenario. ESA has already begun to study how to prevent such impacts. The Earth has been bombarded innumerable times in the past and it is only a matter of time before another large object aims our way again. Because of the uncertainty about when the next impact might occur, and current analyses indicate that none of the known NEOs pose an immediate threat, the present program has a low priority. The danger does not lie in the known NEOs but those that are unknown, and they are certainly out there. For the U.S., NASA is the logical agency to lead this program and obtain needed funds.

### PRIVATIZE NASA PROGRAMS?

The tempting siren call to privatize or commercialize NASA programs has become a chorus. Burt Rutan's success in winning the Ansari X Prize has added volume to the chorus. Recently, NASA Administrator Griffin said he wanted to pursue "nontraditional contractors and contracting" to carry crew and cargo to the Space Station. This was hailed as encouraging news in the trade press.[25] But what do the proponents of turning NASA programs over to the private sector really envision? At present, aside from in-house research, all NASA programs are implemented by aerospace companies or research institutions responding to NASA initiatives. University or other not-for-profit research institutions depend on NASA grants to keep staff employed, purchase or improve facilities, and earn prestige among their peers. During the course of the private sector contracts, and when their deliverables are accepted by NASA, the companies involved have been working to obtain a profit. The only difference in the manner in which this work is done, compared to transactions between two private sector companies, is that NASA contractors perform to a specification provided by the government (sometimes jointly developed) rather than a specification wholly developed by industry.

So when someone complains that NASA should allow private industry to run or manage its space programs, what do they mean? – that NASA should not be involved at all in writing the specifications and overseeing the results for services eventually paid for by the taxpayer? Such an arrangement should never be accepted by the Congress. Who would be held responsible if a critical path product in a major program, contracted with the private sector with free rein to decide how to proceed, were to fail because the contractor did not deliver? As we know, private contractors working for the private sector do not always perform flawlessly. How would a congressional hearing proceed on a failure attributed to a private company? Would congress try to fire the CEO or just put him in jail? Probably what would happen is that the contractor would forfeit an award fee or some other inducement to perform, or go out of business, and taxpayers would be left holding an empty bag. Not a very attractive way to proceed on a high-priority national program. Although not a perfect system, as there are always some failures or costly overruns, NASA programs require the same approach as those funded by DOD. For crucial government programs the agency in charge must be held accountable for failures and must call all the shots.

One could make a convincing argument that some of NASA's recent failures were the result of a lack of NASA contractor oversight. What is the basis for that argument? Over the years NASA staff and in-house research have been reduced, as dictated by personnel ceilings and budgets imposed by OMB and Congress. In Apollo's formative days the question of how large NASA's staff should be was analyzed by Bob Seamans. He came to the conclusion that, in spite of the need to employ thousands of new scientists and engineers for Apollo, there would not be a major increase in the number of civil servants. Rather, staff would "have the skills and capability within NASA to make key decisions."[26] The bulk of the workers needed to carry out Apollo would be hired by private sector contractors. This approach was carefully implemented with subsequent NASA hirings and by the development of in-house expertise that backed up essentially all Apollo contractor work. In addition, NASA Headquarters controlled a contract for a 400-person, dedicated systems engineering capability (Bellcomm) to add depth and expertise to the decision making process. General Electric and Boeing also held Headquarters contracts to directly support the NASA Apollo civil service staff. Those strengths, put in place during Apollo, have deteriorated or disappeared in today's NASA.

Surely advocates of opening up NASA programs to the private sector do not mean that private industry is prepared to fully fund the next generation of space programs, and that the government would be just one of several customers that might be predominantly private companies. Recent overtures by NASA

---

25. As reported in *Aviation Week & Space Technology*, June 27, 2005, "Skin in the Game," Frank Morring, Jr. and Michael A. Taverna, and Editorial "Hats off."
26. Aiming At Targets, Robert C. Seamans, Jr., NASA SP-4106, p. 92.

Administrator Griffin to the private sector to provide services such as a "refueling depot in space" sound intriguing, but it will be interesting to see how many efforts are started. Without a guarantee that the government (NASA) will utilize or pay for a service, how many companies will risk making the large investment needed to develop a service through to application? A guarantee will be required and then the only difference in how services are now provided is that NASA will not dictate all the specifications. Using space for various types of communication and photographic satellites are, and may well be for many years, among the few businesses that promised and delivered a balance sheet in the black. And the technology used, including launch vehicles and launch sites, was developed or heavily subsidized by the government before industry could go it alone.

Experience shows that proposals to privatize NASA programs, many advanced during the Reagan Administration and even earlier, may be well intentioned but are always suspect. A business plan that requires vast sums of up-front money but has no payoff for many years, perhaps decades, such as mining the Moon, will never pass muster with any responsible corporate board. There are always better places to spend a corporation's money in the near-term, and shareholders will demand that they receive a return on their investments.

Without NASA support there would have been few, if any, takers to spend ten years or more designing experiments to be conducted first on Spacelab and then the Station Station, if for no other reason, all the complicated interfaces needed to assure an experiment's success required that NASA be a partner. And if an experimenter had started using just private funding when first extended an invitation to use the Space Station, twenty years later he/she would still be waiting for the first results – not the preferred way to make a profit. In 2002, NASA ended its Space Station commercialization contract with Dreamtime, a Silicon Valley company. Dreamtime said it would raise $100 million for various multimedia projects but never convinced the needed investors to join their endeavor. Examining a more technologically oriented approach, Kistler Aerospace Corporation has balanced on the edge of failure since its inception. Attempting first to capture a private sector clientele and then a NASA customer, the company has had private investors sink hundreds of millions into its launch system with no pay-back. There are several reasons why this venture has come up short of expectations, however regardless of the reasons, a decade of effort has failed to pay off. These are just two examples of those who believed they could start a profitable commercial space venture only to face the truth of the bottom line. Developing and launching spacecraft is complex and expensive and will almost always depend on having the government as a customer, or at the very least as a partner, to provide some services such as communication and tracking, and for space travelers, rescue.

It is always dangerous to suggest an analogy for comparison when opening a new frontier, but consider again what our history has taught. Without extensive government sponsored surveys of the west that took place in the nineteenth century, private investment for ventures, such as mining and railroad construction, undoubtedly would have been delayed. Those surveys, both geological and topographic, provided the foundation for western movement. And even if the government had not undertaken those surveys, eventually the cost of duplicating them would have been within the means of the private sector. Also, opening up the west was successful because it was driven by daring homesteaders with few material resources who took a chance to "go west" and find a better life. "Homesteading," or commercially utilizing the Moon and Mars, will require enormous amounts of initial capital, if they ever comes to pass.

Establishing bases on the Moon to discover if any commercial applications exist would have to start wholly funded or heavily subsidized by the government. Using the Moon as a platform to deploy huge photovoltaic arrays to beam energy back to Earth, or process lunar soil to recover oxygen and hydrogen to power future spaceships, are just a few of the grandiose schemes put forward by space enthusiasts. Mining the Moon for helium-3 to be used as a fuel in fusion reactors, for example, has many proponents. Analyses of lunar soil brought back during the Apollo missions suggests that helium-3 may be present at a ratio of perhaps ten parts per billion. At that ratio, huge amounts of Moon dirt (regolith) would have to be processed to recover useable quantities. Some believe that the ratio might improve at higher latitudes, but regardless, you would still have to process a lot of dirt. A large investment would be necessary just to develop specialized equipment to carry out such a process, not to mention transporting it to the Moon, establishing base operations, and then returning the product.

Today there is no market for helium-3. The design and operation of fusion reactors that could utilize helium-3 to generate sizable amounts of electric power are yet to be demonstrated. Although helium-3 would offer some advantages as a fusion fuel, fusion reactors are not dependent on using helium-3; other fuels, deuterium and lithium / tritium, are more easily available. The promise of having functioning, economically competitive fusion reactors has been just over the horizon for thirty years. It is a tough technological challenge that will not be solved for many years, if ever. All of these proposals, if carefully analyzed, will fail the test of the bottom line or how to spend scarce government resources effectively.

The problem of privatizing space ventures is somewhat analogous to what has occurred with government R&D in energy technologies. As an example, the first modern 100 kW wind turbine was built in 1975 by the Lewis Research Center at its Plumbrook, Ohio facility. NASA went on to design and build larger experimental machines and developed important innovations. Under government contracts, U.S. industry pioneered the manufacture of large, modern wind turbines. Thirty years later, after spending hundreds of millions of U.S. taxpayer dollars, and research in other countries, we are beginning to see acceptance of the technology in suitable areas.

The ugly wind farms that sprouted on the hills east of San Francisco in the late 70s, often shown in photographs by critics trying to belittle the technology, only existed because the government gave large tax breaks to those who would build them. Key considerations, such as spacing, location, operations and maintenance, were ignored as many of the owners did not care if they generated one kilowatt of power; the tax write-off was all they worried about. In the first twenty years after the government began wind turbine R&D, many U.S. wind turbine manufacturers failed to build a business and left the field. It took too long a time and was too large a jump before the technology was accepted. As a result, many of the machines being used today in the U.S. are supplied by foreign manufacturers who found an earlier market in their own countries. In the last decade, as the technology improved, wind turbines have begun to compete in the marketplace and will provide a valuable energy source in the years ahead without subsidies. Wind turbines are just one example of how difficult it can be to leap from a government sponsored R&D program to commercial success.

No matter what technology is used to put a pound of payload in Earth orbit or beyond, it will always be difficult and expensive. There are no easy ways to overcome the laws of physics. Advocating that NASA should hand off some of its programs to the private sector always resonates in some quarters, but serious investors have been hard to find. Relatively easy private sector space-related projects, such as competing for the Ansari X Prize, required a modest initial investment. A financial angel was found willing to bankroll the effort regardless of the outcome. Perhaps Burt Rutan and his associates will capture a niche market and be successful, one can only wish them well. Carrying paying customers on expensive, short suborbital or orbital flights should appeal to a small number of well-heeled enthusiasts. Sending big spenders ($100 million or more per passenger) on a round trip to the Moon by capitalizing on Russian technology would find an even smaller market if it should materialize. Any space program where human lives would be involved, private citizens or astronauts, will require a level of expertise and resources that will only be found in a government-industry partnership. Whatever the next step into space that's totally funded by the private sector might be, it will be expensive. Turning a profit will be difficult.

## — Selected Bibliography —

*The Space Station – An Idea Whose Time Has Come*, Edited by Theodore R. Simpson, Foreword by Harrison H. Schmitt. Published under the sponsorship of the IEEE Aerospace and Electronic Systems Society, IEEE Press, 1985.

*The Space Station Decision, Incremental Politics and Technological Choice*, Howard E. McCurdy, The Johns Hopkins University Press, 1990, 286 pps.

*Together In Orbit, The Origins of International Participation in the Space Station*, John M. Logsdon, Monographs in Aerospace History #11, November 1998, NASA History Division, Office of Policy and Plans, NASA Headquarters, Washington, DC 20546.

*Star-Crossed Orbits – Inside the U.S.-Russian Space Alliance*, James Oberg, McGraw-Hill, 2002, 355 pps.

— INDEX —

Aaron, John W., 24, 36, 45-46, 54-55, 57-58, 121
George Abbey, 128, 155
Advanced Development and Technology Program, 24-25, 32
Advanced Solid Rocket Motor (ASRM), 10, 81, 131
Advanced Technology Advisory Committee, 10, 32-33, 52
Advisory Committee on the Redesign, 123, 126
Aerospace Safety Advisory Panel (ASAP), 10, 86, 155, 180, 185-186
Albrecht, Mark, 89, 114, 205
Aldrich, Arnold D., 93, 115, 117
Aldridge Commission, 211-212, 214
Aldridge, Edward C., Jr., 114, 190-191
Alenia Aerospazio, 168, 174
Allen, Andrew M., 157
Alpha rotary joint, 120
Alpha Station Implementation Plan, 132
Alvarez, Luis W., 38
American Chemical Society, 110
American Geophysical Society, 110
American Institute of Aeronautics and Astronautics (AIAA), 10, 15, 68, 109
American Physical Society, 110
Ames Research Center (ARC), 10, 29, 36-37, 79
Announcement of Flight Opportunities, 84, 91, 118
Ansari X Prize, 218, 220
Apollo, 7, 14, 16, 24, 26-27, 38, 42, 48, 52, 55, 57-58, 62, 71, 81, 85, 92, 97, 108, 111-113, 123, 129, 152, 164, 166, 195, 197, 200-202, 206, 210, 216-219
Apollo-Soyuz Test Program, 117
Ariane, 10, 23, 50, 66, 74, 139, 151-152, 161, 164, 174, 180, 190, 215
Armstrong, Neil A., 38, 50
Associate Contractor Agreement, 84
Associate Contractor Relationships, 82
Assured Crew Return Vehicle (ACRV), 10, 86, 106, 111, 113, 115, 117-118, 124, 129, 132, 135, 139, 140, 144, 151, 154-155, 163
Astronaut Science Support Group, 77
Atlantis (space shuttle), 134, 148, 152-153, 155, 160, 163, 166, 168, 179-180, 186-187, 189, 193
Atlas/Centaur, 65
Attached Payload Accommodation Equipment, 107
Augustine Commission, 104, 212
Augustine, Norman R., 100, 104
Automated Transfer Vehicle, 10, 152, 174, 183, 190
Automation and Robotics Panel, 10, 32
AXAF, X ray Telescope, 10, 169, 172
Baikonur, 135, 149, 153, 172, 177
Banks, Peter, 22, 41, 58
Baseline Configuration, 20, 33, 39, 41, 44-46, 52, 54-56, 59-61, 63, 73, 75, 139-140
Basic Station, 52
Battelle Northwest Institute, 17

Beggs, James M., 9, 14-16, 20, 22-24, 26, 31, 34-35, 42-43, 46, 51, 61, 200-201, 205
Bellcomm, 218
Bentsen, Lloyd, 122
Black, David C., 37
Block 1, 59, 61, 63, 65-66, 68
Block 2, 59, 61, 63
Bloembergen, Nicolaas, 166
Boeing, 15, 17, 32, 34-35, 69, 74, 108-109, 119, 128-130, 133, 140, 142-143, 149-152, 154, 156-158, 160, 163, 167-169, 175, 218
Boland, Edward P. (D-MA), 29-30, 63, 68, 78, 82
Bond, Christopher (R-MO), 163
Booz Allen & Hammond, 17
Bowsher, Charles A., 110
Brazil, 167, 172
Brinkley, Randy, 137-139, 141, 156, 158, 171
Brown, George E., Jr. (D-CA), 123, 126, 133, 135-136, 142, 145, 147, 150, 171, 199
Bryan, Frank, 19
Bumpers, Dale (D-AR), 117, 133, 146, 166, 203
Burns, Conrad, 160
Bus-1, 124, 130, 132, 137, 156
Bush, George H.W., 23, 45, 71, 84-86, 88, 92, 109, 114, 117, 120, 121, 135, 185, 194, 199, 203, 205, 206-208
Bush, George W., 7, 177, 181, 186, 189, 190, 193, 203, 209, 210-211, 213-214
Byrne, Leslie (D-VA), 128
Cabana, Robert D., 134, 175
California Space Institute, 32
Canada, 11, 15, 22-23, 37-38, 49, 71, 81, 83-84, 96, 108, 132, 135, 140, 146, 149, 168
Canadian Space Agency, 10, 126
Carlisle, Richard, 89, 201
Carter, Jimmy, 205
Casey, William, 79
Center for Strategic and International Studies, 193
Centrifuge, 82, 95, 98, 105, 113-114, 119, 151, 155, 157-158, 181-182, 187-188, 196, 211
Chabrow, Jay, 170-172, 177
Challenger (space shuttle), 43-46, 48, 50-51, 53, 55, 57, 59, 62, 64-65, 67-69, 93, 115, 155, 177, 188, 200-203, 212
Chance Vought, 17
Chernomyrdin, Viktor, 129, 135, 157, 161, 170
Chiao, Leroy, 193
China, 134, 215
Clementine, 159
Clinch River Breeder Reactor, 166
Clinton, William J. "Bill", 116, 119-121, 124, 126, 130, 133-134, 136, 146-147, 150, 160, 167, 169, 171, 176-177, 181, 184-185, 191, 203, 208, 210, 212
Cohen, Aaron, 92-93, 114, 185, 208
Cohen, William, 148

Collins, James, 145
Colorado School of Mines, 177
Columbia (space shuttle), 7, 10, 37, 48, 50-51, 64, 134, 153, 155, 166, 177, 186-188, 191, 195, 203, 208-209, 211-212
Columbia Accident Investigation Board (CAIB), 10, 50, 177, 188-189, 191-195
Columbus Laboratory, 179, 188
Columbus Module, 23, 96, 113-114, 140, 150-151, 158, 174, 184
Columbus Polar Platform, 84
Commercial Development of Space Industry Advisory Group, 77
Commercial Space Act of 1998, 175, 178
Configuration/Budget Review Team, 93-94
Congressional Budget Office (CBO), 10, 62, 139, 141-142
Continuous Process Improvement, 116
Cooperative Solar Array, 141
Coopers & Lybrand, 17
Cost Assessment and Validation Task Force (CAV), 10, 170, 172
Cost Containment Team, 57-59, 61
Couvalt, Craig, 53-54
Covey, Richard, 193
Covington, Clark, 55
Crawley, Edward F., 90, 117, 119
Crew Emergency Return Vehicle (CERV), 10, 55, 68, 73, 169, 179, 212
Crew Exploration Vehicle (CEV), 10, 190, 192, 194-195, 215
Crippen, Robert L., 50, 93, 195
Critical Evaluation Task Force (CETF), 10, 54-59, 61, 64, 105
Croomes, Scott, 156
Culbertson, Frank L., 165, 172
Culbertson, Philip C., 9, 14, 15, 23, 24, 31, 33, 34, 35, 37, 38, 39, 40, 41, 42, 43, 44, 45, 46, 59, 87, 108, 198, 200-201, 212
Currie, Nancy, 175
Cutter, W. Bowman, 146-147, 167
Dailey, John R., 119, 185
Daimler Benz Aerospace, 168
Dale, Shana, 211
Darman, Richard, 86
Delay, Tom (R-TX), 123, 186, 194
Delta 3920, 65
DeLucas, Lawrence, 178
Deming, Edward W., 116
Department of Defense (DOD), 16-17, 39, 42, 46, 49, 61, 66, 76, 79, 104, 109, 156, 184, 187, 202-204, 207, 210, 218
Discovery (space shuttle), 50, 86, 134, 152-153, 155, 158, 166, 172, 189, 191, 193-195
Donahue, Thomas M., 21, 178, 195
Dreamtime, 219
Dryden Flight Research Center, 195
Dunbar, Bonnie, 153
Earth Observation System (EOS), 8, 10, 85, 115, 167
Edelson, Bert I., 21

Endeavor (space shuttle), 115, 120, 127, 134, 153, 155, 161, 166, 175-176, 180-181, 186-187, 190, 219
Environmental Control and Life Support System (ECLSS), 10, 34, 51-52, 75, 121, 129, 138
Environmental Impact Statement (EIS), 10, 112
Ernst & Young, 209
European Space Agency (ESA), 10, 15, 22-23, 37-38, 41, 46-47, 49, 55, 59-60, 63, 66, 74, 81, 83-85, 94-96, 104, 108, 113-114, 117, 125-126, 132, 135-136, 139-140, 144, 146, 149-152, 157-158, 161-163, 168, 170, 172, 174, 179-180, 183-186, 188, 190, 211, 217
Executive Technical Committee, 55
Expedited Processing of Experiments to Space Station, (EXPRESS), 11, 89, 119, 135, 144, 167, 182
Expendable Launch Vehicle (ELV), 10, 51, 64, 66-67, 71, 73, 104, 115, 203
Extended Duration Orbiter (EDO), 10, 22, 64, 86, 102, 105, 112, 116
External Maintenance Task Team, 97, 100
External Tanks Corporation, 11, 108
Extra Vehicular Activity (EVA), 11, 35, 40, 42, 46-48, 53, 55-56, 59, 64, 75, 86, 91, 95-101, 104, 106, 108, 140-141, 151-152, 158, 163-165, 175, 187, 191-192, 195
Faget, Maxime "Max" A., 41
Faster, Better, Cheaper, 115, 208
Fettman, Martin J., 157, 181
Jorge E. Feustel-Buechl, 184
Feynman, Richard P., 50
Finarelli, Margaret, 9, 83
Fincke, Michael, 191
Fink, Daniel J., 58
Finn, Terrence T. "Terry", 9, 15-16, 22-23, 31, 201
First Element Launch (FEL), 11, 40, 69, 74-75, 78, 81, 91, 94-95, 100, 105-106, 115, 118, 120, 131-132, 145, 151, 156, 160, 162, 175
Fletcher, James C., 4-5, 48-49, 51-52, 54-55, 57-59, 61-63, 67, 69, 73, 76-79, 82, 86, 88-89, 108, 194, 200-201, 205-206, 213
Flight Telerobotic Servicer (FTS), 11, 57, 60, 69, 79, 82, 86, 88, 91, 101, 105, 107, 109
Foale, Michael, 152, 165
Foley, Theresa M., 62, 134, 202
Free Flyer, 12, 38, 83, 85, 105, 111, 113
Freeh, Louis B., 159
Freitag, Robert F., 15, 22, 45, 201
Frist, William H., 160, 167
Fullerton, C. Gordon, 52-56, 101
Garriott, Owen K., 108
Gazey, Sami, 168
Geller, Margaret, 178
General Dynamics, 15, 17, 36, 42
General Electric, 32, 35-36, 74, 85, 107, 218
General Management Status Review (GMSR), 11, 80, 207
Gerstenmaier, William, 196
Gibbons, John H., 132, 160
Gibson, Robert L. "Hoot", 134
Gidzenko, Yuri, 163, 180-181
Gilman, Benjamin (R-NY), 179

Gingrich, Newt (R-GA), 153
Glicksman, Martin, 162
Goddard Space Flight Center (GSFC), 11, 21, 24, 29, 33, 35, 51-52, 55, 57-58, 74, 85, 101, 105, 107, 201
Goldin, Daniel S., 114-119, 122-123, 126-128, 132-137, 142, 145, 148, 150-151, 154-157, 160, 162-163, 167, 169-172, 174, 176, 178, 184-185, 200, 205, 207-211
Gore, Albert (D-TN), 62, 87-88, 114, 122-123, 130, 134-135, 157, 161, 170
Government Accounting Office (GAO), 11, 63, 67-68, 73, 108, 110, 112, 123-124, 127, 148, 157, 166, 178, 199, 209, 211
Grace Commission, 210
Graham, William R., 42-43, 45-46, 48, 51, 57, 200, 205-206
Gramm, Phil (R-TX), 131
Gramm-Rudman-Hollings Deficit Reduction Act, 34, 62
Green, William (D-NY), 27, 82
Griffin, Gerald D., 24, 43, 45, 51
Griffin, Michael D., 116, 194-196, 205, 209, 211, 215, 218
Gross, Roberta L., 165-166, 180
Grumman, 15, 17, 35-36, 69, 78, 82, 86, 89, 106, 109, 113, 119, 128-129
Guigne Technologies, Ltd., 177-178
Habitation Module (HAB), 11, 17, 23, 54, 60, 129
Hall, Ralph M. (D-TX), 122, 145
Harris, Bernard A., 35-36, 152
Hearth, Donald P., 26-27, 210
Heritage Foundation, The, 16
Herman, Daniel, 9, 201
Hermes, 23
H-II (Japanese rocket), 66
Hinners, Noel W., 8
Hodge, John D., 8-9, 14-17, 19-20, 23, 31, 37-38, 40, 42, 45-46, 48-49, 51-52, 54, 57, 59, 73, 201
Hook, W. Ray, 54-55
Hopson, George, 85
Host Center, 127-129, 133, 138
Hubble Space Telescope, 79, 100, 115, 152, 186, 188, 191, 196
Hughes Aircraft, 32, 35, 114
Hutchinson, Neil B., 24-25, 32, 36, 40-43, 46, 55
Industrial Space Facility (ISF), 11, 41, 76
Initial Operational Capability (IOC), 11, 22, 27, 29, 32, 38, 41, 44-49, 51, 53, 55, 57, 59, 63, 72, 75
Integrated Management System, 209
Integrated Product Teams, (IPTs), 11, 133
International Federation of Professional and Technical Engineers, 127
Isakowitz, Steven, 162
Italy, 15, 22-23, 83, 140, 145, 150, 161
James Webb Space Telescope, 191
Japan, 15, 22-23, 32, 37-38, 46, 49, 62-63, 66, 71-72, 81, 83, 85, 95-96, 104, 108, 116, 125-126, 135, 144, 149, 157-158, 163-164, 168, 179, 182, 184, 188, 195, 211
Japanese Experiment Module (JEM), 11, 38, 59-60, 66, 84, 94-95, 99, 113-114, 125-126, 132, 139-140, 151, 154, 157, 168, 172, 182, 184, 188

Jernigan, Tamara E., 158
Jet Propulsion Laboratory (JPL), 11, 29, 36, 43, 58, 62, 72, 74-75, 111, 122, 183
Johnson, Lyndon B., 204
Johnson Space Center (JSC), 8, 11, 16, 18, 22, 24-26, 29, 32, 34-36, 38, 41, 43, 45-46, 50-55, 57-58, 72, 74, 77-78, 81, 86, 89-90, 92, 94, 96-97, 99, 109, 111-114, 118-121, 123, 125, 127-132, 134, 138, 140-141, 143-144, 146, 148, 152, 155-157, 159, 162, 165, 172, 175, 178, 180, 183-184, 193-194, 201, 208
Joint Vehicle Integration Team, 119, 121
Jones, Thomas D., 158
Kaliningrad, 118, 132, 145
Kazakhstan, 135, 153
Keller, Samuel W., 9, 136
Kennedy Space Center (KSC), 11, 19, 27, 29, 36, 45, 55, 58, 74-75, 80, 82, 86, 91, 99, 102, 109, 111, 113, 115, 119, 141, 143, 145, 148, 156-157, 161, 163, 166, 169, 172-175, 177, 180, 184, 188-189, 192-193
Kennedy, John F., 14, 16, 197, 204-205
Kerrebrock, Jack L., 101-102, 111-113, 138
Kerry, John (D-MA), 62
Khrunichev Enterprise, 147, 150, 156, 158, 160-161
Kiriyenko, Sergei, 170
Kistler Aerospace Corporation, 219
Kohrs, Richard H., 58, 90, 93-97, 102-103, 105, 108, 110-115, 118, 120, 127-129, 148, 156, 208
Koptev, Yuri, 135-136, 153-154, 161-162, 179
Korolev Center, Moscow, 175
Kostelnik, Michael, 188
Kranz, Eugene F. "Gene", 118, 121, 132, 142
Kress, Marty, 121
Krikalev, Sergei, 163, 175, 180-181
Laboratory Module (LAB), 27, 60
Langley Research Center (LaRC), 11, 16, 29, 26, 36, 54, 58, 65, 74, 123
Lanzerotti, Louis J., 108
Lautenberg, Frank R. (D-NJ), 108, 137
Lazutkin, Alexander, 164
Leetsma, David, 132, 142
Lehman, John F., 212
Leinbach, Michael, 193
Leninsk, 153
Lenoir, William B., 67, 90, 93, 97-98, 102, 108-109, 111
Level 0 Management, 79
Level A Management, 24-25, 31, 45-46, 52-53, 55, 57-58, 69
Level B Management, 24-26, 36, 42, 45, 51-52, 54, 81, 201
Level I Management, 70, 75, 78, 94, 118-119, 121
Level II Management, 73, 75, 78, 81, 86, 88-90, 94, 99-101, 103, 111, 119, 121
Level III Management, 73, 78-79, 82, 103, 121
Level IV Management, 79
Levi, Jean Daniel, 150
Lew, Jacob J., 174
Lewis Research Center (LeRC), 8, 11, 24, 29, 33, 35, 51-52, 54-55, 57-58, 74, 79, 105, 109, 120, 129, 141, 201, 220
Life Sciences Strategic Planning Committee, 82, 85

Linenger, Jerry, 163, 165
Lockheed, 15, 17, 35-36, 69, 96, 112, 130, 137, 147, 150, 156, 158
London Economic Summit, 23
Long Duration Exposure Facility (LDEF), 11, 90
Long Duration Orbiter, 11, 116, 124
Luton, Jean Marie, 144
Lyman, Peter, 72
Manhattan Project, 198
Manned Orbiting Laboratory (MOL), 11, 39, 109, 114
Man-tended, 48
Marburger, John, 191
Mark, Hans, 15-16, 42, 205
Marshall Space Flight Center (MSFC), 11, 17, 24, 26, 29, 34, 46, 51-52, 55, 74, 79, 85, 89-90, 94, 113, 115, 118, 125-126, 128-129, 155-156, 178, 180, 182, 201, 203
Martin, Franklin D., 9, 58, 93, 212
Martin Marietta, 15, 17, 32, 34-35, 69, 77, 79, 100, 104, 108
Mathews, Charles W., 37, 42
McClain, Gretchen, 172
McDonnell Douglas, 11, 15, 17, 35, 74, 109, 119, 121, 129, 140, 160
McInerney, Pamela, 149
McLucas, John, 108
Memorandum of Understanding (MOU), 11, 33, 37, 41, 83, 132, 147, 149-150, 158
Microgravity Science Glovebox, 186
Mikulski, Barbara (D-MD), 87, 105, 107, 126, 131, 133-134, 146, 173, 194
Miller, James C., III, 9, 43, 61-62, 79
Miller, John, 117-119
Mir, 7, 11, 29, 69, 108, 117, 129-132, 134, 136-137, 139, 141, 147-149, 151-155, 160, 163-167, 172, 180, 182-183
Mission Associate Administrator, 211
Mission to Planet Earth, 69, 71, 103, 112, 114, 203
Mitsubishi Heavy Industries, 168
Mixed Launch Fleet, 63-64
Mobile Servicing System (MSS), 11, 33, 37-38, 60, 84, 101, 113-114
Mobile Transporter (MT), 75, 186-187
Moore, Jesse W., 43
Moorehead, Robert, 9, 111, 122, 127-128
Moser, Thomas L., 9, 46, 53, 55-56, 58-60, 62, 64, 66, 69-70, 75-76, 79-83, 85, 88
Mueller, George E., 15
Multilateral Coordination Board, 83, 147
Multiple Element Integrated Testing, 11, 169
Multipurpose Logistics Module (MPLM), 11, 174, 183, 186, 195
Mulville, Daniel R., 185, 205
Murphy, Dan, 143, 192
Musgrave, F. Story, 158
Myers, Dale D., 55, 57, 59, 61, 63, 73-75, 84, 88-89, 200-201, 206
NASA / RAKA Program Coordination Committee, 147

NASA Advisory Council (NAC), 12, 21, 58, 64, 71, 77, 82, 85-86, 93, 100, 108, 117, 138, 145, 170, 182, 185-187
NASA Polar Orbiting Platform (NPOP), 12, 86
NASA Strategic Plan, 112, 212
National Academy of Sciences, 62, 110, 138
National Commission on Space, 38, 64
National Microgravity Research Board, 76-77
National Oceanic and Atmospheric Administration, (NOAA), 12, 69, 85, 194
National Press Club, 110
National Research Council (NRC), 12, 21, 62-63, 73, 91, 108, 162, 178, 196
National Science Foundation, 12, 85, 199
National Security Decision Directive-42, (NSDD-42), 12, 51, 76-77, 86, 92, 203
National Space Council (NSC), 12, 88-90, 105, 114, 124, 204-205, 207
National Space Transportation System (NSTS), 12, 64-67, 81, 93, 108
Naval Research Laboratory (NRL), 11-12, 159-160, 162-163, 180
Nelson, Bill (D-FL), 54, 58, 72, 87, 90, 102, 105
Newman, James, 173, 175
Nixon, Richard M., 8, 50, 197, 204
Nodes, 16, 60, 69, 113, 132, 139-140, 145, 151, 157, 160-161, 163-164, 168-169, 172, 175, 185, 188
Noguchi, Soichi, 195
NPO Energia, 117-118, 158
O'Connor, Bryan D., 123-124, 127-128, 133, 136, 143, 154-155, 208
O'Keefe, Sean, 184-187, 191-192, 205, 209-211
Oberfaffenhoffen, Germany (Control Center), 179
Odum, James B., 9, 79-83, 85-88, 90-91, 93
Office of Aeronautics, Exploration and Technology, 12, 107
Office of Commercial Programs, 12, 92
Office of Management and Budget (OMB), 12, 15, 26-27, 31, 33-34, 42-45, 48, 57, 59, 61-63, 66-68, 72, 79, 84-87, 89, 94, 103-105, 111, 113, 116, 118, 121, 128, 171, 174, 184-186, 189, 194, 196, 198-200, 202-203, 206-208, 210, 218
Office of Science and Technology Policy (OSTP), 12, 57, 124, 126, 130, 132, 134
Office of Space Science and Applications (OSSA), 12, 21, 24, 57, 81, 84-85, 91-92, 95, 98, 111, 115, 118, 206
Office of Technology Assessment (OTA), 12, 110
Operations Phase Assessment Team, 118, 121
Orbital Debris, 77, 90, 98, 113, 121
Orbital Maneuvering Vehicle, 12, 19, 26, 49, 81
Orbital Replacement Unit (ORU), 12, 86, 99
Orbital Space Plane (OSP), 12, 187-188, 190
Orbital Transfer Vehicle (OTV), 12, 19
Ostrach, Simon, 162
Oswald, Stephen P., 155
Padalka, Gennady, 191
Paine, Thomas O., 38, 72, 100, 204, 212
Panetta, Leon (D-CA), 87, 121-122
Parker, Robert, 111

Parkinson, Bradford W., 145, 170
Payload Operations Integration Center, 83, 178
Pederson, Kenneth S., 22
Pentagon, 72, 204
Permanently Manned Configuration (PMC), 12, 55-56, 59-60, 64-65, 74, 80, 83, 94-95, 105-107, 109-111, 113, 115
Peterson, Malcolm L., 9, 184
Phase A Studies, 24, 31, 96, 106, 112
Phase B Studies, 21, 24, 26, 29, 31-33, 36, 38-41, 46, 48, 54-55, 61, 69, 74-75, 96, 113, 201
Phase C/D Studies, 33, 40, 44-46, 48, 51, 54, 58-59, 61-62, 68-69, 72-75, 77-79, 85, 101, 111
Phase I, 59, 69, 73, 75-76, 79, 82, 89, 160, 163-165, 172, 206
Phase II, 59, 183, 206
Phased Program Task Force, 59
Phillips, Samuel C., 52, 54, 58, 62, 67, 129, 202, 210
Polar Platform, 60, 83-84, 91, 95
Polyakov, Valery, 7, 182
Powell, Luther, 17-18, 55, 201
Power Tower, 17, 29-30, 41, 43, 53
PriceWaterhouseCoopers, LLP, 209
Principal Investigators (PIs), 12, 111, 157, 166
Priroda, 130
Program Management Council, 51, 207
Program Scientist, 31, 36, 70, 85, 92
Progress, 141, 164, 167, 174, 188-189, 192-193
Progress-M, 130
Project Management Shared Experiences Program, 202
Project Pathfinder, 76
Prometheus Nuclear Systems and Technology, 198
Proton, 136, 163, 179
Proxmire, William (D-WI), 78, 82
Pryke, Ian, 168
Quayle, Dan, 85, 90, 114, 205
Raines, Franklin D., 171
Raney, William P., 9, 77, 90, 94-95, 111, 138
Reagan, Ronald, 7, 9, 14, 16, 22-23, 28, 31, 34, 38-40, 43, 45, 48, 50-51, 63, 76, 84, 86, 88-89, 92, 94, 110, 150, 197-198, 200, 202-206, 210, 219
Reconciliation Act of 1993, 136
Red Team – Blue Team Review, 115-116
Redesign Team, 123-124, 126-127
Reduction in Force (RIF), 195
Reference Configuration, 29-30, 36-37, 39-41
Report of the Task Force on Issues of a Mixed Fleet, 64, 70
Reston, Virginia, 62, 75, 81-82, 89, 103, 109, 111, 119-120, 122, 126, 128
Review Item Dispositions (RID), 12, 83
Ride, Sally K., 69-71, 85, 119, 203, 212
Riegle, Donald W., Jr. (D-MI), 66
Robbins, Frederick C., 82
Robinson, Stephen R., 195
Rockefeller, Jay (D-WV), 167
Rocketdyne, 35-36, 74, 119, 140
Rockwell, 15, 17, 35, 57, 67, 86, 96, 112, 129
Roe, Robert A. (D-NJ), 68, 94, 97
Roemer, Tim (D-IN), 123-124, 127, 153, 156, 181

Rogers, William P., 50
Rogers Commission, 48, 51-52, 63, 67, 71
Rohrabacher, Dana (R-CA), 122, 160-162, 167-168, 175
Romanenko, Yuri, 82
Ross, Jerry, 158, 175
Ross, Lawrence J., 58, 81
Rothenberg, Joseph, 169-171, 181
Russian Aviation and Space Agency (RAKA), 12, 147, 153, 179
Russian Duma, 153
Russian Space Agency, 12, 129, 135-136, 147, 149
Rutan, Burt, 218, 220
St. Hubert, Quebec (Control Center), 179
Salyut, 7, 132, 135
Sasser, James (D-TN), 87
Satellite Solar Power System, 7-8
Seamans, Robert C., 42-43, 63, 218
Senior Interagency Group for Space (SIG Space), 13, 89, 204
Senior Management Council, 207
Sensenbrenner, F. James, Jr. (R-WI), 128, 136, 146, 149-150, 154, 156, 160-163, 165, 171, 174, 176, 178
Service Module (Russian), 13, 136, 139-140, 154, 156-162, 175, 178
Sharipov, Salizhan, 193
Shea, Joseph F., 123
Shelley, Carl, 72
Shepherd, William, 127, 136, 158, 163-164, 180
Shuttle Derived Vehicle (SDV), 13, 65
Shuttle Plume Impingement Experiment, 119
Shuttle Return to Flight Task Group, 193-195
Shuttle-C, 65-66, 125
Silver, Rae, 186
Sisson, James M., 91
Skunk Works, 29
Skylab, 7, 18-19, 101, 108, 111, 141
Small Business Innovation Research (SBIR), 13, 74
Smith, Marcia, 9, 130, 134, 164
Software Support Environment, 13, 69, 78
Solar Dynamic Power System, 105, 129, 136-137, 139
Solid Rocket Booster (SRB), 13, 192
Solovyev, Anatoly, 158
Solovyov, Vladimir A., 153
Soskovets, Oleg, 154
Soviet Union, 7, 71, 150, 159, 215
Soyuz, 112, 117-118, 124-125, 129-130, 132, 135, 137, 139-140, 144, 149, 154-155, 163-165, 168-169, 174, 179-180, 183, 187-189, 192-193, 212
SP-100, 198
Space Applications Board, 21
Space Exploration Initiative (SEI), 13, 92, 109-110, 194, 207, 212
Space Industries Inc (SII), 13, 41
Space Science Board, 13, 21
Space Shuttle Main Engines (SSMEs), 13, 65
Space Station Advisory Committee (SSAC), 8-9, 13, 58, 76-77, 79-83, 85-86, 89-91, 93-96, 98-102, 105, 107-108, 111-114, 116-119, 121, 127, 138, 156, 169, 171, 208

Space Station Alpha, 11, 120, 124, 130-133, 135, 137, 140, 143, 145, 151, 156, 182
Space Station Chief Scientist, 37, 111, 138
Space Station Control Board (SSCB), 13, 36, 42, 83, 147, 163, 172
Space Station Development Plan, 73
Space Station Freedom (SSF), 13, 82, 84-86, 91-93, 98-102, 106, 108, 110-112, 115, 118, 121, 124-128, 131, 133, 137-141, 146, 156, 164, 208
Space Station "Geometries", 17-18, 29
Space Station Management Council, 91
Space Station Operations Task Force (SSOTF), 13, 72-73
Space Station Program Management Council, 51
Space Station Program Office (SSPO), 13, 31, 36, 46, 55, 58-59, 62, 64-66, 70, 76, 78, 81-82, 84-86, 92, 112, 138, 145-146, 148
Space Station Program Requirements Document (PRD), 12, 78
Space Station Science and Application Advisory Subcommittee (SSSAAS), 13, 83, 85, 90-91, 105, 111, 118-119, 138
System Engineering and Integration Task Force, 57-58
Space Station Task force, 9, 14-15, 20, 24-26, 29, 37, 39, 49-50, 74, 201, 205
Space Studies Board, 13, 108-110, 162
Space Transportation System (STS), 12-13, 27, 36, 41, 50, 55, 63-64, 119, 125, 173, 183, 188, 193, 195
Spacehab, 168, 174, 188
Spacelab, 22, 75, 82, 101, 111, 115, 121, 125, 132, 134, 177-178, 219
Spar Aerospace, 36, 168
Spectr, 130, 141
Sputnik, 71
SRI International, 32
Stafford, Thomas P., 109, 162, 193, 212
Stage Integration Facility, 99
Stockman, David, 15
Stofan, Andrew J., 9, 35, 52-55, 57-59, 63, 69, 75-79, 90, 93, 198, 201
Stokes, Louis (D-OH), 123, 133-134
Super Collider, 166, 198
Support Equipment Sharing Agreement, 119
Synthesis Group, 109-110, 212, 214
Talone, John "Tip", 9, 156-157, 169
Tam, Daniel C., 146, 178
Tanner, E. Ray, 85, 88, 90
Task Force on Scientific Uses of the Space Station (TFSUSS), 13, 22, 37, 41, 58
Taylor, William W.L., 111
Technical Management Information System (TMIS), 13, 69-70, 73, 77-79
Teledyne Brown Engineering, 35-36, 108
Teleoperator Maneuvering System (TMS), 13, 19, 27, 100
Thagard, Norman, 149, 151, 153
Thomas, Andrew, 9, 55, 172
Thompson, James R., Jr., 89, 114
Thompson, Tommy, 199
Titan II, III, IV, 65, 86, 215

Tito, Dennis, 183
Toulouse, France (Control Center), 179
Tracking and Data Relay Satellite (TDRS), 13, 41, 83, 92, 114, 143
Trafton, Wilbur C., 137, 141, 143, 145, 148-149, 154-156, 159, 162, 167, 169-170
TransHab, 169
Traxler, Robert (D-MI), 87
Truly, Richard H., 41, 46, 58, 89-90, 92-94, 100, 108-109, 111-112, 114-115, 200-201, 205-207
TRW Systems, 15, 32, 35-36, 114, 170, 186
Tsibliev, Vasily, 163-164
Tsukuba, Japan (Control Center), 179
Tucci, Liz, 134, 137, 143, 145-146
Tug (FGB, also called Zarya), 11, 129-130, 132, 135, 137, 139-140, 147, 150-153, 156-157, 159-161, 163, 168-169, 171, 174176, 178-180, 189
U.S. / Russia Human Space Flight Agreement, 147
U.S. Space Command, 39, 90
U.S. Space Foundation 10th Annual National Space Symposium, 21, 144
Uhran, Mark, 162, 182
United Space Alliance (USA), 158, 186
Unity, 172-173, 175-176, 178-180, 189
User Accommodation Panel, 92
Van Laak, James, 159, 167
Vest, Charles M., 123, 126, 141
Viking, 111, 210
Vision for Space Exploration, 190, 213
Walker, David M., 97
Washington Space Business Roundtable, 137
Webb, James E., 42-43, 207
Weinberger, Casper, 79
Weiss, Stanley I., 113, 117
Wilson, Pete (D-CA), 89
Winslow, Larry, 150
Work Package #1 (WP #1), 34-35, 57, 85, 109
Work Package #2 (WP #2), 34, 57, 98, 109, 120-122
Work Package #3 (WP #3), 35, 85, 101, 105, 107, 109
Work Package #4 (WP #4), 35, 101, 105, 109, 120, 141
Working Group on Russian American Cooperation, 129
X-33, 181
X-34, 181
X-38, 171, 176, 186
Yamano, Masato, 144
Yeager, Charles E., 50
Yellow Book, 26
Yeltsin, Boris, 117, 135, 162, 170
Young, John, 50, 143
Young, Thomas, 10, 138, 184-185, 210
Zarya, See Tug (FGB)
Zenit, 134, 136, 154
Zimmer, Richard (R-NJ), 127, 153
Zvezda, 173, 178, 180, 183, 189

**WITHDRAWN**

JAN 1 2 2009
NOV 2 9 2010
JAN 2 7 2014
MAR 3 0 2015
MAY 1 1 2017
JUN 2 0 2018

TL 797 .B43 2006
Beattie, Donald A.
ISScapades
7/2008